Introduction to Cancer Biology

A Concise Journey from Epidemiology
through Cell and Molecular Biology
to Treatment and Prospects

This concise overview of the fundamental concepts of cancer biology is ideal for those with little or no background in the field. A summary of global cancer patterns introduces students to the general principles of how cancers arise and the risk factors involved. By focusing on fundamental examples of the signalling pathways within cells, the functional effects of DNA damage are explained. Later chapters then build on this foundation to provide a comprehensive summary of the major signalling pathways that affect tumour development. Current therapeutic strategies are reviewed, along with a discussion of methods for tumour detection and biomarker identification. Finally, the impact of whole genome sequencing is discussed, bringing students up to date with key recent developments in the field. From basic principles to insights from cutting-edge research, this book will enable the reader to move into the cancer field with confidence. The online material that accompanies this book can be found at www.cambridge.org/hesketh.

Robin Hesketh has been a member of the Biochemistry Department at the University of Cambridge for over twenty-five years. He has taught at all undergraduate levels from first-year medicine to fourth-year biochemistry on a wide range of cell and molecular biology topics with a particular focus on cancer. His major research area is the development of anti-angiogenic strategies for the treatment of cancer. He is also the author of the popular science book *Betrayed by Nature: The War on Cancer* (2012).

Introduction to
Cancer Biology

A Concise Journey from Epidemiology
through Cell and Molecular Biology
to Treatment and Prospects

Robin Hesketh
University of Cambridge

CAMBRIDGE
UNIVERSITY PRESS

CAMBRIDGE UNIVERSITY PRESS

Cambridge, New York, Melbourne, Madrid, Cape Town,
Singapore, São Paulo, Delhi, Mexico City

Cambridge University Press
The Edinburgh Building, Cambridge CB2 8RU, UK

Published in the United States of America by Cambridge University Press, New York

www.cambridge.org
Information on this title: www.cambridge.org/9781107013988

First published 2013

Printed and bound in the United Kingdom by the MPG Books Group

A catalogue record for this publication is available from the British Library

Library of Congress Cataloguing in Publication data

Hesketh, Robin.
   Introduction to cancer biology: a concise journey from epidemiology through cell and
molecular biology to treatment and prospects / Robin Hesketh.
   Includes bibliographical references and index.
   ISBN 978-1-107-01398-8 (Hardback) – ISBN 978-1-107-60148-2 (Paperback)
   I. Title.
   [DNLM: 1. Neoplasms.   2. Neoplastic Processes.   3. Oncogenes.   4. Sequence Analysis, DNA.   QZ 200]
   616.99′4–dc23

2012015670

ISBN 978-1-107-01398-8 Hardback
ISBN 978-1-107-60148-2 Paperback

Additional resources for this publication at www.cambridge.org/hesketh

CONTENTS

Contents

A colour plate section is between pages 238 and 239.

ACKNOWLEDGEMENTS

No cancer book should fail to acknowledge the vast, anonymous army of those who have confronted cancer and in so doing found the resolution to participate in medical research. Over the years they have made an immense contribution to our knowledge of cancer and, as we move into the era of personal genome sequencing, their role has become even more significant.

The scientific story rests, of course, on the wonderful labours of so many who have built the massive archive that represents our current state of knowledge. Its telling in this book has also drawn on countless hours of discussion with colleagues over many years, as well as on their lectures and published work. In addition I am profoundly grateful to more generations of students than I care to count for their stimulating input, ranging from questions eliciting a long pause, a big gulp and 'Mmm...' to comments on the clarity (or otherwise) of my lecture slides and handouts. The number who deserve acknowledgement thus runs into thousands, represented here, I'm afraid, only by the small number of specific citations I've been able to include.

Finally, it is a great pleasure to thank the following friends and relations for providing photographs or images, for critical comments on all or some of the manuscript or for specific suggestions and general support. Any errors, omissions or gnomic passages that remain despite their efforts are, of course, entirely my responsibility: Dimitris Anastassiou (Columbia University), David Bentley (Illumina), Tom Booth (Cambridge Research Institute), Peter Börnert (Philips Technologie GmbH, Hamburg), Peter Britton (Addenbrooke's Hospital), John Buscombe (Division of Nuclear Medicine, Addenbrooke's Hospital, Cambridge), Jean Chothia (Faculty of English, University of Cambridge), Dan Duda (Harvard Medical School), David Ellar, Ferdia Gallagher (Department of Radiology, Addenbrooke's Hospital, Cambridge, and Cancer Research UK Cambridge Research Institute), Mel Greaves (The Institute of Cancer Research), John Griffiths (Cambridge Research Institute), Brian Huntly (Cambridge Institute for Medical Research), Rakesh Jain (Harvard Medical School), Scott Lowe (Sloan-Kettering Institute), Rahmi Oklu (Massachusetts General Hospital), Richard Sever (Cold Spring Harbor Laboratory Press), Pierre Sonveaux (University of Louvain Medical School), Sir John Sulston (Institute for Science, Ethics and Innovation, University of Manchester), Robert Tasker (Harvard Medical School), Rupert Thompson (Faculty of Classics, University of Cambridge), Sir John Walker (MRC Mitochondrial Biology Unit, Cambridge), Robert Whitaker (Selwyn College), Richard White (Dana Farber Cancer Institute, Children's Hospital Boston), Roger Wilkins, Rick Wilson (Washington University, St. Louis) and, from the Department of Biochemistry, University of Cambridge, Gerard Evan, Richard Farndale, Chris Green, Jules Griffin, Heide Kirschenlohr, Kathryn Lilley, Tom Mayle, Jim Metcalfe and Gerry Smith.

Acknowledgements

My sons Robert (Select English) and Richard (Hillingdon NHS) have been invaluable proof-readers and critics but nevertheless remain two of my best friends. My other best friend is my wife who quite simply makes everything worthwhile and even finds the energy in her *alter ego* as Jane Rogers (The Genome Analysis Centre, Norwich) to be my tutor in genomics.

And, really finally, this book would not have happened without two members of the staff of Cambridge University Press, Katrina Halliday and Hans Zauner. They brought to bear both scientific and publishing expertise as well as the enthusiasm that enabled me to make it to the last page.

FOREWORD

In 1971, President Richard Nixon famously committed the intellectual and technological might of the USA to its great "war on cancer," signing in the National Cancer Act and making eradication of the disease both a national imperative and an international cause célèbre. Other than the space race, few if any peace-time endeavours have consumed such prodigious resources over such a protracted time-scale. Yet 40+ years and billions of dollars later, cancer still kills over one third of all people in what some fondly call the "developed" world. To many, this manifest failure is inexplicable and to a small fringe clear evidence that a cure has been found but is being suppressed for some nefarious end by an international conspiracy of governments and pharmaceutical companies. After all, rumours routinely circulate of natural products or extracts with amazing anti-cancer therapeutic efficacy but whose use is, for some unfathomable reason, shunned by the Western medical elite. The truth, however, is far less sensational but far more intriguing: cancers have emerged as an unexpectedly complex and diverse ensemble of diseases driven by mutations in processes that lie at the heart of the fundamental questions of biology – how cells, tissues and organisms self-build, self-assemble, self-maintain and self-repair. To comprehend cancer is no less daunting a task than comprehending the very organizational principles that underpin biology.

Discussions of cancer are further confounded by the misleading mythology that enshrouds it and endows it with both strategic foresight and murderous intent. Cancers are said to skulk unseen in our bodies - sometimes for decades – waiting to lash out, sicken, invade and kill. Then, as cancers spread and invade they are said to "progress" – a word loaded with directional purpose and carrying the clear implication that the disease is out to get you. In response to treatment, cancers "fight back," developing resistance and then returning with renewed and deadly vigour. All the time, the relationship between patient and cancer is depicted as a battle, literally to the death. Unfortunately, such language is not just counterproductive fearmongering – it is also plain wrong and appalling biology to boot. Cancers are not single entities but diverse, heterogenous ensembles of independently evolving clones. The rules governing cancers are the same as those that govern any evolutionary process – random variation in huge numbers of replicating individuals fuels selection for ever-fitter variants. In the case of cancers, the individuals are tumour cells and fitter means faster growth. A cardinal feature of natural selection is that we see only the winners – we have no idea how often cancers start but die out and the prodigious tumour cell death that chemo or radiotherapy of primary tumours usually triggers counts for little in the face of a small number of resistant residual clones that subsequently regenerate the disease and precipitate relapse.

In this comprehensive new book on the molecular basis of cancer, Robin Hesketh offers us both illumination and demystification. Chapter 1 offers us a lucid and comprehensive survey of worldwide cancer incidence, but immediately confronts us with the still largely unexplained conundrum that the frequencies of so many cancers differ dramatically between cultures and nations. Some insight is offered in Chapter 2, which addresses the general issue of causes of the genetic changes that precipitate cancers. Even setting aside the witting depredations of tobacco and solar UV, the sinister truth is that our bodies and our environment are awash with all manner of carcinogens and mutagens that relentlessly pummel our DNA and epigenome. Mutations are inevitable: and so, as night follows day, is cancer. Chapter 3 lifts the lid on one of life's greatest underlying principles – biology is all about moving information around. Large, long-lived organisms like humans are comprised of a vast, heterogenous colony of self-assembling, self-organizing, self-diagnosing and self-repairing cells, each of which appears to know "who" it is, where it belongs and how it is supposed to behave. The key to unlocking this particular mystery is the processing and transduction machinery within each cell that receives, coordinates and interprets the signals that give cells their instructions. And exactly how this signal processing machinery gets corrupted in cancers is the theme of Chapter 4. Cancers are unique puzzles because they are diseases of misregulation: they arise by mutations in the genes that regulate key cellular processes – growth, proliferation, repair and survival, differentiation, movement and migration resulting in cells that are too many, in the wrong place, doing the wrong thing and at the wrong time. Understanding how such mutations cause cancer offers insights into how normal cells and tissues self assemble, maintain their architectures and regenerate when damaged.

The 1980s and 1990s were heady days in cancer research, when almost every month seemed to herald another dramatic example of the convergence between cell signaling/intracellular regulatory pathways and the proteins encoded by the ever-lengthening list of oncogenes and tumour suppressor genes mutated in various cancers. A fundamental principle of cancer biology with profound explicatory power had, indeed, been uncovered: mutations in various oncogenes and tumour suppressors loosened the shackles of intracellular regulators and brakes, propelling cancer cells into their pathological autonomy. Thereafter, additional traits would gradually accumulate that further augmented their potential to grow and spread. Many of these "hallmarks" affected cell autonomous properties, such as refractoriness to growth inhibitory signals and to programmed cell death and the switch to biosynthetic metabolism and aerobic glycolysis that Warburg had noted back in the 1930s. But oncologists and pathologists had long known that cancers are not merely monocultures of rogue cancer cells but complex, aberrant tissues. In the 1860s, Virchow noted the dramatic infiltration of tumours by leukocytes and other inflammatory cells, opining that cancers resembled "wounds that do not heal," while the remarkable vascular infiltration of many solid tumours led Judah Folkmann in the 1970s to formulate his groundbreaking notions on the necessity of angiogenesis for macroscopic tumour growth and dissemination. More recently, the pivotal role played by both fibroblasts and stroma in the initiation, maintenance, spread and therapeutic responses of cancers has emerged. Such observations highlight an important principle of the biology of multicellular organisms.

Foreword

To crudely paraphrase the 17th century satirist John Donne: "no cell is an island, entire of itself." Somatic cells may appear physically discrete down the microscope but everything about their biology is obligatorily social – their proliferation, differentiation, migration, metabolic state and, indeed, very survival are all dictated by extracellular signals. And while cancer cells have, indeed, lost some of these social constraints, they remain deeply reliant on the signals from, and activities of, neighbouring normal cells both for sustenance and to remodel tissue stroma and vasculature. It is therefore appropriate that Chapter 5 takes the time to consider not only what defines a cancer cell but also what defines a tumour.

The final chapters, 6, 7 and 8 make an especially intriguing trio. Having dealt in Chapter 3 with the nuts and bolts of how information is conveyed across cells and tissues, Chapter 6 takes us further down the rabbit hole to the deeper understanding that such signaling "pathways" are organized into networks – complex, self-regulating and self-correcting ensembles more akin to the internet than to telephone cables. Biologists understand such integration as an inevitable consequence of evolution – the potently creative mix of variation and selection painstakingly incorporating any tweaks, fixes, patches, feedback and integration with other processes that confers greater fitness. Ascertaining how such networks "function" is daunting for two reasons. First, they tend to respond as an integrated whole rather than as discrete but interconnected functional units. Where information is relayed in such a diffuse, iterative and parallel manner, where even the distinction between output and input is blurred, it can prove impossible to pin down cause-and-effect relationships. Second, evolution has a predilection for employing functional networks as general purpose signaling machines, redeploying them in a variety of other roles in other tissues. Perhaps of greatest significance to cancer therapy, however, is that such networks typically exhibit remarkable robustness, stability and uniformity of output despite perturbation, damage and variability and inconsistency in input. Given this, it is hardly surprising that even our newest cancer drugs, notwithstanding the unparalleled specificity and efficacy with which they inhibit their targets, frequently elicit unforeseen side effects and often lose efficacy as cancers progress and patients relapse. Finally, Chapter 8 is our ticket on the post-genomic roller coaster, a ride at once both entrancing and terrifying. State of the art technologies can now garner and catalogue extraordinary masses of data: whole cancer genomes can be sequenced and every dent or scratch in the DNA documented; the entire gamut of genes turned on or off at any juncture can be defined, compiled and presented in glorious, multicoloured ontogenies; and every protein present in a cell or tissue can be itemized, characterized and quantified. The devastating conclusion we distill from recent detailed analyses of individual cancers is that each person's cancer is different from every other, each with its own peculiar ensemble of mutations and epigenetic changes. Worse yet, recent studies reveal that a tapestry of interwoven diverse clonal lineages resides even within each single tumour. This is daunting, yes, but not altogether surprising. After all, the engine that powers the evolution of cancers is random heritable variation at (we now appreciate) both the genetic and epigenetic level and such variation, fueled as it is by intrinsic and environmental forces, is an inescapable fact of life. Our task now is find productive ways to use this information to improve cancer therapy.

Foreword

The most obvious strategy is to personalize cancer treatment – to take advantage of the speed and plummeting costs of state-of-the-art genomic technologies to define the molecular architecture of each patient's cancer and build a treatment regimen around that architecture – is personalized therapy. In so doing, it will be essential to identify where the robustness and functional redundancy resides in oncogenic pathways and networks, and to map within those networks how and where resistance arises in response to targeted therapies. Ultimately, to avoid the evolution of drug resistance and the all-too-familiar spectre of patient relapse we may need to develop strategies that specifically target functionally non-redundant processes, since these cannot be circumvented by adaptive compensation or evolution. Here, we may be heartened by one of the conclusions from Chapter 6 – that most of the diverse mutations in human cancers in the end converge on a relatively small number of core pathways (e.g. Ras, Myc, E2F/RB1, p53, NF-$\kappa\beta$, STAT3, TGFβ). It seems that underneath the glittering diversity of cancers there lies a deep commonality – and maybe the potential for novel therapeutic strategies in the future.

Gerard Evan, Ph.D., FRS, F.Med.Sci
Head of Department and Sir William Dunn Professor
Department of Biochemistry
University of Cambridge, UK
and
Adjunct Professor
Department of Pathology and Helen Diller Family Comprehensive Cancer Center,
University of California San Francisco, USA.

INTRODUCTION

The aim of this book is to provide an introduction to the science of cancer for those coming to the topic for the first time – be they students or graduates or post-doctoral scientists moving into the field of oncology. That is, to paint a picture of what we think happens to cells and molecules in the making of cancers, how it bears on diagnosis and prognosis, and where the science is taking us in terms of treatment. To scientists and non-scientists alike, cancer can seem almost the ultimately daunting subject. It's true that cancers are the most complex diseases that afflict us and it is arguable that no two cases are identical, if all the biochemical changes involved are identified. However, while at some point we will face that problem – indeed we shall see that its very complexity may offer some advantage in the therapeutic battle against the disease – it's now clear that the underlying principles by which cancers arise are remarkably consistent and conceptually straightforward.

By keeping that in mind it is possible to advance through the story by the following relatively simple steps:

1. Consider some facts about the frequency with which cancers arise across the world and what the distribution patterns of different types of cancer, that is the epidemiology, tell us about the underlying causes.
2. Review the major risk factors. Some of these are beyond our control: for the others we consider possible measures to reduce the risk of getting the disease. This takes us from cancer statistics to the underlying science, and a discussion of why the causes can sometimes be difficult to confirm and why there is such variation in treatment efficacy between and even within countries.
3. The fundamental feature of cancer is the perturbation of normal cellular control, most critically of the machinery that regulates cell growth and division – the cell cycle – and we consider next how cells respond to environmental cues, focusing on one major signalling pathway.
4. Mutations are the driving force in cancer progression and the critical targets are the central pathways that regulate the cell cycle. In this chapter we consider mutations: the nature of the alterations that can occur in DNA and how they can affect protein and, hence, cellular function. Although many genes may be targets, there are only three basic mutational mechanisms and we use a minimal number of examples as illustrations.
5. These mutational changes convert cells from normal to aberrant, cancerous behaviour. In this chapter we define what makes a tumour cell in terms of the characteristics that differentiate them from normal cells. These emerge as eight major features associated with the development of primary cancers and the progression to a malignant phenotype. Of these, the most critical is the capacity to metastasise,

that is, for cells to migrate from primary tumours through the body to colonise secondary sites.

6. This leads to a more comprehensive view of the aberrations that can occur in intracellular signalling pathways to promote cancer. The initial focus is on a core of pathways, parts of which are abnormal in virtually all cancers. The discussion then expands to consider all the major signalling routes from the plasma membrane to the nucleus and the associations that some of these have with specific types of cancer. This reveals a complex interacting 'information network' of proteins and raises the question of how cells make sense of such complex signalling inputs to enable them to mount a discrete response.

7. The penultimate chapter reviews the progress in both diagnosis, monitoring and therapy in the latter part of the twentieth century and the remarkable advances that have already occurred in the first decade of this century, highlighting the extraordinary scientific diversity that is being brought to bear on cancer.

8. The final chapter focuses on the impact of the revolution in DNA sequencing that has seen the birth of 'personalised medicine' and has already wrought drastic changes in our approach to cancer research, diagnosis and treatment.

The intention therefore is to provide a book that will be the ideal companion to most student courses on cancer. The starting level required is not much more than knowing what a molecule is and having a general concept of a cell. The chapters that introduce cell signalling and mutations are designed to be as easy as possible by describing only sufficient specific examples to illustrate the key points. Similarly, the chapter discussing the behavioural changes associated with the switch from normal to tumour concentrates on the cell biology, keeping specific mentions of genes and proteins to a minimum. The emphasis therefore is on the principles, with diagrams and photographs to help convey the key points without becoming submerged in detail, as can so easily happen in this field. To avoid appearing to over-simplify, Chapter 6 comprises a basic summary of the central molecular defences against cancer but then reviews the major signalling pathways involved in some detail. This confers a reasonably comprehensive perspective and it also opens the way to discussing therapeutic strategies, a field that is considered in the following chapter in a review that takes us from the introduction of the first specific drugs to the current situation. The last chapter deals with genomics and the sequencing revolution that has taken us into the world of 'personalised medicine'. This has dramatically enlarged the scope for therapeutic intervention and we end with a summary of the prospects for cancer treatment as we move into this new era.

In addition to a glossary (the terms therein being emboldened on first appearance in the text), there are five appendices that provide supporting reference sources: Appendix A describes how tumours are graded and staged; Appendix B provides a list of molecular targets for currently available anti-cancer drugs; Appendices C and D summarise the main classes of oncoproteins and tumour suppressor genes; and Appendix E summarises the principal features of the ten types of cancer that predominate in terms of global mortality.

Introduction

The encouraging message is that if we limit the amount of detail we attempt to absorb, the essential principles of cancer become easy to grasp – in contrast, say, to the problems our physicist colleagues have in trying to explain theories of quantum gravity. The enduring message is that, although cancer remains as fascinating as ever from a scientific point of view, the progress that has been achieved has replaced at least some of its mystery with effective treatments and that each passing year sees an increase in the proportion of individuals who are afflicted with cancer and yet triumph.

The online materials that accompany this books include chapter-by-chapter multiple-choice questions, short-answer questions and essay titles for students. Answers and suggested key points for essays are available for instructors. These can be found at www.cambridge.org/hesketh.

Gene nomenclature

The HUGO Gene Nomenclature Committee (HGNC: www.genenames.org/index.html) assigns unique symbols to human genes. Gene names are written in italicised capitals; the protein that they encode is non-italicised: *EGFR* (gene)/ EGFR (protein). They are pronounced phonetically when possible (SRC is *sarc*, MYC *mick*, ABL *able*). Viral forms are prefixed by v- (e.g. v-*src*). For some genes that have commonly used informative names both are shown (e.g. *SLC2A1*/ GLUT1 and *SLC2A5*/GLUT3).

Lessons from epidemiology

We begin by looking at cancer patterns worldwide. These are of interest because they show marked variations in the forms of the disease that afflict different populations. These differences indicate the importance of environmental factors that include lifestyle – for example, what we eat and tobacco use – in determining both the type of cancer and the frequency of occurrence. Although there is variation in cancer types, there is a broad trend of rising incidence across the world, for which a major driving force is increasing longevity. In the developed world lung, breast, bowel and prostate cancers head the mortality table. Taking all cancers together, the last 30 years has seen a gradual increase in the five-year survival rate, although there remain significant variations between nations and even within some countries. For the developing world the outlook is more depressing: not only is the annual number of new cases rising but inadequate screening programmes often mean that diagnosis is delayed until tumours have spread to secondary sites in the body and therefore become very difficult to treat. Analysis of cancer mortality in different age groups revealed many years ago that the additive effects of about half a dozen discrete events drives cancer development – the first direct evidence that the accumulation of mutations is the underlying cause.

Incidence

Every year over 12 million people worldwide are diagnosed with cancer. Europe and North America together contribute about 40% of this figure with just over half of all new cases (54%) arising in developing countries. Of the various types of cancer that contribute to these figures, lung cancer heads the list with 1.4 million new cases annually (12.5%) followed by breast (1.2 million, 10.6%) and colorectal (1 million, 9.4%) cancers (Fig. 1.1). The names of these three cancers will be familiar to American and British readers because they are also in the top four of their national figures for both incidence and deaths due to cancers. The other member of the Big Four in those countries, and in most of the developed world, is prostate cancer (679,000, 6.3%). However, in the world rankings prostate is pushed down to sixth on the list by stomach and cervical cancer (934,000 and 692,000 cases annually, respectively).

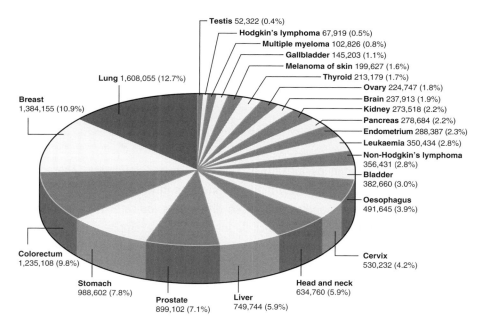

Figure 1.1 **Incidence of major cancers worldwide, 2008.** Shown in rank order with number of cases and percentage of the total number (12,662,554).

Incidence varies widely around the world being highest in the USA for both men and women. For all forms of cancer the lowest rates are about five times less than the USA figure (e.g. in Gambia), although some specific cancers show much greater variation (e.g. there are about 300 new skin cancer cases in some parts of Australia for every one in Kuwait).

In the UK in 2008 there were 309,527 new cancer cases: for the USA in 2012 the estimate is that there will be 1,638,910 new cases. For women breast cancer is the most common: 48,788 British women were diagnosed in 2009, and one in six Americans will get it, which means over 226,800 cases in 2012, about the same percentage in the two populations.

Deaths

Collecting numbers about the distribution of disease is called epidemiology and it has a fascination of its own but, as we shall see, when you get to the scale of the major cancers it can be very informative. Every year over 7 million people die from the disease – that's 13% of the total of 56 million deaths in the world from all causes each year (Fig. 1.2). Unsurprisingly, the ten most common cancers worldwide are also among the leading causes of mortality. Bladder and **non-Hodgkin's lymphoma**, however, are displaced in the mortality table by pancreatic cancers and **leukaemias**. In part this reflects the fact that there are no effective treatments for tumours of the pancreas and it may be noted that this cancer joins the Big Four in the leading causes of USA and UK cancer deaths.

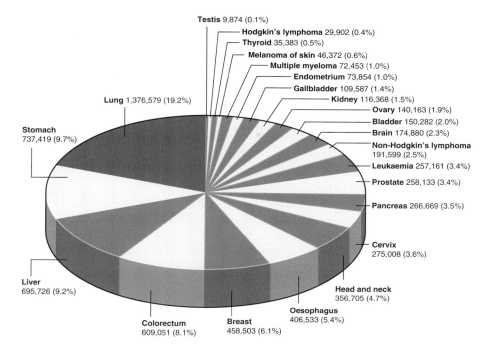

Figure 1.2 Deaths from major cancers worldwide, 2008. Shown in rank order with percentage of the total number (7,564,802)

To put cancer in its place we might note that mankind's biggest scourge is heart disease, which claims over 13.5 million deaths per year (24%) – this total comprises deaths caused by ischaemic heart disease, cerebrovascular disease and hypertension. After heart disease, the next biggest killers all fall some way behind cancer, namely, HIV/AIDS (4.9%), tuberculosis (2.7%) and malaria (2.2%). Much less well known is the figure of 2.2 million children (3.9%) who will die from diarrhoea this year – largely caused by ingesting germs through living in insanitary conditions.

Gradual shifts have occurred in the worldwide cancer pattern over the last century due to changes in lifestyle. The most dominant factor is that we are living longer, which gives us more time to develop the disease, and this together with other factors, has led the International Union Against Cancer to conclude that the number of new cases each year will rise to more than 16 million by 2020.

Patterns around the world: how many cancers and what sort?

UK and USA

Unsurprisingly, the death toll that results from the avalanche of new cases also has lung cancer at the top of the list with 1.2 million (17.5% of all cancer deaths). However, the next biggest worldwide cancer killers are stomach (0.7 million), liver (598,000) and colon (529,000), accounting for 10.4%, 8.9% and 7.9% of all cancer deaths, respectively. In 2009, the UK weighed in with a contribution of 156,090 to the 7 million. In 2010, the USA provided 299,200 male and 270,290 female cancer deaths. We've already noted that on the UK and USA incidence lists prostate displaces stomach and so it's unsurprising that it does the same on the killer lists, with stomach cancer being sixth in the UK and thirteenth in the USA (Fig. 1.3). All the same, 'twas not ever thus: go back 80 years in the USA and stomach cancer was the biggest cancer killer, albeit that in 1930 the total death toll was only ~120,000. Nowadays nearly two-thirds of stomach cancers occur in the developing world, to which we will return in a moment, but there too the incidence is declining, as it has done in the USA, for reasons that are not entirely evident but may include the increasing availability of fresh fruit and vegetables and of meat preserved by refrigeration rather than by salting. Nevertheless 10,540 will die of stomach cancer in the USA in 2012.

Despite having the second highest incidence, breast cancer is only fifth on the mortality list with 458,503 deaths in 2008 (11,633 in the UK in 2010; 39,840 in the USA in 2010). This figure of about half a million represents 1.6% of all female deaths but again there is an imbalance – this time in the other direction. In rich countries 2% of all female deaths are due to breast cancer but in poor countries the figure is only 0.5%. Here the difference roughly reflects incidence – about five times lower in developing countries where the average age of populations is younger. The distribution pattern of cancer mortality between men and women in the USA is closely mirrored in the UK (Figs. 1.4 and 1.5). The two notable exceptions are liver cancer, comparatively common in the USA, and stomach cancer, which is more significant in the UK. The latter may reflect dietary differences in at least some regions of Britain and high protein and cholesterol diets may be a factor in the USA liver rates.

Europe

The pattern across European nations (Fig. 1.6) shows the overall cancer death rates to be generally higher in eastern Europe (Hungary is the highest with 234 per 100,000) with the lowest being in Finland (138 per 100,000). Even so, for women over the period 2000–2004, Denmark, Scotland and Hungary were the top three (141, 123 and 132/100,000, respectively). One might expect that the variations between countries in total cancer mortality would be strongly dependent on differences in the toll of the major cancers. Indeed the pattern of breast cancer, for example, broadly follows that of total cancer deaths, although there is one notable exception: Poland has the third highest rate of total cancer deaths but one of the lowest breast cancer rates. Despite these generalities it should be borne in mind that breast cancer and other major cancers, such as **melanoma** and prostate cancer, vary markedly in national impact.

Russia has by far the highest population of any European nation (148 million) but in terms of cancer it remains something of an enigma. In the period after 1960 there was a gradual decline in Russian cancer death rates but this reversed in the early 1980s. The major contributors to the reversal were male lung and female breast cancer. In the 1990s, however, the trend reversed again, mainly due to a drop in cancer deaths among the elderly. Broadly similar trends have also occurred in the Ukraine (52 million),

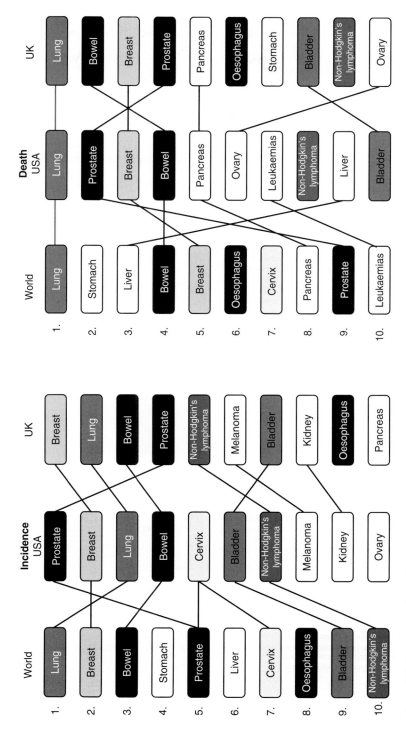

Figure 1.3 Top ten world rankings for incidence and death compared with UK and USA positions.

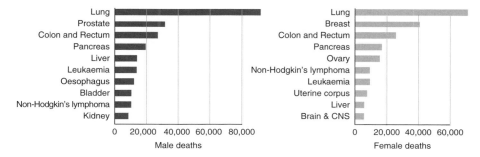

Figure 1.4 The ten most common causes of cancer death estimated for the USA for 2008.

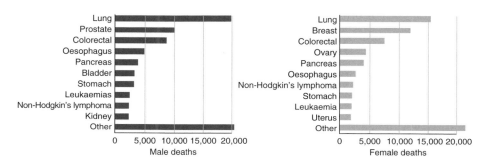

Figure 1.5 The ten most common causes of cancer death in the UK in 2008.

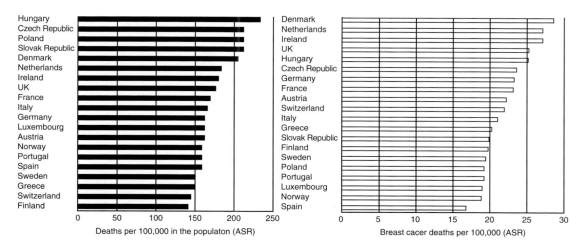

Figure 1.6 Total cancer deaths and breast cancer deaths in 20 European countries in 2006, shown as the **Age Standardised Rate (ASR)** per 100,000 of the population.

which was, of course, part of the Soviet Union until 1991. Analysing Russian data is particularly problematic partly because the quality of the information has varied over time and with the area from which it has been collected. There has also been a competitive effect from large rises in other causes of death, notably heart diseases and accidents. As in other countries, improvements in health care have also begun to make an impact. In addition, the turbulence of the twentieth century has produced marked birth cohort effects, a major example being those who reached their early teens during the Second World War when the living conditions for Russians were even more severe than usual. One can only guess whether history has been a driving factor but smoking in Russia is strikingly prevalent with about 40 million lighting up – that's 63% of men and 12% of women. There's no national anti-smoking campaign and Russia has not signed the World Health Organization (WHO) Framework Convention on Tobacco Control. To make the outlook even worse, a recent survey of doctors and nurses revealed that they smoke even more than the general population. Historically Russia has had a relatively low rate of breast cancer, the early age of first childbirth being a contributory factor but, in a contrary shift, the trend in the last decade has seen the mortality rate rise to 17.3/100,000. Russia is also slightly at odds with general worldly wisdom in producing one survey showing that a bottle of vodka a day keeps breast cancer at bay – that is, heavy drinkers have a lower breast cancer death rate. Unsurprisingly, this level of drinking greatly increases overall mortality, so that whether the effect is real or not, it is of little relevance to strategies for cancer therapy.

In the UK about 100 people die each day from lung cancer and another 33 from breast cancer. Within the European Union there is a diagnosis of breast cancer every 2.5 minutes and a woman dies from the disease every 7.5 minutes. Altogether in Britain there are over 400 cancer deaths every day. That's roughly the number of people in a typical commuter train with all seats taken and the standing space full – dying. Every day. For students of twentieth century European history the train analogy may have chilling resonances but perhaps that's appropriate given, as we shall see, that cancer is a bit of a mixture, in part due to things beyond our control but with a hefty leavening of what human beings do to themselves and each other.

The rest of the world

We've already seen that there are big contrasts in the prevalent types of cancer across the world, and stomach and liver cancer are not alone in this respect. In fact most forms of cancers show marked variation between countries that differ significantly in what might broadly be called lifestyle. Thus, for example, as recently as 2002 the chances of a woman developing or dying from breast cancer were about three times lower in Japan than in Britain or the USA (Fig. 1.7). Lung cancer, highest incidence, biggest killer in the developed world, is rare in East Africa.

We know that these differences reflect lifestyle because of what has happened when significant numbers of people have emigrated. Thus, for example, after the Second World War a considerable number of Japanese people moved to the USA: within a generation the women had acquired the statistical profile for breast cancer of American women. Similar shifts have been observed in the incidence and mortality of colon

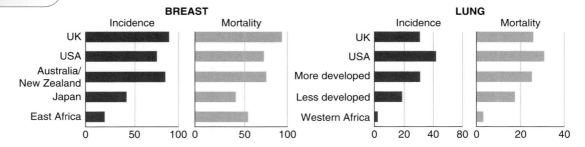

Figure 1.7 **Worldwide variation in breast and lung cancer.** The numbers are rates per 100,000 of the population. The total figures (2008) for incidence were 1,384,155 (breast) and 1,608,055 (lung) and for mortality 458,503 (breast) and 1,376,579 (lung).

cancer following group migrations. Even so, the gap in breast cancer incidence between the USA and Japan is closing, presumably as Western habits swamp traditional Japanese lifestyles.

The picture is, however, complicated by the fact that such differences occur not only between countries but also within national populations. Thus, for example, we noted that in the USA stomach cancer is thirteenth on the list of killers but in 2009, for every white person that this disease claims, 2.3 African Americans will die despite 66% of the population being white and only 14% African Americans. A further 15% of the USA population is Hispanic or Latino, originating mainly from Mexico, Puerto Rico, Central and South America, Cuba and Dominica, 60% of these having been born in the USA. For this group, overall cancer death rates are lower than for non-Hispanic whites (419 versus 574/100,000 in the period 2000–2003). This is because the rates for the major cancers (prostate, breast, colorectal and lung) are lower although Hispanic rates are higher for stomach, liver, cervix, acute lymphocytic leukaemia and gallbladder cancers. This difference appears to be more sustained than occurred in the cohorts of migrants mentioned earlier, but the cancer rates in descendants of Hispanics are nevertheless approaching those of non-Hispanic whites.

Latin America

For the Caribbean (Fig. 1.8) the overall cancer mortality rate is similar to that of South America (around 120 per 100,000) but both are significantly greater than in Central America (92). Taken together the average cancer mortality for Latin America is somewhat below that of North America (Canada and the USA being very similar). As might be anticipated, the rates vary markedly between countries. For men, Uruguay, Argentina and Chile have the highest mortality and for women it is Colombia and Chile. A significant factor is tobacco use, which varies widely across the region but is high in Uruguay where the male lung cancer rate is the highest in Latin America. In Argentina male lung, prostate and colorectal cancers are high but in Chile, Colombia, Ecuador and

Patterns around the world: how many cancers and what sort?

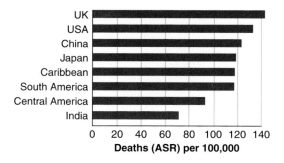

Figure 1.8 Cancer mortality rates in the Americas, China, Japan and India compared with the UK.

Peru stomach cancer is the major male cancer and, for females in those countries, it is also one of the most common, together with cervical cancer. The latter, together with breast cancer, are major causes of female death throughout South America. There is some evidence that stomach cancer deaths may be declining across Latin America as food quality improves. However, the effect of smoking together with the trend towards excess weight and obesity in the entire region is a significant factor in the increasing cancer mortality evident in most countries, Chile being an exception where there has been a slight fall since 1996.

China

In the most populous country in the world, with over 1,300 million people spread over a vast area, the difficulty in obtaining cancer figures is even greater than for elsewhere. Notwithstanding data collection problems, it is clear that by the 1990s cancer had become the second most common cause of death in China, as in so many other countries (Fig. 1.8). In 2008 the major cancers were of the lung, stomach, liver and bowel for both sexes, together with oesophageal cancer for men and breast cancer for women. The rates are rising, there being, for example, 130,000 more deaths from breast cancer in 2005 than there were in 2000 and this is the trend for leukaemia and cancers of lung, liver, bowel and prostate in men, and for breast, lung, liver, bowel and cervical cancer in women. At the same time the incidence of some cancers is declining – notably stomach, oesophagus and nasopharynx – as is mortality from cervical cancer. These improvements are due to the dramatic changes occurring in China as sanitary conditions and diets improve and screening programmes are introduced. The latter are extensive for employees of the government and major companies who provide annual health checks that may be beginning to have an impact on mortality rates for some cancers. On the other hand, as in the Western world, these changes are increasing lifespan and, together with the continuing widespread use of tobacco, the overall result is that cancer rates in China are rising by about 3% each year. Currently lung cancer kills over 1,200 Chinese every day and the WHO estimates that by 2025 there will be a million new cases a year. By then annual tobacco deaths in China will exceed three million.

India

With a population (1,140 million) only slightly smaller than that of China, India is also undergoing a transition to being a 'developed' nation (Fig. 1.8). This means that there is migration from country to city, a general change in lifestyle and a rising life expectancy. Overall cancer rates are lower than in more developed countries (by about three times compared with the USA, for example) despite the rates for oral and oesophageal cancers being among the highest in the world (chewing betel is generally held to be responsible). The most common Indian cancers in men are lung, oesophageal, stomach and larynx; for women they are cervical, breast, ovarian and oesophageal and the total cancer death figure is currently about 600,000 per year. Unfortunately the socio-economic changes taking place mean that, like many other developing nations, the numbers of people dying from non-communicable diseases such as cancer and heart disease are growing.

Africa

We have noted the prediction of 16 million new cancer cases in 2020, 70% being in the developing world. Sub-Saharan Africa will contribute over 1 million of these but, as might be predicted, the patterns across the different regions of Africa show wide variation (Fig. 1.9). The major forms are liver and prostate in men and breast and cervical cancer in women, although AIDS has propelled the otherwise rare **Kaposi sarcoma** to the top of the table in some countries (Uganda, Swaziland, Malawi and Zimbabwe). Seventy per cent of cervical cancers arise from infection by human papillomavirus (HPV). Liver cancer, the major cancer in males, is caused by hepatitis viruses, mainly hepatitis B virus (HBV), which is carried by about 12% of the population. More than 2 billion people, mostly in Africa and Asia, are infected with HBV, accounting for a good deal of the 600,000 deaths a year from liver cancer.

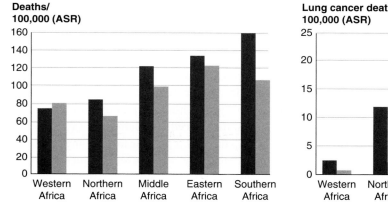

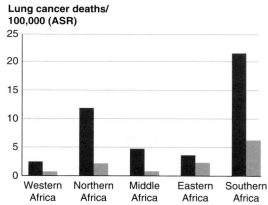

Figure 1.9 Total cancer deaths in regions of Africa and lung cancer deaths in the same regions (men and women).

Although there are now effective vaccines against both HPV and HBV, these have not yet been used on a significant scale in Africa, where the expenditure on health per head of the population is minute compared with developed nations, with the result that when cancers are detected they have usually advanced beyond the treatable stage. As yet there is no hepatitis C virus (HCV) vaccine, and existing therapies are limited, often poorly tolerated and frequently ineffective.

The global data raise the general question of why the incidence of different cancers is so variable. We've noted several factors (tobacco, diet, economics, etc.), to which we will return later, but yet another is infection. Cervical and liver cancers dominate the global picture but are less prominent in the developed countries because most cases occur as a result of infection by viruses.

Trends

One of the most interesting questions arising from all the global cancer data is how are things changing? Most countries in the developed world have produced a gradual decline in their total cancer mortality rate (Fig. 1.10). Nevertheless, Japan has maintained its position of having one of the lowest rates and the UK remains significantly worse than the USA.

One of the most dramatic changes has been brought about by Finland. Up to the 1970s Finland had the world's highest death rate from cardiovascular disease and was also high in the cancer league – mainly due to the factors we'll come back to (heavy tobacco use, high-fat diet and low vegetable consumption). The effect of a national effort to improve these aspects of lifestyle has been to reduce heart disease in men by at least 65% and to reduce cancer mortality to one of the lowest rates in the world, thereby extending the average life expectancy of its citizens by over six years.

Age

Part of the explanation for the worldwide differences we've noted is how long we live. The longer we're around the more likely we are to develop cancer (Fig. 1.11), even though we generally equate longevity with an idyllic lifestyle. For the latter, according to the US Census Bureau, the place to be is Andorra where you'll live to be nearly 84

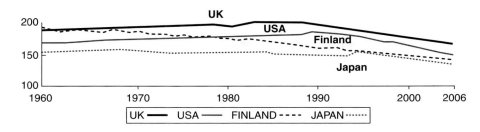

Figure 1.10 Cancer mortality rates from 1960 to 2006 in the UK, USA, Finland and Japan.

Lessons from epidemiology

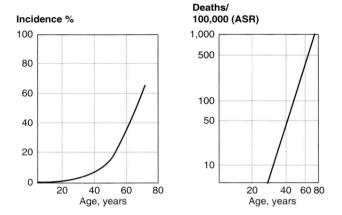

Figure 1.11 The (global) relationship between age and cancer incidence (left) and a double logarithmic plot of cancer deaths versus age (right). (From Nordling, 1953.)

(on average, that is); whereas if you're a male living in some parts of Glasgow 54 is the cut-off age and in Swaziland the average life expectancy is under 32 years. So you would guess, therefore, that not many Swazis die of cancer – and you'd be right: fewer than 5% compared with about 30% of UK denizens. You might also guess that their spectrum of cancers is a bit odd and indeed we've already noted that in countries where the AIDS epidemic is out of control, Kaposi sarcoma is the most common cancer.

Both heart disease and cancers are diseases of old age. Men over the age of 50 who have never smoked are more likely to die of heart disease than anything else and more than two thirds of cancers are first identified in individuals who are more than 65 years old. In general, the average age of populations is steadily rising both in the developed world and in developing countries. In the Bronze Age the average lifespan was 18 years; the average worldwide is now 66 – and that has risen from about 35 in the early part of the twentieth century. So, of course, the heart disease and cancer figures are rising and the prediction that the figure of 10 million new cancer cases in 2000 will rise to 16 million in 2020 is perhaps not so surprising. By then there will be 30 million individuals with the disease, 70% of them in the developing world. Other factors are contributing to this progression, notably trends in smoking and unhealthy lifestyles. Of course, not all cancers are so protracted in development. The entire category of childhood cancers, by definition, only appears in the first 14 or so years of life. In addition, most adults will know of someone who has succumbed to a particularly aggressive cancer in early life. So cancers do strike the young but these are mercifully rare manifestations and, from an overall statistical point of view, cancers are indeed one consequence of getting old.

When either incidence or mortality figures are plotted against age as a double logarithmic plot they give straight lines (of the form $y = a^n$ where a is age). The index n is the gradient of the log-log plot and (n + 1) represents the number of specific events that drive the development of the disease (Fig. 1.11). Between different types of cancer the number varies from four to seven but is typically six, indicating that, on average, acquisition of a major cancer-promoting event requires more than ten years. This type of deduction from cancer statistics was made 60 years ago and was the first evidence that a cancer cell contains a number of mutations that have accumulated over extended periods. The log-log plot shown is for USA white males but similar linear plots can be derived for all major cancers, regardless of country, sex and race. This indicates that although, as we have seen, the incidence of different types of cancer varies widely around the world, similar molecular events drive the development of all cancers.

We now know that cancers are the result of the cumulative effects of mutations in genes that control cell division and repair damaged DNA such that cells are released from the normal controls governing their proliferation and location.

So the major conclusion from this kind of finding is that, to a considerable extent, cancers are self-inflicted in that we have a degree of choice in the way we lead our lives: they are a consequence of our lifestyle – where we live and how we live. The problem with 'lifestyle' is that it is a bit tricky to define what is good and what is not. That notwithstanding, in the context of cancer susceptibility there are one or two basic guidelines, namely, be lucky enough to have a decent standard of housing and sanitation, eat well and keep reasonably physically fit. It's also a good idea to live where airborne pollution isn't an unseen hazard. That may be easier said than done, particularly given the considerable evidence that links auto emissions to breast and lung cancers. Living near a railway is probably OK – just don't get a job maintaining any diesels that might run on it because at least one study shows that significantly increases your risk of lung cancer.

Where do we stand and is there any good news?

It has to be admitted that on a quick glance there doesn't seem to be much to be cheerful about. The predictions for the worldwide increase in cancer rates are almost beyond comprehension and, as we've noted in passing, there isn't anywhere to hide. Certainly not in the UK where in 2008 there were 309,527 new cancer cases, an increase of 54,000 compared with 1994. In particular, there was a 15% increase in breast cancer and sharp increases in bowel and prostate cancer. Of course, part of the reason for the increasing trends is that we are living longer and, as we saw earlier, cancers are mainly diseases of old age. Even so, cancers also afflict the young. The ~156,000 cancer deaths/year makes up ~25% of all UK deaths but that proportion rises to 36% for all deaths under the age of 65 and it's ~50% for deaths in women under 65. This reflects sub-sets of cancer (e.g. childhood cancers and aggressive forms of breast cancer, as we noted earlier) that are relatively rare but have a big statistical impact on an age group that, by and large, doesn't tend to die from other causes.

Some encouragement may be had from the progressively earlier detection of the disease through improved screening programmes (Fig. 1.12) together with the development of effective drug treatments. This has been particularly true for cervical

Lessons from epidemiology

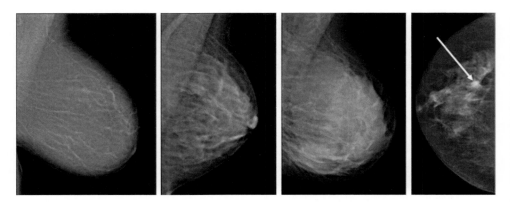

Figure 1.12 Imaging breast tumours. These tumours can be difficult to detect by mammography, that is using X-rays, because the organ is a mixture of fatty tissue, through which X-rays pass relatively easily (dark areas), and denser tissue that absorbs X-rays (white). The three left-hand images show density variation in normal breasts. The right-hand image shows a tumour (arrow) partially obscured by adjacent dense breast tissue, showing how difficult it can be to detect a small tumour by mammography. (Images kindly contributed by Dr Peter Britton, Consultant Radiologist, Addenbrooke's Hospital, Cambridge, UK.)

and breast cancers and has meant that the developed world has seen a steady improvement in the five-year survival rate for breast cancer. Overall in the richest countries about 50% of cancer patients survive the disease. However, in the developed world you are more than twice as likely to be diagnosed with cancer than elsewhere, with the result that in developing countries 80% of initial diagnoses are of late-stage, incurable tumours.

Thus generally the world needs to improve the screening programmes referred to earlier, but we could also do much more by way of prevention. In developed countries only about 8% of cancers arise from infections but elsewhere almost one quarter are caused by agents such as HBV and HCV (liver cancer), HPV (cervical and ano-genital cancers) and bacteria. Thus, for example, 80% of deaths from cervical cancer are in developing countries. We noted the efficacy of HBV and HPV vaccines and it is to be hoped that these will become sufficiently available to make a major impact on the incidence of the cancers promoted by these viruses. Despite these wonderful contributions by science, it is difficult to be optimistic about controlling cancer in Africa, given the economic and logistical problems. The setting up in 2008 of AfrOx to bring UK expertise to bear on the African cancer problem is at least one encouraging step.

In the UK the picture is somewhat mixed. Better diagnosis and treatment has produced an annual decline in cancer deaths since 1983. The long-term survival rate is ~40% although the figure varies widely across the ~200 distinct forms of cancer (Fig. 1.13). Of these, as we noted earlier, four types (breast, bowel, prostate and lung) account for ~50% of adult cancer deaths in the UK. In the period from 1984 to 2004 the five-year percentage survival rates for the first three of these cancers improved, prostate from 42 to 74%, breast from 59 to 81% and bowel from 40 to 50%. For lung cancer the rate remained unchanged at 5%. The most dramatic improvements of all have been for testicular cancer and some childhood cancers where the five-year survival rate has gone from being very low 20 years ago to >90%. These large

Where do we stand and is there any good news?

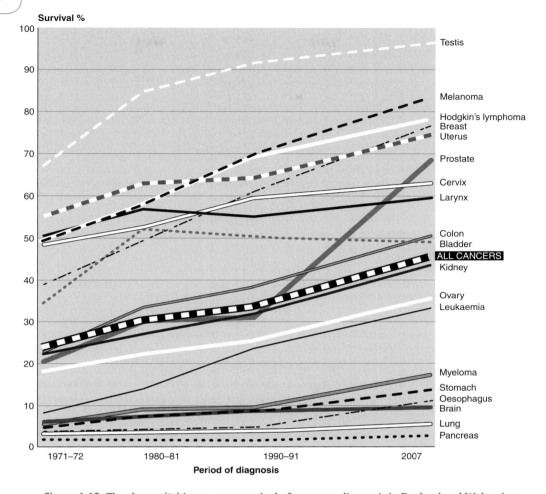

Survival %

- Testis
- Melanoma
- Hodgkin's lymphoma
- Breast
- Uterus
- Prostate
- Cervix
- Larynx
- Colon
- Bladder
- **ALL CANCERS**
- Kidney
- Ovary
- Leukaemia
- Myeloma
- Stomach
- Oesophagus
- Brain
- Lung
- Pancreas

Period of diagnosis

Figure 1.13 The change (%) in ten-year survival after cancer diagnosis in England and Wales since 1971. Figures for 1971 to 1991 are actual survival; those for 2007 are predicted. (Coleman, Cancer Research UK, http://www.cancerresearchuk.org/cancer-info/cancerstats/survival/latestrates/#Ten [accessed August 2012])

differences in the survival rates and in the impact that medical research is having arise from the fact that although, as we shall see, there are some basic features that apply to most if not all cancers, distinct molecular mechanisms drive the diseases and they arise in different types of cell. For these reasons no single therapeutic regime is likely to be fully effective even against one cancer type yet alone against all. The one advantage of this diversity from the point of view of devising therapeutic strategies is that cancers present many potential targets for slowing or reversing their progress.

So there have been some advances in the UK, but in terms of detection and treatment there remain some major problems that have more to do with the health system in Britain than with the biology of cancer. Despite the improvements mentioned above, UK five-year survival rates for common cancers are worse than the European average

Lessons from epidemiology

by 5 to 15%. Considering breast cancer, this gulf is even greater for deaths within six months of diagnosis. There are three possible explanations for this: (1) later diagnosis; (2) more aggressive forms of the disease in Britain; (3) lower standard of treatment. It seems likely that all three play a part and tellingly, a 2001 review delicately concluded that 'it is difficult to refute' the suggestion that breast cancer care in the UK has been non-uniform and sometimes inadequate. Comparison with the USA for breast cancer tells a similar story that is mirrored by other cancers. Thus the prostate rate of death in the USA is falling four times faster than it is in the UK: the real difference may be less than this due to differences in ways of attributing the cause of death but nevertheless there is a significant difference.

Perhaps the most important point behind all the incidence and death figures is how well people do after diagnosis of cancer. In terms of five-year survival after diagnosis the rates for all cancers in Europeans are significantly worse than in the USA by about 47% versus 66% for men and 56% versus 63% for women, respectively.

Equally perturbing is the fact that where you live in the UK bears significantly on your cancer risk. The National Cancer Intelligence Centre has produced a Cancer Atlas that compares incidence and death rate from the 21 most common cancers in different counties of the UK. The differences reflect levels of smoking, drinking, poor diet and social deprivation and show that regions of northern England and Scotland are cancer 'hot spots'. Their estimate is that if the worst areas could be converted to the best there would be 25,000 fewer new cases and 17,000 fewer deaths a year: with about 156,000 cancer deaths per year that would represent an 11% decrease.

One of the problems, of course, is that patients from poor backgrounds are more likely to be diagnosed with cancer at a later stage and thus adversely prejudice the efficacy of treatment. This is almost certainly the reason for the disparity between the breast cancer death rates of blacks and whites in America (Fig. 1.14). However, more

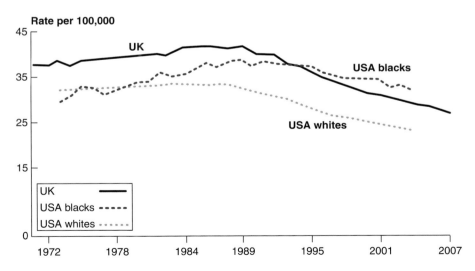

Figure 1.14 Decrease in US and UK breast-cancer mortality rates from 1970 to 2007 (Age Standardised).

subtle factors can also be involved. Thus a report of 2001 (CancerBACUP) noted that, although most breast cancer treatment centres screen tumours for the presence of oestrogen receptors, the methods and interpretation of the data varied so widely that a significant number of women were receiving unsuitable drug treatment. These problems are a strong argument for specialist treatment centres.

Conclusions

The cancers together with heart disease are responsible for two thirds of all deaths worldwide. In developed parts of the world the last three decades have seen a general trend to longer five-year survival times. However, this reflects substantial improvements in the detection and treatment of some types, notably breast cancer, and masks the fact that the prognosis for other forms, particularly lung and pancreatic cancers, remains extremely poor, notwithstanding the application of immense scientific effort, ingenuity and money. As the world population and average life-span rises, so too does the cancer burden. In less developed regions infection by hepatitis and papillomaviruses continues to contribute to the huge number of liver and cervical cancers while tobacco use, causing the majority of lung cancers, is predicted to carry on rising. Inevitably in poor regions screening programmes remain limited so that by the time of detection there are usually few treatment options.

This review of the numerical background has already touched on some of the major causes of cancer. With the awe-inspiring statistics in mind, we'll now look at these in a little more detail to identify what it is we're trying to outwit and whether we can exert a measure of control over any of them. We should bear in mind that the analysis of cancer deaths in age groups revealed the accumulation of distinct events – mutations in DNA – as being the critical driving force. However, we know that there are subtle differences between each human being in their genetic make-up so we should prepare ourselves for the revelation that establishing cause and effect is often far from straightforward when the 'effect' is cancer.

Key points

- There are nearly 13 million new cases of cancer every year and these diseases kill nearly 8 million people worldwide every year.
- There will be 16 million cases by 2020.
- The continuing upward trend in the cancer burden is predominantly driven by increasing average life-span but unhealthy lifestyle is also a major factor.
- The five major forms of cancer contributing to worldwide mortality are lung, stomach, liver, colorectal and breast.

Lessons from epidemiology

- In the developed world substantial increases in the five-year survival rates have been achieved over the past 20 years for, in particular, breast and prostate among the major cancers.
- For other cancers, notably that of the lung, very little therapeutic progress has been made.
- Prognosis varies widely across the world, a major factor being the stage at which cancers are first diagnosed.

Future directions

- Despite the gradual decline in cancer deaths over the last 20 years in most developed countries, the global cancer burden is rising dramatically. As early detection and surgery remains the most effective treatment for most cancers, improved screening programmes, especially in Africa and other developing regions, are perhaps the most attainable of aims.
- The variation in treatment outcome in different regions of, for example, the UK suggests the desirability of concentrating resources into major centres of expertise, as has been proposed for cardiovascular disease.

Further reading: reviews

PATTERNS AROUND THE WORLD: HOW MANY CANCERS AND WHAT SORT?

Coleman, M. P., Quaresma, M., Berrino, F. and the CONCORD Working Group. (2008). Cancer survival in five continents: a worldwide population-based study (CONCORD). *The Lancet Oncology* 9, 730–56.

Jha, P. (2009). Avoidable global cancer deaths and total deaths from smoking. *Nature Reviews Cancer* 9, 655–64.

UK AND USA

Jemal, A., Siegel, R., Ward, E. *et al.* (2008). Cancer statistics, 2008. *CA: A Cancer Journal for Clinicians* 58, 71–96.

Kort, E. J., Paneth, N. and Vande Woude, G. F. (2009). The decline in U.S. cancer mortality in people born since 1925. *Cancer Research* 69, 6500–5.

EUROPE

Verdecchia, A., Francisci, S., Brenner, H. and the EUROCARE-4 Working Group. (2007). Recent cancer survival in Europe: a 2000–02 period analysis of EUROCARE-4 data. *The Lancet Oncology* 8, 784–96.

Further reading: reviews

CHINA

Qin, X.-J. and Shi, H.-Z. (2007). Major causes of death during the past 25 years in China. *Chinese Medical Journal* 120, 2317–20.

INDIA

Anderson, S. R., McDonald, S. S. and Greenwald, P. (2003). Cancer risk and diet in India. *Journal of Postgraduate Medicine* 49, 222–8.

AFRICA

Lingwood, R. J., Boyle, P., Milburn, A. *et al*. (2008). The challenge of cancer control in Africa. *Nature Reviews Cancer* 8, 398–403.

TRENDS

Purushotham, A. D., Pain, S. J., Miles, D. and Harnett, A. (2001). Variations in treatment and survival in breast cancer. *The Lancet Oncology* 2, 719–25.

Cancer statistics websites

GLOBAL

CancerBACUP: www.jamkit.com/Clients/Nonprofits/CancerBACUP
Center for Communications, Health and the Environment: www.ceche.org/
International Agency for Research on Cancer (IARC): http://globocan.iarc.fr/
Union for International Cancer Control (UICC): www.uicc.org/
World Health Organization (WHO): www.who.int/cancer/en/
WHO, Global Infobase, Country Profiles: www.who.int/infobase/report.aspx?rid=126

UK AND USA

American Cancer Society: www.cancer.org/
Cancer Research UK: http://info.cancerresearchuk.org/cancerstats/
National Cancer Institute: www.cancer.gov/

EUROPE

European Cancer Observatory: http://eu-cancer.iarc.fr/

LATIN AMERICA

The Pan American Health Organization (PAHO): http://new.paho.org/

2 | Causes of cancer

The major causes of cancer fall into two categories: those over which we have some control and the rest. The latter includes radiation from the Earth that has been a background to human evolution and about which we can do nothing, although we could take more effective steps to limit the accumulation of radon in houses in regions where there is high, localised emission of the gas. The factors that we can control also fall into two groups: those exerting major effects for which the epidemiological evidence is overwhelming and those for which the data are inconclusive and therefore controversial. Of the former, the most familiar is the use of tobacco, which is responsible for 90% of lung tumours. By contrast, there are a number of prominent agents that may have weak tumour-promoting effects. For these the epidemiology is generally unpersuasive and direct experimental evidence has not been forthcoming, as exemplified by the continuing debate over the risks associated with the use of mobile phones. The two most effective measures we could take that would reduce the global cancer burden by at least one third would be to abolish the use of tobacco and to limit the consumption of red meat.

Introduction

In this chapter we'll look at the major external factors that have been incriminated as causes of cancer. Several of these – radiation, tobacco and alcohol – are so well known that, rather than repeating the basic statistical evidence, we'll say a little about the mechanism and other aspects of their involvement. For magnetic fields, however, which have been the subject of much media publicity, cause and effect remains unproven and it's interesting to consider why it has been so difficult to come up with a clear answer about a connection. In some respects the matter of what we eat has been equally intractable. Defining a 'good diet' is easy, but pinning down precisely what is 'bad' has turned out to be tricky and we'll say a little about the difficulties associated with large-scale epidemiological studies. It turns out to be a bit like the chap who is said to have eaten 25,000 'Big Macs': the worst of his problem is likely to be the fruit and vegetables he's missed out on rather than the polar bear's weight of fat he's eaten.

Let us, though, commence with a rather neglected cancer cause – infection – and we'll come back to one of its consequences in more detail in Chapter 5.

Infection

In the previous chapter we noted that the majority of cervical and liver cancers are initiated by viruses. In fact nearly 20% of all cancers worldwide are caused by infection, either by viruses, bacteria or other microorganisms. In developing countries this figure rises to about one in four. While viral transmission between individuals is a major factor, living conditions and especially sanitation also bear on cancer, partly because chronic infection is likely to decrease the efficiency of the immune system. Quite how the immune system targets tumour cells at the molecular level remains controversial but it is unquestionably one of our anti-cancer defences and its suppression can release one of the brakes on tumour development.

The impact of microbial infection is most clearly established for stomach cancer in response to *Helicobacter pylori*, the bacterium mentioned earlier that gives rise to chronic infection and ulcers in the stomach. It is mainly spread by the consumption of food or water contaminated with faecal matter and it can double the risk of developing gastric cancer although the most potent strains may increase the risk by 30-fold. About a quarter of us pick up *H. pylori* and there are effective drug combinations for treating this infection. Given that most of those infected show no symptoms and stomach cancer is an uncommon outcome, we're clearly quite good at keeping it under control ourselves.

Another bug, the tuberculosis (TB) bacterium (*Mycobacterium tuberculosis*), not only kills two million people a year but is present in latent form in about one in three of us. Effective drug therapy is available and the mortality rate is less than 5%. It's been known for many years that patients with TB have an increased frequency of lung cancer, possibly as a result of inflammation, and recent evidence from mice indicates that chronic infection can indeed initiate the development of malignant lung cancer.

Although, as we shall see, cancers essentially arise from genetic defects, microbial or viral infection can act as an initiating event and we will consider some of the molecular mechanisms by which this occurs in Chapters 5 and 6.

Radiation

Everything that we see in the world around us comes, quite literally, in the form of electromagnetic radiation, an oscillating electric and magnetic field that travels at the speed of light (Fig. 2.1). The electromagnetic spectrum is the range of all the frequencies (or wavelengths) of electromagnetic radiation. This includes a tiny region that our eyes can detect (the visible spectrum) as well as radio waves and X-rays. At shorter wavelengths is 'ionising radiation', which includes gamma rays, produced by subatomic particle interactions (e.g. in radioactive decay), and X-rays that result from high-speed electrons colliding with metal. Alpha and beta particles are also forms of

Causes of cancer

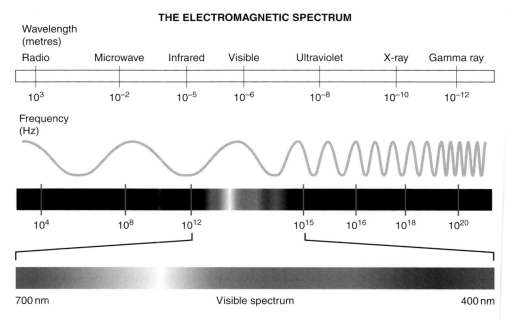

Figure 2.1 The electromagnetic spectrum. The wavy line shows the relationship between frequency (ν) and wavelength (λ). One complete oscillation of the field (called the period) is the distance between two adjacent peaks. The wavelength of a complete cycle is related to the frequency by the speed of light, c ($\lambda = c/\nu$). Ultraviolet (UV) radiation (i.e. 'beyond' violet light in the sense of being of shorter wavelength and hence higher energy) occupies the wavelength bands from 400 to 315 nm (UVA), 315 to 280 nm (UVB) and 280 to 100 nm (UVC). 1 nm (nanometre) $= 10^{-9}$ m. (See plate section for colour version of this figure.)

ionising radiation released when unstable isotopes undergo radioactive decay. Alpha particles are made up of two protons and two neutrons; beta particles are electrons emitted by nuclei.

Ionising radiation

Ionising radiation and ultraviolet (UV) radiation are important because they can damage living tissues, most significantly by causing mutations in DNA that can lead to cancer, as we shall discuss later. We have evolved against a background of natural radioactivity that comes from radioactive elements in the earth and rocks, trace amounts of which are also present in food and water, and in radiation from space. On average, about 87% of the ionising radiation that hits humans comes in these unavoidable forms. The rest is artificial radiation of one sort or another, including X-rays used in medicine.

By way of reassurance before confronting the more extreme forms of radiation, it is instructive to spend a moment considering potassium, which is important for generating the electrical signals relayed by nerve cells (an **action potential**).

Table 2.1 Doses from medical radiation sources.

Procedure	Effective dose (mSv)
Single radiograph (X-ray): limb	0.06
Single radiograph (X-ray): chest	0.06
Single radiograph (X-ray): abdomen	0.7
Intravenous pyelogram (kidneys: 24 images)	2.5
Barium meal (11 images)	3.0
Computed tomography (CT) (head)	2.0
Computed tomography (CT) (abdomen)	10.0
Coronary angiogram	5.0–15.0
Mammogram	0.13
Brain scan (^{99m}Tc)	7.0
Bone scan (^{99m}Tc)	4.4
Tumour scan (^{67}Ga)	12.2
Tumour scan in pregnancy (^{67}Ga): foetal dose	18.0

Effective doses are given in millisieverts (1 mSv = 1/1000 of a Sv).
Radiation dose is also sometimes measured in rem (1 Sv = 100 rem).
^{99m}Tc (technetium-99m) and ^{67}Ga (gallium-67) are radioactive isotopes.

Potassium is the major radioactive emitter in our bodies and there are three forms (isotopes) that occur naturally: ^{39}K and ^{41}K (together comprising over 99.9%) and ^{40}K. The first two are stable but ^{40}K is radioactive and has a **half-life** of 1.3 billion years – so it's a good job that it only makes up 0.012% of our potassium. We can't avoid it because it's in the earth – though we increase the amount we eat by using fertilisers that contain nitrogen, phosphorus and potassium, which adds several thousand curies a year to US soil. This means that the fruit and vegetables considered essential in a good, cancer-protecting diet contain lots of potassium, including the radioactive fraction. Because cows eat grass, if you drink milk you will be downing about 74 Bq in every litre (or 2×10^{-9} Ci).

Radiation exposure can also occur in diagnostic medical procedures – mainly X-rays and computed tomography (CT). In a conventional X-ray the radiation is directed at part of the body and what passes straight through is collected on a photographic film, which when developed gives a two-dimensional image of the sort familiar to almost anyone who has ever broken a bone. Computed tomography also uses X-rays to acquire two-dimensional images but, from a large number of such images taken as the radiation beam moves through the body, a three-dimensional picture can be pieced together.

The biological damage caused by radiation is measured in units called the sievert (Sv). A typical chest X-ray requires a dose of ~0.04 mSv; for a corresponding CT scan the figure is 8 mSv (Table 2.1). These might be compared with our annual dose of 'unavoidable' natural radiation, which is about 3 mSv. So, in general, it is probably safe to say that these typical medical exposures are not going to have harmful effects. Nevertheless, we should bear in mind that some conditions may require multiple scans. Currently in the USA there are about one million CT scans and worldwide about two billion medical exposures every year.

Radiation is, of course, one form of cancer therapy, the strategy being to target and kill the cells of the tumour. For this, much higher exposures are required, which has driven the development of increasingly sophisticated machines that can target precisely the contours of the tumour.

The first major case of accidental exposure to radiation came in the years following the First World War as a result of the commercial use of radium to make luminous dials for watches and instruments. Women employed by the US Radium Corporation to apply the radium-containing paint were instructed to lick their brushes to maintain their shape (of the brushes, obviously). As the dial painters began to fall ill and indeed die it became clear that the cause lay in their working conditions. Eventually five 'radium girls' sued US Radium. The high-profile case focused public attention on the hazards associated with radium, a surprising necessity given that the element had been discovered as long before as 1898. It was in fact radium radiation that was to kill one of its discoverers, Marie Curie, in 1934. Radium (^{226}Ra) is a decay product of uranium, has a half-life of 1,602 years, and decays to a number of short-lived products, one of which is the gas radon. Despite all this, and quite amazingly, concoctions of radium were avidly marketed in the USA in the early 1900s as a treatment for everything from high blood pressure to stomach cancer, and only ceased to be sold in 1931.

The most extreme example of exposure to artificial radiation came, of course, from the atomic bombs dropped on Hiroshima and Nagasaki in 1945. These caused immediately about 120,000 deaths from the explosions together with the extremely high levels of nuclear radiation released in the form of gamma rays and (fast) neutrons. Although this killed almost everyone within one kilometre of the epicentre, the radiation level subsequently fell so rapidly that within one week it had returned to normal. This dramatic fall may in part account for an unexpected long-term consequence of the atomic bombings, namely that in those who survived because they were more distant from the epicentre but had been transiently exposed to high levels of radiation, the overall incidence of radiation-induced cancers was very small.

Of 48,000 atomic bomb survivors who were exposed to at least 5 mSv, some 5,900 died between 1950 and 2000 as a result of developing solid tumours but of these only 480 (~8%) were attributable to radiation. Furthermore within the exposed group of survivors there has not appeared significant evidence for mutations caused by radiation being passed to off-spring (that is, for the acquisition of germ line mutations). A much higher percentage of cancers of the blood (**leukaemias**: Fig. 2.2) was attributable to radiation (~50%) but the number of such cancers was very small (<4% of the total). All of which suggests that although transient exposure to high levels of radiation may cause mutations, the body is quite good at repairing the damage.

The worst nuclear accident so far has been the explosion of a reactor at the Chernobyl power plant in 1986, which blew a large amount of radioactive material into the air – 1.1×10^{19} Bq in fact, some 400 times that released by the Japanese bombs. Although 60% fell on Belarus, radioactivity from the cloud was subsequently detected in most countries of Northern Europe and as far away as North America. To this day sheep from some 330 farms in North Wales have to be scanned for radiation before they can be dispatched to market. The main products of nuclear fission released were caesium-137 (half-life 30 years), iodine-131 (half-life 8 days) and

(a)   (b)   (c)   (d)

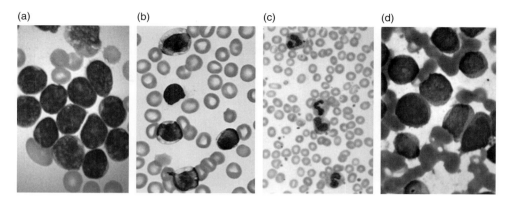

Figure 2.2 **Leukaemic cells.** Leukaemias arise in the blood-forming tissues, the bone marrow and lymphatic system, giving rise to abnormally high levels of circulating cells. These are usually leucocytes (white blood cells). (a) Acute lymphocytic leukaemia (ALL). The cells stained purple are white blood cells (lymphocytes) at a very immature stage (called blasts): they have very large nuclei and little cytoplasm. The majority of the cells are normal red blood cells. ALL is the most common type of childhood cancer. (b) Chronic lymphocytic leukaemia (CLL). The slide shows relatively high numbers of mature lymphocytes (purple nuclei). CLL most commonly occurs in older adults. (c) Chronic myeloid leukaemia (CML): marked leucocytosis (i.e. raised white blood cell count) with increased numbers of precursor neutrophils. (d) Acute myeloblastic leukaemia (AML): a bone marrow smear showing a high proportion of blast cells (immature precursors of granulocytes) of large size and with visible nucleoli. Acute erythroblastic leukaemia is a variant of AML in which high levels of immature red blood cells are produced. (Source: http://commons.wikimedia.org/wiki/) (See plate section for colour version of this figure.)

strontium-90 (half-life 29 years). These decay by beta and gamma emissions and thus constituted a health hazard to humans on the receiving end of the fall-out. There's a degree of irony in this because all have been used in medicine: iodine and strontium isotopes specifically in the treatment of cancer. Iodine is taken up by the thyroid gland (and used to make hormones, e.g. thyroxine): because thyroid cells are the only ones in the body that can absorb iodine, patients with thyroid cancer are given radioactive iodine orally after surgical removal of the gland to kill any residual cells and hence ensure the cancer does not recur (they require hormone tablets thereafter to compensate for the loss of their thyroid). The problem with the aftermath of Chernobyl was the uncontrolled uptake of iodine-131 that caused thyroid tumours to develop in a large number of young children (Figs. 2.3 and 2.4). The comprehensive reports (Chernobyl Forum) on the accident found that just over 50 people were killed directly and that by 2005, of the ~600,000 people most exposed to fall-out, there had been some 4,000 cancer deaths attributable to radiation, most of these being thyroid tumours. These reports also conclude that there has been no increase in inherited birth defects or in other types of solid cancers as a consequence of the fall-out, a result consistent with the Japanese atomic bomb data. They are also consistent with studies of wildlife in the vicinity of the reactor, specifically comparing populations of voles with those in uncontaminated areas. These showed that a lifetime's exposure to

(a) (b)

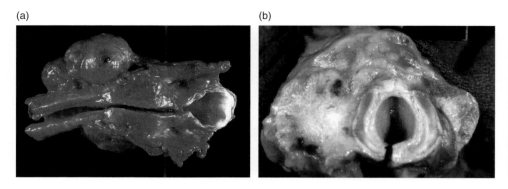

Figure 2.3 The thyroid gland. The normal thyroid gland has two small lobes that wrap around the trachea. (a) Shows an opened trachea with a goitre, an enlargement of the thyroid most frequently caused by iodine deficiency. (b) Shows a cross-section through the thyroid and trachea of a metastatic **carcinoma**. (© Dr Peter Anderson, University of Alabama at Birmingham, Department of Pathology.) (See plate section for colour version of this figure.)

(a) (b) (c)

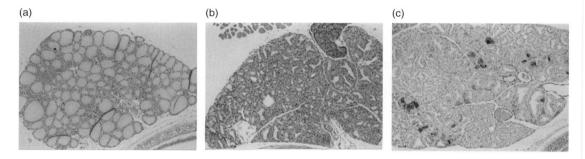

Figure 2.4 A tissue section of a normal and an enlarged thyroid gland (a mouse goitre). Goitres can be caused by low iodine levels and they increase the chance of developing thyroid cancer. (a) and (b) have been stained with haematoxylin and eosin. (a) Normal thyroid architecture with a portion of the trachea. (b) A chemically induced goitre, showing diffuse thyroid hyperplasia with part of the parathyroid. (c) A section of a goitre that has been treated with an anti-vascular drug, which has caused localised damage manifested as multiple thrombi that stain positive for von Willebrand factor (brown). (Griggs *et al.*, 2001.) (See plate section for colour version of this figure.)

about 10 mSvc of radiation a day caused no detectable DNA damage in the animals. The simplest explanation for these somewhat surprising findings is that exposure to low levels of radiation may activate DNA repair mechanisms and thus confer protection.

Radon

In most parts of the world the largest source of ionising radiation is radon, a product of radium. Radon-222 is an inert gas that has a half-life of four days and arises from radioactive decay of uranium-238, which is present throughout the Earth's crust.

The average radon concentration in British homes is $21\,Bq/m^3$ and the estimate is that 1,110 lung cancer deaths per year (about 3.3% of lung cancer deaths) are caused by this background source of alpha particles (Fig. 2.5). In the European Union the figure is 20,000, in the USA 21,000 and worldwide one million deaths are caused by radon, with smokers being more vulnerable. Significant radon concentrations normally accumulate only inside buildings but this can readily be prevented by installing a sealed membrane at ground level. Although there is a statutory level at which this preventative measure must be applied ($200\,Bq/m^3$), the Gray report (Gray *et al.*, 2009) estimates that 85% of deaths caused by radon arise from lower levels for which preventative measures are not legally required. Much higher concentrations of radon can, however, be encountered in some thermal springs, notably those in Boulder, Colorado and Lurisia in Northern Italy.

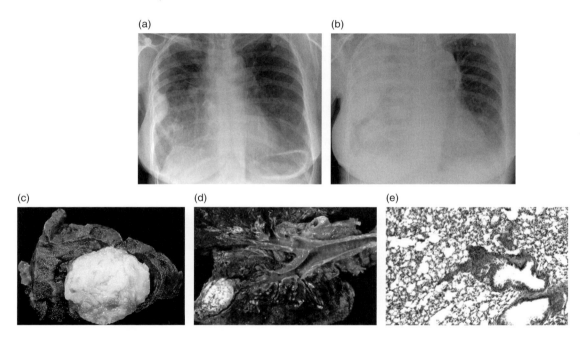

Figure 2.5 **Lung tumours.** (a) and (b) Two chest X-rays of the same female patient showing (a) a tumour at an early stage of development and (b) at an advanced stage. The tumour is a **metastasis** to the pleural space from a gynaecological primary. Images kindly contributed by Dr Ferdia Gallagher, Department of Radiology, Addenbrooke's Hospital, Cambridge and Cancer Research UK Cambridge Research Institute. (c) A pulmonary hamartoma, the most common form of **benign** lung neoplasm. The yellow and white tissue of the well-circumscribed hamartoma is fat and cartilage, respectively. (d) Lung primary carcinoma (at base of left lower lobe). Many tumours metastasise to the lung and secondary growths are more common than primary lung tumours. (e) Lung section stained with haematoxylin and eosin (H & E) showing numerous alveoli and two bronchioles. In the centre is a nest of tumour cells that have metastasised from a blood vessel. These have the characteristic dark purple colour of hyperchromatic cancer cells stained with H & E. (© University of Alabama at Birmingham, Department of Pathology.) (See plate section for colour version of this figure.)

Ultraviolet radiation

The energy carried by the UV region of the electromagnetic spectrum is ~4 **electron volts** (eV), which means that it is capable of breaking covalent bonds and thus damaging biological molecules either directly or indirectly (by generation of free **radicals**). Most of the radiation reaching the Earth from the Sun is absorbed by the ozone layer, and nearly all the UV radiation that reaches us is UVA (98.7%). Nevertheless, UVB is important because it is responsible for producing vitamin D in the skin, a deficiency of which causes a high proportion of premature deaths and has been linked to many diseases including cancers.

Thus UV radiation is necessary for good health but it is also a **carcinogen**, so its effects become a matter of balance. As a contribution to this equilibrium we make melanin pigments in our skin that absorb UV. The cells that do this are melanocytes (they're also present in the eye and the bowel). They make both black eumelanin and a second type of pigment, phaeomelanin, that is red. The relative amounts of these two pigments determine our hair and skin colour. When we get sun tanned it's because our skin makes eumelanin as a protection against UV light. Dark-skinned people make more eumelanin and are thus better protected against skin cancer than their brethren in northern climes whose lighter skin may have evolved as a response to lower levels of sunshine, thereby boosting their vitamin D production.

Melanocytes have come to prominence in the cancer field in recent years because their uncontrolled growth gives rise to a malignant tumour called a melanoma (Fig. 2.6).

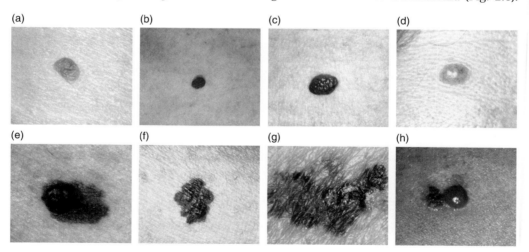

Figure 2.6 Moles and melanomas. (a) to (d) Four normal moles. These are benign birthmarks, a category that also includes café au lait spots, light brown areas that may be associated with the genetic disorder neurofibromatosis type I. (e) to (h) Melanomas ((e) to (g)) and an advanced malignant melanoma (h). Melanoma is a malignant tumour of melanocytes and usually arises in moles or in skin of normal appearance. The most common type of skin cancers are basal cell carcinomas, one of the two major groups of non-melanoma skin cancers (the other being squamous cell carcinoma). (Photographs courtesy of the Skin Cancer Foundation with the permission of the National Cancer Institute.) (See plate section for colour version of this figure.)

Melanoma is a relatively rare form of skin cancer – the non-melanoma skin cancers basal cell **carcinoma** and **squamous** cell carcinoma are the most common – but it is the most serious. It causes 75% of skin cancer deaths of which worldwide there are 48,000 per year with over 160,000 new cases. In the USA there are over 68,000 new malignant melanoma cases and 8,700 deaths every year, the corresponding UK figures being 8,000 and 1,800, respectively. In Britain and the USA it is the second most common cancer in young people (aged 15 to 34) and the incidence is rising by 1% per year in both countries. White Americans are almost 30 times more likely to develop melanoma than African Americans. For non-melanoma skin cancers UV exposure appears to be a major cause, and having had this condition increases the risk of malignant melanoma. Because malignant melanomas develop from moles on our skin they are the easiest cancers to detect at an early stage and if you have any such mark on your skin that changes colour, size or shape you should seek medical advice. The encouragement to do so is that, if identified early, melanomas can be treated by surgery alone and the prognosis is then excellent (five-year survival >95%).

Low-frequency magnetic fields

In most countries the electrical power systems run at a frequency of 50 Hz and at about 230 V (60 Hz and 120 V in the USA). This means that most of us, whenever we are indoors, are surrounded by a network of wires carrying alternating currents that induce corresponding alternating electric and magnetic fields. Over the last 25 years a question that has frequently surfaced in the media is whether exposure to these electromagnetic fields (EMFs) can contribute to the development of cancers.

Concern over EMFs was aroused in 1979 by a series of epidemiological surveys that attempted to assess the strength of domestic fields and concluded that people living in houses with higher field strengths were more likely to develop cancer. The first such survey was American and the influential successors were predominantly Scandinavian. The emotional temperature of the ensuing debate was undoubtedly elevated by the fact that the main cancer for which EMFs were incriminated was childhood leukaemia. Because childhood leukaemia is a very rare condition – there are about 500 new cases a year in the UK and about 2,200 in the USA – it is difficult to associate with causative factors. A second problem is that there is no known mechanism by which low-energy fields can affect biological processes. That is, unlike X-rays or even UV radiation, the energy imparted by EMFs is too low to change the structure of molecules – it cannot directly cause mutations in DNA. The most plausible mechanism advanced so far is that magnetic fields could alter the lifetime of free radicals – highly reactive species that are generated as part of the normal metabolism of living cells. Free radicals do have the potential to initiate tumour development because they can mutate DNA. However, although magnetic fields have been shown to perturb free radical formation in chemical reactions, there is no evidence that this happens in biological systems. Moreover the field strengths required to influence chemical reactions are about 1,000 times higher than those we experience from environmental EMFs.

Public concern has led to the establishment of independent bodies both in the USA (EMF RAPID Program) and the UK (The EMF Trust) whose remit is to support and coordinate high-quality research into EMF effects and to promote independent efforts to replicate any promising findings. Thus far there is no convincing experimental evidence that has been independently replicated showing that EMFs cause cancer.

High-frequency magnetic fields (mobile phones)

A similar concern arising from radiation of higher frequencies has developed with the widespread use of mobile phones. These devices work in the ultra-high frequency (UHF) radio frequency (RF) range of the electromagnetic spectrum (300–3,000 MHz, wavelength 10–100 cm). Even at these wavelengths the energy of radiation is only about seven millionths of an electron volt (7 μeV). Because it requires ~ 1 eV to break the weakest chemical bonds in DNA it seems out of the question that such radiation has direct effects on genetic material.

The major concern over mobile phones has been the suggestion that their use increases the risk of two types of brain tumour, **acoustic neuroma** (Fig. 2.7) and glioma, and that youngsters are particularly susceptible. As with childhood leukaemia, a major problem is that these cancers are rare in adults (one in 100,000 for acoustic

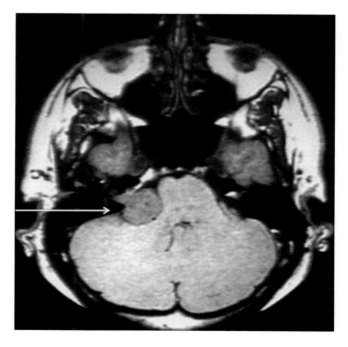

Figure 2.7 A vestibular schwannoma visualised by magnetic resonance imaging (MRI). Often called an acoustic neuroma, these benign tumours arise in the internal auditory canal but may expand until they compress the brainstem, as shown in this image (arrowed spherical mass). The tumours grow slowly and are often treated conservatively by annual MRI monitoring. Tumours can be removed by surgery or radiation treatment. (© Slice of Life and Suzanne S. Stensaas Peir digital library.)

neuroma and one in 30,000 for glioma) and even rarer in children. The events that cause these tumours therefore occur infrequently and it is difficult to show that using a mobile phone increases their probability. The ideal statistical way to tackle this sort of problem is to study a large number of people. The biggest survey so far, from Denmark, was of 420,000 people and concluded that there was no link between mobile phone use and any kind of cancer. Those who had used mobiles for over ten years showed no increased incidence of brain tumours and there was no trend of cancer development with time after first subscribing to a phone. Another Scandinavian study showed no significant increase in the incidence of brain tumours between 1974 to 2003 (when there were 59,984 cases), despite the fact that from the mid-1990s on the Nordic populace took to mobile phones in a big way.

More recently still the biggest survey yet to be published came from The INTERPHONE Study Group (2010). Covering 13 countries it was set up by the International Agency for Research on Cancer (IARC) to address the question of whether mobile phones increase the risk of brain tumours within the first 10 to 15 years of use. The authors concluded that 'Overall, no increase in risk of glioma or meningioma was observed with use of mobile phones'. The large cohorts in these three studies give considerable weight to their conclusions by comparison with surveys that *have* shown a link between mobile phones and cancer, in which the number of cases was generally very small (mostly fewer than 40).

It is appropriate to leave the last word on the risks of mobile phones to the exhaustive UK Government-commissioned Stewart report (IEGMP, 2000), an analysis by an independent group of experts of both the epidemiology and the experimental evidence relating to biological effects of mobiles. The overall conclusions were that the experimental evidence from studies on cells and animals does not suggest that mobile phone emissions, within existing International Commission on Non-Ionizing Radiation Protection (ICNIRP) guidelines, have any adverse effects on health. Furthermore the epidemiological evidence does not suggest that mobile phone emissions cause cancer nor is there any risk to people who live near base stations. Nevertheless, the period over which cancers may develop is long compared with that for which mobile phones have been in widespread use and there *is* evidence that mobile phone emissions may interfere with the electrical activity of the brain. For these reasons the report recommended continued research into this question.

Tobacco

These days, at least in America and Europe, we are all aware that to keep cancer at bay avoiding tobacco smoke is as important as sensible eating. The estimate for the USA is that 30% of cancers are caused by smoking, with poor diet accounting for 25%. Ninety per cent of lung cancers are attributable to smoking and the World Health Organization has estimated that in the twentieth century tobacco-associated diseases killed 100 million people (Fig. 2.8).

The seminal UK study that conclusively linked smoking to lung cancer was published in 1950 by Richard Doll and Austin Bradford Hill. That and subsequent reports

Causes of cancer

(a)   (b)

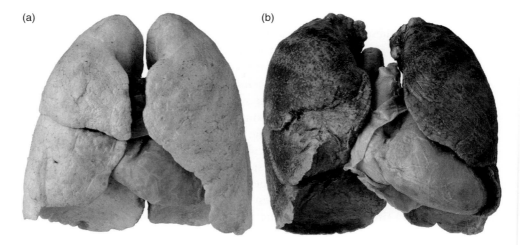

Figure 2.8 **The effect of smoking on the lungs.** (a) Non-smoker's lungs. (b) Smoker's lungs. (Gunther von Hagens' Bodyworlds, Institute for Plastination, Heidelberg, Germany, www.bodyworlds.com.) (See plate section for colour version of this figure.)

certainly influenced attitudes, leading in the ensuing 50 years to the progressive prohibition of smoking in public places and a gradual decline by about half in the number of UK smokers. In the USA things are a bit more idiosyncratic with some states having banned smoking in all enclosed public places but others having no state-wide prohibition.

Perhaps the most depressing aspect of the smoking saga is that the first statistical evidence linking lung cancer and cigarette smoking was published not by Doll and Hill in 1950 but 90 years ago in the 1920s. By 1935 the German physician Fritz Lickint felt able to write there was 'no longer any doubt that tobacco played a significant role in the rise in bronchial cancer' and to coin the term 'passive smoking'. Because those pre-war studies were carried out in Germany and published in German they have tended to be ignored.

The gloomy reading provided by the WHO figures for twentieth century deaths due to smoking is as nothing compared to their predictions. The figure of 5.4 million a year that tobacco use kills now (that's one every six seconds) will rise to over 8 million (a year) by 2030. Of the more than one billion smokers, over 80% live in low- and middle-income countries. Currently there are 200,000 tobacco-related deaths in Africa and it is there and in other under-developed regions that cigarette smoking is being heavily promoted by the manufacturers. It is the people of these regions who will, if this trend continues, contribute 80% of the 8 million dead.

The effects of smoking are not confined to the lung: almost all the major cancer types are between two and six times more likely to develop in smokers than in non-smokers (i.e. mouth, pharynx, larynx, bladder, oesophagus, pancreas, stomach, liver, cervix, kidney and myeloid leukaemia). The carcinogenic effects of tobacco arise because specific chemicals in nicotine can cause mutations that disable the function of critical genes. As well as initiating tumours, tobacco smoke can also act as a **tumour promoter** by causing chronic inflammation.

The fact that tobacco smoke is carcinogenic and the problem of involuntary (passive) inhalation, recognised so many years ago by Lickint, has been confirmed by numerous studies and has now prompted legislation in many countries banning smoking in public places. The most convincing evidence on passive smoking comes from long-term studies of non-smokers living with smokers, for which pleasure their lung cancer risk goes up by 20 to 30%. Although it is clearly not possible to quantify how much carcinogen involuntary smokers inhale, the evidence is that exposure to cigarette smoke at work can increase the risk by up to 20%.

It is perhaps surprising that most studies have found no link between smoking and breast cancer, or indeed prostate or endometrial cancer of the uterus. Surprising because chemicals in tobacco smoke have been shown to cause breast cancer in rodents and these compounds have also been detected in both breast tissue and in breast milk. The most persuasive of such studies compared 58,515 women with the disease with 95,067 who were disease free, a data set large enough to separate effects of smoking from those of alcohol. Almost inevitably, there are other surveys that suggest there may be a link. The California Environmental Protection Agency (2005) concluded that, in women who were mainly pre-menopausal, passive smoking could be associated with breast cancer and the US Surgeon General's (2006) report described the evidence as 'suggestive but not sufficient'. The age at which women start smoking does appear to be very significant in that smoking within five years of the first menstrual cycle almost doubles the risk before menopause. This may be because teenage breast tissue that is still developing is more sensitive to smoke carcinogens.

Obviously, it is best never to start smoking but if you have succumbed all is not lost. Yet another major survey conducted by Richard Doll and his colleagues has shown that giving up smoking will improve your chances of avoiding lung cancer. The earlier the better, of course, but even those who cease after the age of 50 reduce their risk by over 60%.

Alcohol

Alcohol is produced when the sugars from fruits or cereals are fermented by yeasts to release CO_2 from sugar ($C_6H_{12}O_6$) giving ethanol (C_2H_5OH). When we drink alcohol 20% is absorbed by the stomach and the small intestine from which it passes into the bloodstream and hence to all tissues and organs in the body (though it's not taken up by adipose tissue because it doesn't dissolve in fat). Eventually most gets broken down in the liver (90%) by the enzymes alcohol dehydrogenase (producing acetaldehyde) and aldehyde dehydrogenase. Alcohol consumption worldwide is estimated to cause about 4% of all cancers and there is evidence for specific association with most of the major cancers, including breast, mouth, oesophagus, pharynx and larynx, stomach, bowel, lung, liver (Fig. 2.9), ovary and prostate.

The mechanism by which alcohol promotes cancer is not clear although acetaldehyde may be mutagenic and alcohol itself acts as a local anaesthetic. Nevertheless, consumption is clearly linked to an increased risk of developing breast cancer. The Collaborative Group on Hormonal Factors in Breast Cancer (2002), referred to earlier in

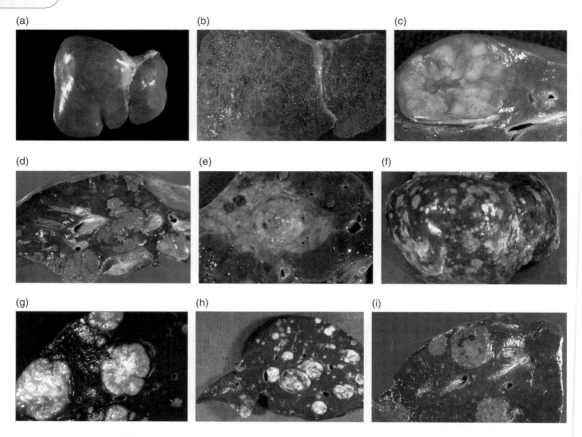

Figure 2.9 Human liver. (a) A normal liver. There are four lobes of unequal size and shape. The gallbladder can also be seen. (www.aafp.org/afp/2006/0901/p756.html) (b) Cirrhosis: external view of micronodular cirrhosis. The result of damage, this condition causes hardening of the organ due to the formation of non-functional scar tissue and surface nodules, here ~3 mm in diameter. The most common causes of cirrhosis are alcoholism and infection with hepatitis viruses. The condition can lead to liver cancer (hepatocellular carcinoma). (c) Liver adenoma showing the natural colour in a close-up view of a well demarcated lesion. (d) Hepatitis: inflammation of the liver characterised by the presence of inflammatory cells. Regions of atrophy can be seen caused by chronic blood vessel damage. (e) Hepatocellular carcinoma (malignant hepatoma): a primary malignancy of the liver. (f) Metastatic carcinoma from a primary stomach tumour on the liver surface. (g) Metastatic carcinoma showing necrosis in the centre of tumour masses. (h) Metastatic lesions in the liver from a primary stomach tumour ranging in diameter from <1 mm to several cm. (i) Magnified view of metastatic lesions in the liver from a primary tumour. (© University of Alabama at Birmingham, Department of Pathology.) (See plate section for colour version of this figure.)

the context of tobacco, also showed that about 4% of the breast cancers in developed countries are attributable to alcohol. Compared with non-drinkers, women who consume one alcoholic drink a day have a very small increase in risk. Those who have two to five drinks daily have about one-and-a-half times the risk of women who drink no alcohol.

The American Cancer Society recommends that women limit their consumption of alcohol to no more than one drink per day. One plausible explanation of its effects on the breast is that it elevates oestrogen production, which in turn increases cell growth.

Tea and coffee

Tea and coffee are the most common hot drinks in the world and it is well known that they contain the drug caffeine (a xanthine alkaloid) that can cross the blood–brain barrier and act on the central nervous system. So one might well ask, with all this drug abuse of our throats, is there any evidence that drinking tea or coffee gives us cancer? The most comprehensive analysis so far (of pooled data from nine case-control studies) shows, perhaps slightly surprisingly, that there is an inverse association between drinking caffeinated coffee and the risk of cancers of the mouth and pharynx. In other words drinking coffee protects against these cancers and the more you drink, at least up to four cups per day, the greater the protection. For laryngeal cancer, however, drinking caffeinated coffee has no effect on the risk. The data on decaffeinated coffee are less solid but suggest that at least it does not increase risk. Tea drinking showed no association with head and neck cancer. The leaves of the *Camellia* tea plant are a particularly rich source of anti-oxidants in the form of polyphenols, and on a per serving basis coffee provides even more anti-oxidants. Substantial proportions appear to be absorbed by the body so it is possible that, through their anti-oxidant effects, they can inhibit mutagenic events.

Thus far, although we've mentioned alcohol, tea and coffee, we haven't considered diet – one of the most obvious potential sources of carcinogens. The consensus on this matter, to the extent that there is one, has emerged from the integration of a large number of epidemiological studies and, by way of background, Box 2.1 summarises the basic approaches.

Diet: epidemiological studies

The Nurses' Health Study started in the USA in 1976 to investigate risk factors for cancer and other diseases in women and has now involved over 90,000 nurses. Thus far among its major findings are that the risk of bowel cancer is increased by eating a lot of red meat and reduced by taking folic acid in multi-vitamin supplements. Breast cancer incidence is unaffected by fat and fibre intake but is increased by one third in response to moderate amounts of alcohol, as we noted.

Bowel cancer has been intensively studied in the context of diet and it is worth considering the results of some of the major surveys to illustrate why this is such an intractable problem. Apart from the fact that it is one of the three major cancer killers, colorectal carcinoma is much studied because it is a well defined, multi-step process in which, without treatment, adenomas develop into full carcinomas (Fig. 2.10). In the early stages polyps form on the wall of the bowel that can be surgically removed (polypectomy). Individuals so treated have been particularly studied to determine the effect of diet on adenoma recurrence.

Box 2.1 Epidemiology

Epidemiology is the study of the occurrence and distribution of diseases, and the application of the data to the control of diseases and other health problems. In this context it is particularly concerned with the evaluation of risk factors for cancer. Broadly speaking such studies are either 'observational' or 'controlled'. In case studies and case series, individual or small groups are studied to provide qualitative information about an illness-associated factor. Cross-sectional studies (or cross-sectional analysis) collect data from a single group (sub-population) at a given time. These may relate a disease to a potential contributory factor and the association may be quantified by calculating an odds ratio. Longitudinal studies make repeated observations of the same variables over protected periods to establish correlations.

Observational studies also include case-control and cohort studies. Case-control studies compare two groups, one with disease (the 'case' group), the other being controls who are disease free (the 'control' group), in terms of possible causes. The control group should be matched as closely as possible for factors that are not under study (e.g. age, race, etc.). The study is a 'retrospective' survey of both groups for potential exposures. The association may be calculated in the form of an odds ratio (OR) from the numbers of exposed cases (A), exposed controls (B), unexposed cases (C) and unexposed controls (D):

$$OR = (A/C)/(B/D) = A \cdot D/C \cdot B.$$

Cohort studies fall into two categories, being either 'prospective' or 'retrospective'. Prospective studies start with healthy individuals who are followed for exposure to potential factors and subsequent development of disease. Exposed and unexposed sub-groups are compared for disease rates. The follow-up period may be short (a few days) for acute diseases, but for cancer, cardiovascular disease and other chronic conditions may be several decades. Cohort studies differ from experimental studies in that the exposure status is observed rather than determined. Thus, for example, you might use two cohorts, one of smokers the other non-smokers, to estimate the incidence of lung cancer over time. In retrospective (or historical cohort) studies data are collected from the records of individuals to establish the contribution of potential factors to disease development.

Cohort studies generate the relative risk or risk ratio (RR), the probability of disease in an individual from the exposed group divided by that for one in the unexposed group ($RR = p/q$ where p = the estimated probability of the event occurring in the exposed group and q that in the control group. $OR = p(1-q)/q(1-p)$).

For example, if 80 lung cancers develop in a group of 1,000 smokers and 5 occur in 1,000 non-smokers:

$$RR = (80/1,000)/(5/1,000) = 16.$$

$$OR = (80/920)/(5/995) = 17.3.$$

Box 2.1 cont'd

Randomised controlled trials (RCTs) are a form of prospective cohort study in which the participants are randomly divided into groups that receive, for example, a specific therapeutic regimen or diet. The main advantage is that random distribution into the groups reduces selection bias and the first part of the analysis determines whether the two groups are comparable in demographic and other variables, as they should be after random allocation. Randomised controlled trials have been extensively used to determine associations between dietary factors and cancers, the main problems being the absence of control over how well participants stick to the dietary rules and the fact that cancers take many years to manifest themselves.

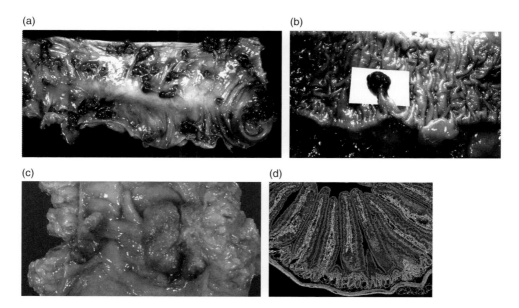

(a)

(b)

(c)

(d)

Figure 2.10 Colon cancer. (a) A section of colon with multiple polyps (typically 5 mm in diameter); (b) a colon polyp showing also the convoluted structure of adjacent normal colon; (c) malignant colon carcinoma (~5 × 7 cm). (© University of Alabama at Birmingham, Department of Pathology.) (d) Mouse intestine section imaged with wide-field multi-photon microscopy, showing the villi that provide a large area for digestion and absorption. Actin (green), lamin (red), nuclei (blue). (Thomas Deerinck, NCMIR, UCSD.) (See plate section for colour version of this figure.)

Folate

Three familiar dietary players have been to the fore in this context: folate, calcium and fibre. Folate (there are several naturally occurring forms) is a B vitamin that helps to shuttle carbon atoms around when the building blocks of DNA and proteins are being made. It is therefore important in DNA replication and in the repair of damaged DNA, so it's easy to see that if you were short of folate you might be prone to cancer. A synthetic form of folate – folic acid – is used as a food supplement. A number of both retrospective and prospective studies suggest that folate has a protective effect, consistent with its importance in maintaining the integrity of DNA. Furthermore, other studies have measured the levels of folate in plasma and serum and found an inverse association with bowel cancer – that is, the more folate you have the better – consistent with the Nurses' Health Study.

As usual in this field, there are conflicting reports, some of which conclude that diet supplementation can actually increase the risk. For folate you need enough to maintain your DNA in a healthy state, thereby minimising your susceptibility to cancer, but if you have too much it may block DNA repair and thus help to drive carcinogenesis. So, supplementing diet with folic acid might provide protection against colon carcinoma for those whose normal circulating levels of folate are low but might be a very unhelpful thing to do for individuals with higher levels. All of which indicates the dangers of giving diet supplementation to general populations.

Calcium

As with folate, there is a general view that plenty of calcium is good, not only because it gives you strong bones but also because you are less likely to get bowel cancer. Again there are prospective studies that support this view but, as ever, there are others, including one of 36,000 cases, showing that calcium supplementation has no effect. These findings scarcely make the case for calcium supplementation and, as with folate, may be confounded by individual variation – that is, additional calcium intake will not benefit those on healthy diets who have normal levels of calcium.

Fibre

Dietary fibre, sometimes called roughage, is the stuff we eat that can't be digested but that does an important job in taking up water and generally helping our insides to work. The well publicised advice is that eating plenty of fibre helps to prevent colon cancer and there are many supporting studies. Notable among these is the European Prospective Investigation into Cancer and Nutrition (EPIC) study showing that 35 grams per day reduced the risk by 40% compared with 15 grams per day. This is a particularly powerful contribution because it involved over half a million (520,000) people from ten European countries. However, there are, of course, other studies (e.g. the Polyp Prevention Trial and the Wheat Bran Fiber Trial: National Institutes of Health, 2000) that show no protective effect and at least one that concludes men are much better protected than women. In addition to the pitfalls mentioned earlier, a further cause of confusion may be the variation in study duration and the fact that follow-up periods are generally short relative to the many years over which cancers usually develop.

It is probably obvious that, when faced with a considerable number of independent studies about a specific factor with conclusions spanning all categories, it might be worth putting them together and seeing if a more focused message emerges. This is what epidemiologists call a 'meta-analysis' and it's really a way of increasing your sample size and hence the statistical power of the data. It's not absolutely straightforward because there's no point in including a study you consider so badly designed as to be worthless. Thus, an element of judgement was required to combine data from 13 separate studies to show that there was no effect of a high fibre diet on the risk of developing colon cancer, a conclusion that was, of course, completely at odds with the EPIC finding.

It is now 30 years since Richard Doll and Richard Peto (Doll and Peto, 1981) estimated that diet may contribute to one third of all cancers. While this figure is still widely accepted, we've seen that there is very little direct chemical evidence linking the action of specific food components to cancer. Part of the problem is that many foods contain agents that, either directly or indirectly, can exert DNA damaging (genotoxic) effects. The link between eating red or processed meat and bowel cancer, shown by a number of large and seemingly well conducted studies, arises from the fact that cooked meat can release agents in the stomach that, on reaching the circulation, undergo enzymatic conversion to carcinogens. Compounds added to give colour and flavour and to stop *Clostridium botulinum* growing can have similar effects. It is scarcely surprising therefore that epidemiological studies of the association between diet and cancer are often inconclusive and contradictory. Fortunately, making sense of the data has been considerably helped by the establishment of the Cochrane Collaboration (www.cochrane.org), a non-profit consortium dedicated to analysing and summarising the literature on healthcare interventions. Its aim is to explain, briefly but in 'plain English', why a question is being asked, how the trials were set up, the main results and the conclusions. From its files the most recent conclusions for dietary calcium are that, although there is evidence that calcium supplementation might make a modest contribution to the prevention of bowel cancer, there is not sufficient evidence to recommend its general use. A corresponding report on dietary fibre concludes that there is currently no evidence to suggest that increased intake will reduce the incidence or recurrence of bowel cancer within a two- to four-year period. The collective advice of the World Cancer Research Fund and the American Institute for Cancer Research on food and cancer is to eat a 'balanced diet' with plenty of fibre, vegetables, fruits, lentils, beans and whole grains such as brown rice and wholemeal pasta, less than 500 grams per week of red meat, 'very little if any' of that being processed (i.e. preserved by smoking or by the addition of salt, sugar, nitrate or nitrite) and, of course, no alcohol.

Obesity

Over 300 million people in the world are obese, that is, are more than 25% overweight. Obesity dramatically increases chances of developing a wide range of life-threatening conditions including diabetes (in the USA more than one person in three ends up with adult-onset diabetes), arteriosclerosis, hypertension, heart disease, age-related degenerative disease, sleep apnoea, gallstones and some cancers. Obesity specifically promotes cancers of the colon, kidney, liver, oesophagus, pancreas, endometrium and breast and there is evidence that it may also contribute to gallbladder and ovarian cancers. The risk of breast cancer is particularly significant in post-menopausal women. Before menopause the ovaries produce most of a woman's oestrogen, and fat tissue makes the rest. After menopause the ovaries stop making oestrogen and it comes mainly from fat tissues. Having more fat tissue after menopause can increase oestrogen levels and thereby the likelihood of developing breast cancer. The risk is greater for women who put on weight as adults by comparison with those who have been overweight since childhood. Furthermore, there are differences between fat cells in different regions of the body so that waist fat appears to be worse than hip or thigh fat in terms of breast cancer risk. Despite the fact that obesity is linked to breast cancer, it is also associated with a lower density of breast tissue, whereas increased tissue density is a significant risk factor for breast cancer.

While the upsurge in obesity is a relatively recent trend, there is evidence dating back 100 years for a link between the amount one eats and cancer. In rats and mice dietary restriction, which means being fed between 10% and 50% fewer calories, reduces the incidence of at least some types of cancers – and extends lifespan. Note that it's total calorie intake rather than the nature of the food that counts. Does this apply to humans? That's a bit more difficult to be sure about but certainly some human tumour cells when grown in mice are very sensitive to dietary restriction. The explanation may be that when diet is restricted the levels of key metabolic hormones decrease, particularly insulin. As well as being a major metabolic regulator, insulin is also a very potent promoter of cell growth – and hence, potentially a driving force for cancer. Sustained low levels of such a factor may therefore be protective.

Stress

One other factor that may be associated with cancer is stress. It is possible to put a number on stress, so to speak, by measuring the amount of cortisol in blood or saliva. Cortisol is a steroid hormone released from the adrenal gland in response to signals from the brain. Normally secretion is maximal early in the morning and declines thereafter during the day (diurnal variation) but it is increased by food, fasting, exercise or stress. Its role is to provide energy when required by stimulating sugar production and the breakdown of lipids/fat and proteins, resulting in raised blood sugar levels and blood pressure. However, cortisol can also indirectly increase appetite and promote fat deposition. Studies of breast cancer patients have shown that about two-thirds may have abnormal cortisol profiles (higher, relatively constant or maximal at abnormal times of day) and this group survive for significantly shorter times (3.2 versus 4.5 years) than those with normal cortisol profiles. The reason may be suppression of the immune system by raised cortisol levels because these patients have reduced numbers of **natural killer cells** in their blood. A second strand of cortisol association is the evidence that night shift work, which disrupts normal diurnal rhythms, is associated with increased incidence of breast cancer. This has been attributed to melatonin suppression but may well be due to perturbed cortisol variation. Despite all this, in the confusing way that cancer has, cortisol-type steroids can suppress the growth of some tumours and have been used as therapeutic drugs.

Conclusions

Ideally the demonstration that a specific agent causes cancer requires both strong epidemiological evidence of association between the two and knowledge of a mechanism by which the effect can be achieved. These conditions have been met for a number of well-known causes – chronic infection, radiation and tobacco use. All of these can be mutagenic, although for infection the effect is usually indirect as a consequence of long-term inflammation (discussed in Chapter 5). For alcohol the association is overwhelmingly established although, beyond its action as a local anaesthetic that perturbs cell signalling, the mechanism remains obscure. Despite considerable media attention, the evidence for any association between electromagnetic fields and cancer, either from power lines or mobile phones, remains insufficient for any of the regulatory authorities to have recommended increased stringency in the current exposure guidelines. As with alcohol, the epidemiological evidence linking obesity to enhanced risk of a range of cancers is clear. The mechanism is less well established but is likely to involve hormonal imbalance, especially of oestrogen, a powerful promoter of cell growth proliferation. Poor diet predisposes to a variety of types of cancer. However, many foodstuffs contain a mixture of essential nutrients and compounds that have carcinogenic potential. Thus the most informed advice is to eat a balanced diet and avoid supplements unless they are medically prescribed.

Key points

- The major exogenous factors that can promote cancer are infection, radiation, diet and tobacco smoke.
- In principle we can exert a substantial measure of control over all of these.
- In the developed world chronic infection causes about 8% of all cancers; in developing countries the figure is 23%, suggesting the influence of poor sanitation and insufficiently pure drinking water.
- Viral infection accounts for about 15% of human cancers worldwide, of which human papillomaviruses (HPVs) and hepatitis B virus (HBV), causing cervical cancer and liver cancer, respectively, are the major contributors.
- There are limits to what can be done about natural radiation from the Earth and in sunlight. However, the most important source of ionising radiation to which humans are normally exposed is the gas radon, the major cause of lung cancer in non-smokers. The burden of lung cancer deaths in which radon plays a role could be reduced by lowering the threshold for which building insulation is required.
- Exposure to UV radiation in the form of sunlight is a risk factor for non-melanoma skin cancers that can be reduced by appropriate protection. Although the evidence that sunlight can cause malignant melanoma is equivocal, this serious condition does have an association with previously having had non-melanoma skin cancer for which UV radiation is a predisposing factor.

- Public concern has promoted considerable research into the questions of whether low-frequency magnetic fields (power lines and domestic electrical circuits) or high-frequency magnetic fields (mobile phones) cause cancer. Low-frequency fields remain classified by the World Health Organization as a 'Possible human carcinogen', the weakest of three categories used to classify scientific evidence on potential carcinogens. The other two higher categories are 'probably carcinogenic to humans' and 'is carcinogenic to humans'. No adverse health effects have been established for mobile phone use.
- Tobacco smoke is the biggest cause of preventable death worldwide, being a major risk factor for cardiovascular disease and cancers. In addition to lung cancer, tobacco use increases the risk of cancers of the respiratory tract, bladder, kidney, pancreas, stomach and cervix as well as acute myeloid leukaemia.
- Alcohol consumption increases the risk of respiratory tract, breast, liver, stomach and large bowel cancers.
- The effects of combined tobacco and alcohol use are synergistic for cancers of the mouth, throat and oesophagus. The relative risk is approximately 6-fold for one or the other but is 35-fold for both, relative to the risk for those who neither smoke nor drink.
- Epidemiological studies show that poor diet can increase the risk of cancer. The World Cancer Research Fund and the American Institute for Cancer Research recommend eating vegetables, fruits, lentils, beans and whole grains such as brown rice and wholemeal pasta with less than 500 grams per week of red meat. Together with limiting the amount of salt and alcohol consumed, this will provide a balanced diet with no requirement for food supplements.
- Obesity increases the risk of cancers of the colon, kidney, liver, oesophagus, pancreas, endometrium and breast. In addition to the above diet, 30 minutes physical exercise a day is a World Cancer Research Fund and the American Institute for Cancer Research recommendation.
- Stress, which is reflected by abnormal levels of the steroid hormone cortisol in the circulation, may contribute to some forms of cancer.

Future directions

- Despite extensive epidemiological surveys that generally indicate the association of 'good, balanced diets' with reduced cancer risk, the critical biochemical ingredients remain undefined.
- A major initiative is underway, promoted by the United Nations, to inform the public worldwide of the major risk factors for the leading 'non-communicable' diseases (cardiovascular disease, cancer, diabetes and respiratory disorders), namely unhealthy diet, tobacco, lack of exercise and abuse of alcohol. While this is laudable, the widespread dissemination of information in, for example, the UK on the dangers of smoking and obesity has not had a noticeable impact on behaviour.

Further reading: reviews

INFECTION

Rook, G. A. W. and Dalgleish, A. (2011). Infection, immunoregulation, and cancer. *Immunological Reviews* **240**, 141–59.

IONISING RADIATION

Jeggo, P. and Lavin, M. F. (2009). Cellular radiosensitivity: how much better do we understand it? *International Journal of Radiation Biology* **85**, 1061–81.

ULTRAVIOLET RADIATION

Narayanan, D. L., Saladi, R. N. and Fox, J. L. (2010). Ultraviolet radiation and skin cancer. *International Journal of Dermatology* **49**, 978–86.

LOW-FREQUENCY MAGNETIC FIELDS (POWER LINES AND DOMESTIC ELECTRICAL CIRCUITS)

Kheifets, L., Renew, D., Sias, G. and Swnason, J. (2010). Extremely low frequency electric fields and cancer: assessing the evidence. *Bioelectromagnetics* **31**, 89–101.

HIGH-FREQUENCY MAGNETIC FIELDS (MOBILE PHONES)

The INTERPHONE Study Group. (2010). Brain tumour risk in relation to mobile telephone use: results of the INTERPHONE international case-control study. *International Journal of Epidemiology* **39**, 675–94.

Independent Expert Group on Mobile Phones. (2000). Report of the Group (The Stewart Report). *Mobile Phones and Health.* www.iegmp.org.uk/report/index.htm [Accessed 4 June 2012].

DIET: EPIDEMIOLOGICAL STUDIES

The Cochrane Collaboration: www.cochrane.org/

OBESITY

Calle, E. E. (2007). Obesity and cancer. *British Medical Journal* **335**, 1107–8.

STRESS

Wek, R. C. and Staschke, K. A. (2010). How do tumours adapt to nutrient stress? *The EMBO Journal* **29**, 1946–7.

3 | Signalling in normal cells

A universal feature of the membranes that form the boundary of all cells is their capacity to permit the transfer of both chemical entities and information. Critical to the latter are trans-membrane receptors, of which the receptor tyrosine kinase family form a particularly important category in the context of regulating growth and proliferation. Through being able to respond to ligand binding by stimulating multiple pathways of intracellular proteins, activated receptors signal to the nucleus to regulate gene expression. Components of these pathways, including the receptors themselves, are frequently mutated in cancers, giving rise to aberrant proliferation control. To illustrate the principles of signalling we will consider one major example: the RAS-MAPK pathway activated by epidermal growth factor (EGF) through its receptor (EGFR), a member of the receptor tyrosine kinase family. The critical result of such signalling is the activation of transcription of key genes required for cells to enter the division cycle, a master regulator being MYC. Steroid hormones that diffuse across membranes before binding to their receptors also play significant roles in cancer and modulation of their effects has been an important therapeutic strategy.

Introduction

A fundamental property of all cells is their capacity to respond to external signals. Mammalian cells receive a vast number of such signals in the form of chemical messengers, many of which are hormones – often called **growth factors** or **mitogens** (or **cytokines** if they've been secreted by cells of the immune system). The intracellular pathways that are activated are complex and their upshot is a pattern of gene expression that determines the critical decisions that cells have to take – whether to proliferate or remain quiescent, to change their function (**differentiate**) or even to commit suicide. However, only four basic mechanisms are used to transduce these signals. Three of these utilise proteins that span the plasma membrane (Fig. 3.1). These cell surface receptor proteins comprise: enzyme-coupled receptors, G-protein-coupled receptors (GPCRs) and ligand-gated ion channel receptors. The fourth system uses receptors within the cell to respond to steroid hormones.

In this chapter we shall consider each of these four in turn, focusing on their actions in normal cells. The greatest emphasis will be on enzyme-coupled receptors because

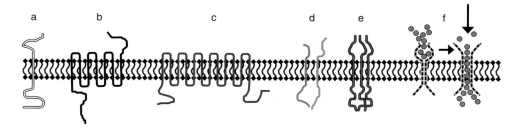

Figure 3.1 Representation of the structures of trans-membrane signalling receptors. (a) RTK; (b) GPCR; (c) 12-trans-membrane domain protein; (d) integrin; (e) tumour necrosis factor receptor (TNFR); (f) monovalent ion channels. These proteins are classified according to the orientation of their N- and C-termini. Types I, II and III are single pass molecules, Type IV are multiple pass molecules (IV-A: N-terminus cytosolic; IV-B: N-terminus extracellular). The trans-membrane domains are all alpha helices. (a) Receptor tyrosine and receptor serine/threonine kinases have a single trans-membrane domain and are activated by receptor cross-linking. (b) The GPCR family have seven trans-membrane domains. (c) The 12-trans-membrane domain protein Patched (PTCH) is the receptor for Hedgehog proteins (see Chapter 6). (d) Integrins are heterodimers of two trans-membrane subunits, a and b. (e) The tumour necrosis factor receptor (TNFR) superfamily comprises receptors for proteins of the tumour necrosis factor (TNF) family that induce a controlled programme of cell death. The best characterised is the Fas receptor that controls cell death in the immune system. TNF proteins function as a trimer of three identical proteins: their binding induces receptor (TNFR) trimerisation. Fas activates caspase 8 to induce cell death. (f) Ion channels. One other form of trans-membrane domain is known, the beta barrel, and occurs in some mitochondrial and bacterial proteins.

they play the major role in regulating proliferation and both the receptors themselves and components of the intracellular signal pathways they drive frequently behave aberrantly in cancer. To illustrate the basic principles we will mainly consider the mitogen-activated protein kinase (**MAPK**) pathway and take up the detail of how this and other pathways are perturbed in cancer in Chapters 5 and 6.

Each of these signal systems differ in mechanism but they are really just a variant on the theme of transmitting information carried by a chemical signal to the cell so that it adjusts its lifestyle accordingly. The four key features of cellular signalling are (1) specificity conferred by ligand–protein and protein–protein interactions; (2) signal amplification generated when activated enzymes are components of pathways; (3) the capacity for pathway convergence (two signals affect the same messenger) and/or divergence (multiple pathways emanate from a single component); and (4) signal termination (by decrease in the concentration of the activating ligand together with de-**phosphorylation** and/or receptor internalisation).

Enzyme-coupled receptors

In mammals there are four types of enzyme-coupled receptors that have intrinsic enzymatic activity within a single trans-membrane protein (Table 3.1): those in which the enzyme phosphorylates the amino acid tyrosine (receptor **tyrosine kinases** or RTKs (Fig. 3.2)),

Signalling in normal cells

Table 3.1 Classes of mammalian enzyme-linked receptors.

Class	Activity (stimulated by ligand binding)	Example (ligand)
Receptor tyrosine kinases	Tyrosine kinase or association with intracellular tyrosine kinase	Epidermal growth factor T-cell receptor ligand
Receptor serine/threonine kinases	Serine/threonine kinase	Transforming growth factor β
Receptor guanylyl cyclases	cGMP	Atrial natriuretic peptide
Receptor-like tyrosine phosphatases	De-phosphorylate tyrosine residues in intracellular proteins	Switch off tyrosine kinase signalling and function in cellular recognition

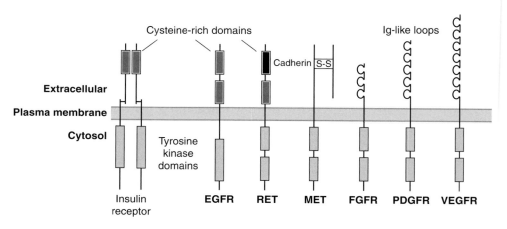

Figure 3.2 Major members of the receptor tyrosine kinase family. The insulin receptor family includes insulin-like growth factor 1 receptor (IGF1R), which has the same $\alpha_2\beta_2$ structure as the closely related insulin receptor. EGFR (epidermal growth factor receptor), RET, MET (also known as the hepatocyte growth factor receptor, HGFR), FGFR (fibroblast growth factor receptor) and PDGFR (platelet-derived growth factor receptor) are oncoproteins (Chapter 4). The platelet-derived growth factor receptor (PDGFR) family includes KIT (the receptor for stem cell factor) and CSF1R/FMS (colony stimulating factor-1 receptor). VEGFR is the receptor for **vascular endothelial growth factor.** FGFRs, the two forms of PDGFR (A and B) and VEGFRs have the same general structure as the EGFR but with two differences: (1) the extracellular domain is composed of immunoglobulin-like loops – a common feature of the superfamily of lymphocyte cell surface proteins; (2) the tyrosine kinase domain is split with the lysine required for adenosine-5′-triphosphate (ATP) binding separated by a 'kinase insert' sequence from the catalytic domain.

receptor serine/threonine kinases, receptor guanylyl cyclases and receptor-like tyrosine phosphatases. In a smaller group of enzyme-coupled receptors the cytosolic domain does not have intrinsic enzyme activity but instead recruits and activates cytoplasmic tyrosine kinases to relay the signal. There is also a family of histidine-specific protein kinases present mainly in bacteria.

The RTKs are the largest class of enzyme-coupled receptors and we will focus on them because of the major roles they play in controlling cell proliferation and the fact that they frequently signal abnormally in cancer (Fig. 3.2).

Receptor tyrosine kinases (RTKs)

In the case of signals that bind to cell surface receptors, binding of the hormone/growth factor to the extracellular part of the receptor promotes a conformational change that is transmitted to the intracellular domain, via the trans-membrane bridge. This is often facilitated by the capacity to bind to two receptors at once (so-called 'cross-linking'), which, in effect, draws the two receptors together.

The intracellular domains of these receptors are kinase enzymes that can add phosphate groups to specific target proteins, including the receptor's own intracellular domains (Fig. 3.3). Hormone binding to the ectodomains juxtaposes the cytosolic domains of two receptors, promoting a conformational change that results in catalytic activation. Activated receptors phosphorylate other receptors of the same type (*trans-phosphorylation*), rather than tyrosine residues within the same molecule but the critical point is that a set of phosphorylated tyrosine residues are created to act as docking sites for intracellular signalling proteins (Fig. 3.4).

In the example shown in Fig. 3.3 the hormone is epidermal growth factor (EGF), one of the first growth hormones to be identified, its name reflecting the fact that it

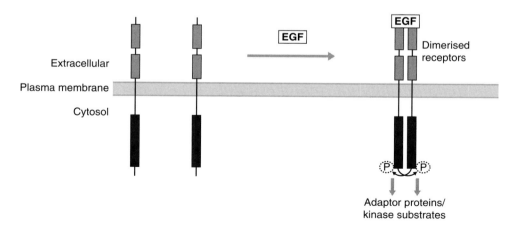

Figure 3.3 Tyrosine kinase receptor activation involves *trans*-phosphorylation. EGF binding activates the cytoplasmic tyrosine kinase domain of EGFR promoting *trans*-phosphorylation followed by the recruitment of signalling molecules to phosphorylated tyrosine residues.

Signalling in normal cells

stimulates the growth of epidermal (skin) cells in animals. Epidermal growth factor differs from most other RTK ligands in being monomeric. It therefore binds to EGFR with a 1:1 stoichiometry but the conformational change induced in the ectodomain promotes receptor dimerisation with the two tyrosine kinase domains taking up an asymmetrical structure stabilised by the juxtamembrane region of the receptor. There are four members of the EGFR family (Fig. 3.5) to which a variety of ligands, in addition to EGF, can bind with differing affinity, the exception being ERBB2, which has no direct ligand, although it forms signalling heterodimers with the other members of the family.

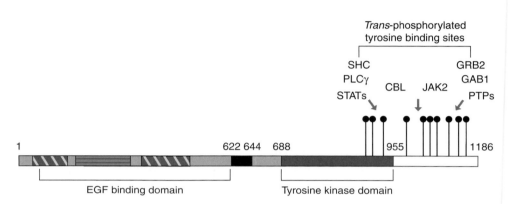

Figure 3.4 The epidermal growth factor receptor, EGFR. The phosphorylated tyrosine residues in the cytosolic part of the protein (black discs) provide multiple binding sites for cellular proteins via SH2 and PTB domains. These include adaptor proteins (SHC, GRB2, GAB1), transcription factors (STATs) and enzymes (phospholipase Cγ, PLCγ) and tyrosine phosphatases (e.g. PTPs). The numbers refer to amino acids: 622–644 is the trans-membrane region.

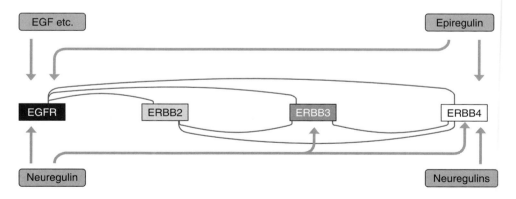

Figure 3.5 EGFR family receptors (EGFR, ERBB2, ERBB3 and ERBB4) and ligands. All possible heterodimeric complexes can exist (shown by the connecting brackets), as well as homodimers. EGF and other ligands interact specifically with the EGFR. Epiregulin and other ligands bind ERBB4 and also EGFR. A family of neuregulins bind with differing affinity to EGFR, ERBB3 and ERBB4. ERBB2 is an 'orphan receptor' with no known direct ligand.

Intracellular signalling from activated tyrosine kinase receptors

A wide variety of intracellular proteins bind with high affinity to activated RTKs via interaction domains that recognise phosphorylated tyrosines within specific peptide sequences (Fig. 3.6). The broad family of proteins that recognise phosphorylated tyrosines include 'adaptors' and enzymes, some of which also function as adaptors. Adaptor proteins generally do not have enzymatic activity but interact specifically with other proteins to mediate the formation of protein complexes or to draw proteins to specific locations. In essence they act like bits of molecular Velcro. Two major types of domain are responsible for protein binding to phosphorylated tyrosines within specific peptide sequences: *Src homology 2* (SH2) domains and *phosphotyrosine binding* (PTB) domains. The SH2 designation derives from homology with one of the domains of the SRC protein that contains two other domains, SH1 (the tyrosine kinase domain) and SH3 (see Chapter 4). Other binding domains include *Src homology 3* (SH3), which attaches to proline-rich amino acid sequences and pleckstrin homology (PH) domains, which bind to the charged head groups of **phosphoinositides**. In addition to adaptors and enzymes, some activated RTKs can recruit docking proteins that also undergo tyrosine phosphorylation to provide an additional 'scaffold' of multiple binding sites. The most familiar example of docking proteins is the insulin receptor substrate (IRS) family that contribute to signalling from insulin receptors.

An important example of adaptors are the three isoforms of the SHC (Src homology 2 domain containing) family that contain single SH2 and PTB domains and act in signal pathways from a variety of receptors including RTKs, antigen receptors, cytokine receptors, G-protein-coupled receptors and integrins. Activated SHC proteins have been associated with cancers and there is evidence that two forms of SHC are highly expressed in aggressive breast tumours.

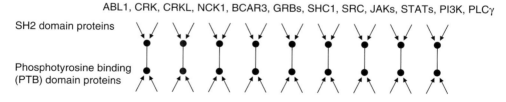

ABL1, CRK, CRKL, NCK1, BCAR3, GRBs, SHC1, SRC, JAKs, STATs, PI3K, PLCγ

SH2 domain proteins

Phosphotyrosine binding
(PTB) domain proteins

Figure 3.6 The EGFR interaction network. Genome-wide profiling using protein microarrays representing essentially every SRC homology 2 (SH2) and phosphotyrosine binding (PTB) domain encoded in the human genome reveals over 150 interactions with phosphorylated tyrosine residues in the EGFR. As the concentration of the binding sites is increased, corresponding to the over-expression of EGFR that occurs in many human cancers (Chapter 4), the number of interactions increases (Jones *et al.*, 2006). The black discs represent phosphorylated tyrosine amino acids (Fig. 3.4). Similarly complex interaction patterns occur with other members of the EGFR family although fewer proteins interact with ERBB3.

Signalling in normal cells

Enzymes (other than the receptors themselves) are recruited to signalling pathways in exactly the same way as adaptor proteins – by having appropriate high-affinity binding sites. Thus, for example, signals activated by growth factors may be terminated by the cytosolic protein-tyrosine phosphatase PTPN6, which has two SH2 domains that permit binding to activated RTKs, permitting the dephosphorylation of adjacent tyrosines. The role of tyrosine phosphatases in suppressing tumour progression is not well understood but another of these cytosolic enzymes, PTPN12, does contribute to some types of primary breast cancers in which its loss of function relieves constraint of a number of kinases, including EGFR and ERBB2.

In signal propagation, enzymes may be directly recruited to activated RTKs or form part of the downstream signalling pathways that are turned on. Phosphatidylinositol-specific phospholipase C gamma 1 (PLCG1) contains two SH2 and one SH3 domain in addition to its enzymatic domain and can be directly activated by RTKs to promote inositol 1,4,5-trisphosphate-mediated calcium signalling (Fig. 3.7).

Like SHC proteins, PLCG1 is also highly expressed in some tumours. Whether directly activated or part of a downstream pathway, because enzymes are catalysts, once activated they continue carrying out their specific reaction until they are switched off. This feature means that when there is an enzyme in a pathway the signal is amplified at that point. Inbuilt amplification is a mechanism for increasing the sensitivity of the response and signal pathways often have a sequence (a cascade) of successive enzyme steps. The upshot of such pathways is the movement into the nucleus of phosphorylated proteins that can affect transcription.

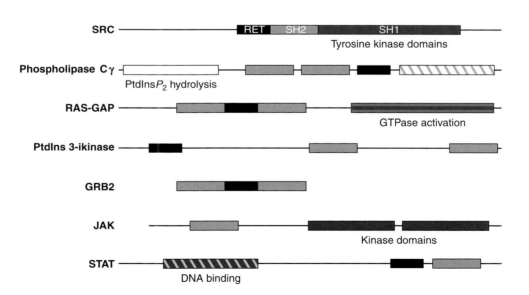

Figure 3.7 SH2 family signalling proteins.

Activating RAS and MAPK

One of the central pathways in eukaryotic cell signalling is that linking activated RTKs to RAS and the *mitogen-activated protein kinase* (MAPK) pathway (Fig. 3.8). An essential component of this sequence is an adaptor protein, GRB2 (*growth factor receptor-bound protein 2*). Made up of an SH2 domain flanked by two SH3 domains, GRB2 is recruited to activated RTKs and its N-terminal SH3 domain binds to the guanine exchange factor SOS. Activation of RAS requires the dissociation of GDP, which is achieved by SOS that, in addition, draws RAS to the plasma membrane. These SOS-mediated effects change the conformation of RAS to promote interaction with the next member of the chain, RAF1. There is no evidence implicating SOS mutations in cancer but they do occur in some hereditary conditions, notably Noonan syndrome, which gives rise to a form of dwarfism.

Three subsequent protein associations cascade to the nucleus: RAF1 (activated by RAS) interacts with MAPKK, MAPKK with MAPK and, finally, MAPK phosphorylates **transcription factors** (TF in Fig. 3.8; e.g. ERK1). After phosphorylation by MAPK, transcription factors can enter the nucleus and bind directly to DNA.

Box 3.1 Intracellular signalling: basic principles

The hormone signal (a growth factor) binds to its receptor and activates a protein relay (A → B → C → etc.) that signals to the nucleus. The receptor is shown as two globular proteins at either end of a wiggly line – the trans-membrane domain. Each circled P represents a phosphate group attached to the internal part of an activated receptor: they act as initiation sites for cell pathways comprised of relays of interacting proteins (A → B → C → etc.) that ultimately activate transcription factor (represented by G) that can enter the nucleus and regulate gene expression.

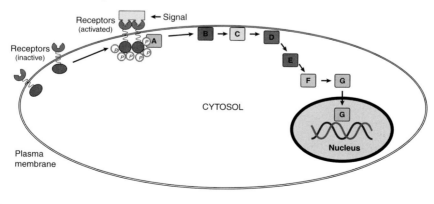

Signalling in normal cells

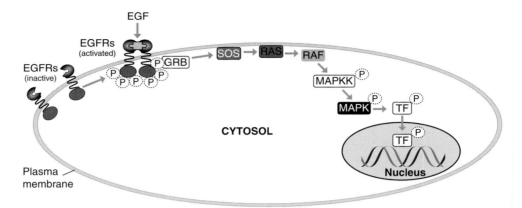

Figure 3.8 A major signalling pathway from plasma membrane to nucleus. In this example the initiating hormone is epidermal growth factor (EGF) that binds to its receptor (epidermal growth factor receptor, EGFR) and activates a sequence of proteins that carry its signal from the membrane to the nucleus. Each circled P represents a phosphate group attached to an amino acid. In the cytosolic domain of an activated receptor tyrosine kinase (RTK) the phosphorylated amino acids are tyrosines. A major target for activation by RTKs is the RAS-MAPK (mitogen-activated protein kinase) pathway. The downstream enzymes in this pathway (RAF1, MAPKK and MAPK) are serine/threonine kinases. The EGFR is one of 58 RTKs encoded by the human genome. There are also 32 non-RTKs (i.e. cytosolic enzymes) that are involved in signalling. In addition to tyrosine, two other amino acids can accept a covalently attached phosphate group (i.e. undergo phosphorylation): serine and threonine. There are ~125 serine/threonine kinases (STKs) that are particularly important in regulating metabolism.

RAS is the prototype of a superfamily of molecular switches comprising more than 50 members in five sub-families (Table 3.2). We look at how the functional effects of mutations in RAS can promote cancer in the next chapter and will meet several other members of the superfamily of small GTPases playing roles in both proliferation and cell death in Chapter 6.

The MAPK cascade that is activated by RAS is also conserved across all organisms as a central mechanism for signal transduction, and there are several families of MAPK pathways that respond to different types of agonist (hormones, cytokines, stress signals, etc.) (Fig. 3.9).

These enzymes are further examples of molecular switches: they are switched on by phosphorylation and they remain activated, independent of contact with their upstream activator, until they are de-phosphorylated by the action of a phosphatase. The sequential interactions between MAP3Ks, MAP2Ks and MAPKs define linear routes leading to the activation of transcription factors. In addition, 'cross-talk' can occur between some components of distinct pathways. Mammalian cells have a variety of other signalling pathways in addition to the MAPK family and which of these are activated depends not only on the hormones present in the circulation but on the specific receptors and signal components expressed by the cell at any time. In effect, the pattern of different receptors carried defines cell type (e.g. a liver cell or an **epithelial cell**).

Table 3.2 The RAS superfamily.

Sub-family name	Principal roles	Representative members
RAS	Proliferation, differentiation	HRAS, KRAS, NRAS
RHO/RAC/CDC42	Actin cytoskeleton, adhesion, secretion, invasion and metastasis	RHOA, RAC1, CDC42
RAB	Intracellular protein transport	RAB1A, RAB2A
RAN	Nuclear protein import and RNA export	RAN/TC4
RAD/GEM	Receptor-mediated signal transduction	RRAD, GEM

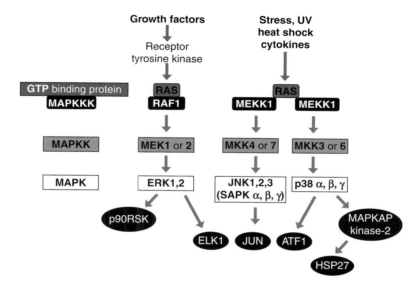

Figure 3.9 Conserved MAPK pathways in mammalian cells.

Sustained versus transient activation of RAS and MAPK

RAS-MAPK signalling has been extensively studied in the PC12 cell line, derived originally from an adrenal tumour. These cells proliferate in response to EGF but they are a useful model for neuronal differentiation because treatment with nerve growth factor (NGF) halts their proliferation and induces the formation of sympathetic neurons. Both responses are signalled via RAS-MAPK but with different requirements from two members of the RAF family, RAF1 and BRAF (ARAF is the third family member). Transient activation of RAF1 and BRAF promotes proliferation and it is only when BRAF activation is sustained that differentiation occurs (Fig. 3.10). These cells have been subjected to numerous manipulations of receptors and signalling proteins with the consistent result that transient signalling leads to proliferation, whereas sustained BRAF activation promotes differentiation. Both receptors (EGFR and TRKA)

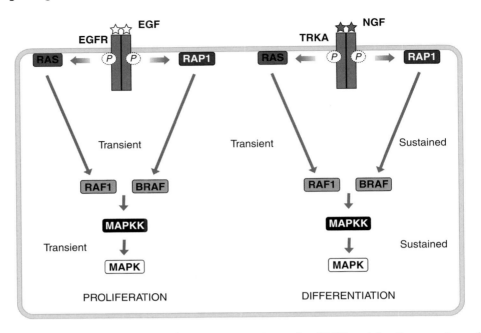

Figure 3.10 Sustained versus transient ERK activation: role of RAF proteins. Two members of the RTK family engage different combinations of adaptor and scaffold proteins. While both activate RAS and RAP1 (also a member of the RAS family) transient activation of RAF1 and BRAF promotes proliferation in response to EGF whereas sustained activation of BRAF leads to cell differentiation.

are RTKs, contrasting signals being generated through subtle differences in the activated coupling proteins.

The tissue dependence of signalling can be illustrated by considering two types of tumour. In the lung, tumour development is driven by RAF1 and BRAF can be dispensed with, at least in mice. However, in melanocytes (pigmented skin cells from which melanoma develops) NRAS activates BRAF and MAPKK.

The underlying point is that, in principle, a mutation in *any* component of an intracellular signalling pathway that results in abnormal activity has the potential to convert a normal cell into one with the capacity to become cancerous. We will see in the next chapter how the EGFR and RAS in particular frequently acquire such activating mutations, thereby making them major players in human cancers. For reasons that are not clear, mutations in MAP2Ks and MAPKs appear to be very rare events.

Mitogenic activation of cell cycle progression

The consequence of signalling pathway activation by growth factors and mitogens is, of course, the initiation of cell proliferation. Activated receptors propagate primary mitogenic signals, triggering the transition from G_0 into G_1 (Fig. 3.11 and Box 3.1).

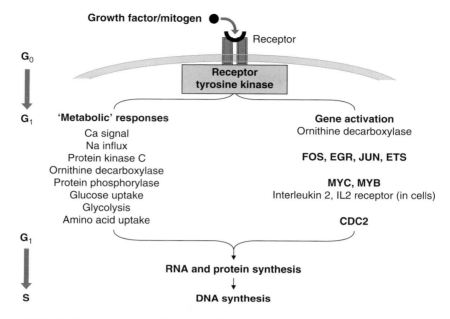

Figure 3.11 Biochemical events during proliferation in eukaryotic cells. The interaction of growth factors with their receptors on the cell surface causes quiescent, somatic cells to leave G_0, traverse G_1 and enter S phase, whereupon cells are normally committed to at least one round of the cell cycle. Following the generation of primary signals, a sequence of 'metabolic' events occurs and, in parallel with and independent of these events, the coordinated transcription of ~100 genes is activated within six hours. These include **ornithine decarboxylase** and *JUN*, *FOS* and *MYC*.

Irrespective of cell type, the primary signals generate three responses that occur in parallel: (1) metabolic activation, characterised by increased uptake of essential precursors for RNA and protein synthesis and of glucose to produce adenosine-5′-triphosphate (ATP); (2) activation of RNA and protein synthesis; and (3) activation of a programme of gene expression that begins with the 'immediate early response genes' within the first few hours. Central to cell proliferation among these genes is *MYC*, the expression of which is switched on by a variety of growth factors that activate RTKs.

MYC is a central regulator of cell growth and proliferation

MYC is essential for normal cell proliferation and high levels of expression accelerate growth, although it is not expressed in dividing germ cells. MYC is a transcription factor that regulates the expression of ~15% of human genes and it may therefore be considered a 'master' cell regulator. The MYC protein family (MYC, MYCN and MYCL) each contain **basic helix–loop–helix** and **leucine zipper** domains that characterise a major class of transcription factors (Fig. 3.12). Each forms heterodimers with the protein MAX (which also contains all three motifs) that bind specifically to DNA as

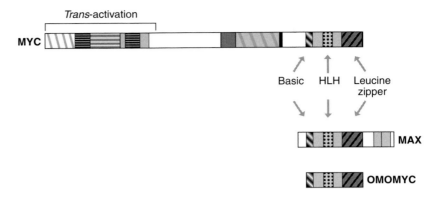

Figure 3.12 **Structure of MYC and MAX.** The conserved sequences MYC box 1 and 2 lie within the *trans*-activation domain (i.e. that which binds to DNA to promote transcription of specific genes). MYC also contains a nuclear localisation signal. Thr^{58} and Ser^{62} are phosphorylated by MAP kinases and their phosphorylation modulates *trans*-activation by MYC.

trans-activating complexes (i.e. complexes that activate the transcription of other genes). MAX expression is independent of MYC and MAX may thus regulate gene transcription independently of MYC. Other leucine zipper proteins (the MXD family, MLX, MXI1 and MNT) form heterodimers with MAX that repress transcription, giving rise to a complex regulatory network.

MYC and proliferation

MYC exerts a dominant role in cell proliferation through its capacity to *trans*-activate a number of key cell cycle genes, notably cyclin D1 (*CCND1*) and *CDC25A*. Cyclin D1 controls progression through G1 and into the S phase of the cell cycle in which the genomic content of DNA is duplicated. CDC25A is a critical regulator of progression into the mitotic phase (see Box 3.2) and its expression pattern during the cell cycle closely resembles that of MYC. It is also aberrantly expressed in some types of cancer and it is easy to see how abnormal regulation of a key step in the cycle could contribute to uncontrolled proliferation. MYC also regulates all three RNA polymerases and thus exerts an indirect effect on the transcriptional expression of the entire genome. In addition to these actions as a transcription factor, MYC plays a direct role in DNA replication by interacting with the pre-replicative complex. Consistent with these central roles in cell division and the duplication of DNA, *MYC* is repressed when cell growth is arrested, for example, in response to DNA damage.

The importance of MYC in regulating key cellular processes suggests that control of its expression is critical. Normal MYC abundance is indeed regulated by the rate of transcription into mRNA, and by the stability of both its mRNA and of MYC protein itself. The de-regulation of MYC expression in cancers can also occur via multiple

Box 3.2 The cell cycle

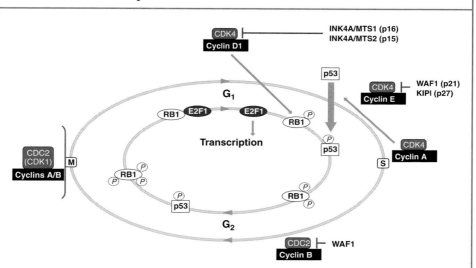

The cell division cycle comprises two growth phases (G1 and G2 – G for 'gap') that separate the S phase (when DNA *synthesis/replication* takes place) from *mitosis* (M phase) when one cell becomes two. The main driving force for the cycle is the sequential action of a number of enzymes (kinases) that phosphorylate key targets. Kinase activity depends on the regulated appearance and breakdown of partner proteins called cyclins (cyclins D, E, A and B are shown) so that they are 'cyclin-dependent kinases' (CDKs). (The yeast homologue of CDK1 is Cdc2, 'cdc' referring to 'cell division cycle'). RB1 is a negative regulator controlling progression from G1 to S phase. When expressed in response to stress, p53 also arrests the cell cycle and may promote **apoptosis** (programmed cell death). The sequential activation of kinases not only promotes the next step but can inhibit proteins involved in earlier stages, ensuring that the cycle is irreversible. The basic control processes are therefore phosphorylation, both by and of specific CDKs, and the processive synthesis and breakdown of cyclins.

CDC25 proteins are phosphatases that control entry into and progression through phases of the cycle by removing inhibitory phosphate residues from target CDKs. Entry into mitosis (G2 to M transition) is specifically regulated by the WEE1 kinase that phosphorylates and inactivates cyclin B1-complexed CDK1. Thus M phase entry is regulated by the balance between the inhibitory WEE1 and the removal of a key phosphate by CDC25A that permits mitotic progression.

An additional layer of control is provided by cyclin-dependent kinase inhibitors (CDIs) that inhibit CDKs to induce cell cycle arrest. INK4 (*inhibitor of CDK4*) family (A, B, C and D) inhibit cyclin D dependent CDKs. WAF1, KIP1 and KIP2 are less specific CDIs.

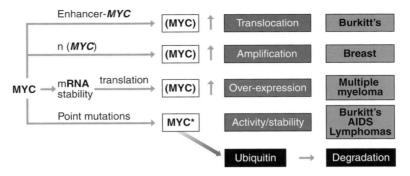

Figure 3.13 Mechanisms regulating MYC expression. Normal MYC abundance is regulated transcriptionally, post-transcriptionally and post-translationally. The de-regulation of MYC expression in cancers can also occur via multiple mechanisms.

mechanisms (Fig. 3.13). Normal MYC undergoes ubiquitin-mediated proteolysis, which confers a short half-life (~20 minutes). A number of mutations, arising for example in **Burkitt's lymphoma**, inhibit this degradative pathway and substantially increase the lifetime of MYC. **Amplification** and/or over-expression of *MYC* commonly occurs in a wide range of tumours. Mutations in the protein sequence are not necessary to render MYC oncogenic although when they do occur they may enhance its tumour-promoting capacity.

Cytokine receptors

The large and diverse family of cytokine signalling proteins includes lymphokines, interleukins and chemokines and the members play major roles in the immune system and hematopoietic cell development. The major classes of cytokine receptors also signal via tyrosine kinase activation, but for these the enzyme is a Janus kinase (JAK) – a distinct protein that associates non-covalently with the receptor (Fig. 3.14). Receptor dimerisation activates JAKs to *trans*-phosphorylate each other and to phosphorylate the C-termini of the receptors, to which STAT (*s*ignal *t*ransduction and *a*ctivator of *t*ranscription) proteins bind before becoming phosphorylated. STATs then dimerise and translocate to the nucleus to direct gene transcription. STATs may form stable homodimers or heterodimers that are active as transcription factors.

In addition to cytokines, growth factors (e.g. EGF) can also activate the JAK/STAT signalling pathway through the recruitment of the SRC tyrosine kinase to the EGFR. Activated SRC can then phosphorylate both JAKs and STATs. Conversely, a number of cytokines (e.g. IL3 and granulocyte-macrophage colony stimulating factor) can activate both JAK/STAT and MAPK pathways and the phospho-tyrosyl moieties of some activated cytokine receptors can provide binding sites for the regulatory subunit of **phosphatidylinositol 3-kinases** (PI3Ks: produce PIP, PIP_2 and PIP_3 from PI), as do EGFR family members.

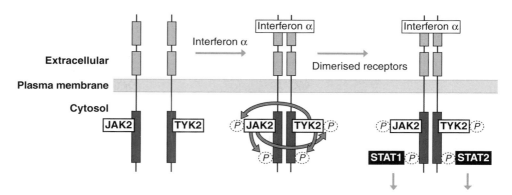

Figure 3.14 Cytokine signalling. Receptor activation by a member of the interferon alpha family. Other cytokines (e.g. interleukins, interferons, erythropoietin, thrombopoietin, growth hormone, prolactin and granulocyte colony stimulating factor) signal via receptors bound to other members of the JAK family (JAKs 1, 2, 3 and TYK2).

G-protein-coupled receptors

G-protein-coupled receptors (GPCRs), (also known as seven-transmembrane domain receptors, 7TM receptors, heptahelical receptors, serpentine receptors and G-protein-linked receptors (GPLR)), comprise a superfamily of over 1,000 genes that contribute to the control of most aspects of cellular behaviour. On the basis of shared sequence motifs and functional similarity they are grouped into four classes:

Class A (or 1) (Rhodopsin-like)
Class B (or 2) (Secretin receptor family)
Class C (or 3) (Metabotropic glutamate/pheromone)
Class D (or 4) (Fungal mating pheromone receptors)
Class E (or 5) (Cyclic AMP receptors)
Class F (or 6) (Frizzled/Smoothened)

GPCRs activate intracellular signal pathways via trimeric G proteins (Fig. 3.15). Ligand binding induces a conformational change in the receptor that promotes guanine nucleotide exchange from an interacting G protein (when the α subunit dissociates from the complex of β and γ subunits). This results in the activation of one of four classes of G_α subunits ($G_{\alpha s}$ (stimulatory), $G_{\alpha i}$ (inhibitory), $G_{\alpha q/11}$ or $G_{\alpha 12/13}$).

The two major signal pathways that are activated as a result are cyclic adenosine monophosphate (cAMP) and phosphatidylinositol (PI). G_s proteins activate adenylyl cyclase, which catalyses the production of cAMP from ATP, thereby activating protein kinase A (PKA), a major regulator of cell metabolism. $G_{\alpha i}$ inhibits PKA.

GPCRs activate PI signalling when the receptor binds to a G_q subunit: this activates phospholipase Cβ causing the hydrolysis of phosphatidylinositol 4,5-bisphosphate (PIP$_2$) into two second messengers, inositol 1,4,5-trisphosphate (IP$_3$) and diacylglycerol (DAG). The release of calcium from the endoplasmic reticulum raises the cytosolic

Signalling in normal cells

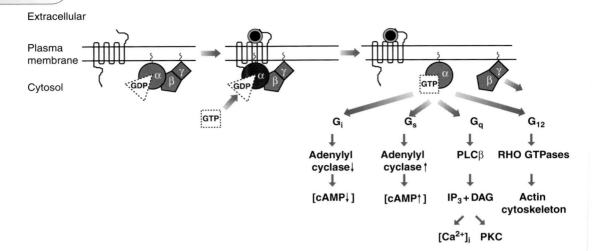

Extracellular

Plasma membrane

Cytosol

Figure 3.15 Hormonal activation of G-protein signalling via a G-protein-coupled receptor (GPCR). Hormone binding promotes exchange of GDP for GTP bound to the α subunit of trimeric G proteins. The α and γ subunits are attached to the plasma membrane by lipid anchors. Some of the activities transduced by representative members of the mammalian α subunit family are shown. IP_3: inositol 1,4,5-trisphosphate; DAG: diacylglycerol; $[Ca^{2+}]_i$: intracellular free calcium; PKC: protein kinase C.

concentration of calcium, which, together with DAG, activates the serine/theonine kinase protein kinase C (PKC).

Ligand-gated ion channel receptors

Ligand-gated ion channel receptors (LGICs) are trans-membrane pores, typically highly selective (e.g. for Na^+, K^+, Ca^{2+} or Cl^-), that are opened or closed in response to the binding of a ligand (usually a neurotransmitter). These ion channels are either an integral part of the receptor molecule (ligand-gated ion channels) or are linked to the receptor through a G-protein-mediated mechanism (ion channel-linked receptors). Ligand binding induces a conformational change in the channel-forming protein that increases ion flux across the membrane.

There are three LGIC superfamilies: cys-loop receptors (e.g. $GABA_A$ receptor, nicotinic acetylcholine receptor), ionotropic glutamate receptors (e.g. NMDA receptor) and ATP-gated channels (e.g. P2X) and each of these channels can be switched on and off rapidly. There is no evidence that abnormal LGICs contribute to cancers although there is evidence that altered expression of some *voltage*-gated ion channels is associated with the invasive capacity of some tumour cells.

These two classes of receptors thus play a less prominent role in cancer than the RTKs. Nevertheless, many forms of GPCR, some of their ligands and the coupling G proteins are aberrantly expressed in a range of cancers. Their activation may be autocrine (affecting the cells in which it is produced) or paracrine and promoted by tumour cells themselves (e.g. transforming growth factor α, insulin-like growth factor) or by stromal cells in the vicinity of the tumour (Chapter 5). Some tumour cells use detection of the level of the ligand stromal-derived-factor-1 (SDF1) by CXCR4 (a GPCR sometimes called fusin) as a guide to secondary sites during **metastasis. Endothelial cells** that form the inner lining of blood vessels are characterised by strong expression of S1PR1, which when activated by its ligand, sphingosine 1-phosphate, switches on cAMP, RHO and RAC GTPases and phospha-tidylinositol signalling pathways, acting as an important mediator of angiogenesis. An emerging complexity is cross-talk between GPCR pathways and the RTK network. Thus, for example, S1P receptors are *trans*-activated by RTKs and the prostaglandin receptor EP2 regulates the activity of the EGFR.

Steroid hormones

Having noted that most signalling molecules, that is hormones, are proteins and don't need to get into cells to deliver their message, we should mention a second group that work in a completely different way. These are the steroid hormones familiar to all because they include the sex hormones testosterone and oestrogen. All steroids are synthesised using cholesterol as a precursor, also familiar because, we are often told, too much of it is a bad thing (Fig. 3.16). Cholesterol itself is a rigid ring with a fatty (acid) molecule attached to it. That suggests cholesterol might be a membrane com-ponent and indeed in plasma membranes there's roughly one cholesterol for every phospholipid. Because of its rigid ring, cholesterol has the effect of limiting the flexibility of the phospholipid fatty acid chains – in other words cholesterol determines the fluidity and the permeability of membranes.

The various modifications that give rise to the steroid hormones convert cholesterol from an essential structural component of cells to molecules that can act as signals. The major steroid hormones are vitamin D, cortisol (or hydrocortisone, the major human glucocorticoid), oestrogens, progesterone and testosterone (which is a member of the androgen family). All these steroids are pretty insoluble in water, which means that to be carried around the circulation they need to attach to something that *is* soluble, the most common carrier being the protein serum albumin (which accounts for about 60% of the protein in plasma, the fluid in which blood cells are suspended). When these complexes come into contact with cells the steroid hormone can leave its carrier and diffuse into the lipid environment of the plasma membrane. By this means the hormone's signal is delivered to the cell and, in contrast to the protein hormones, the messenger actually enters the cell and completes the journey to the nucleus to direct gene expression. They can do this because mammalian cells make specific proteins to which steroid hormones bind.

These receptors are members of the nuclear receptor superfamily of transcription factors. Steroid hormone receptors may function as monomers or dimers but all contain a conserved DNA binding domain that recognises consensus DNA sequences

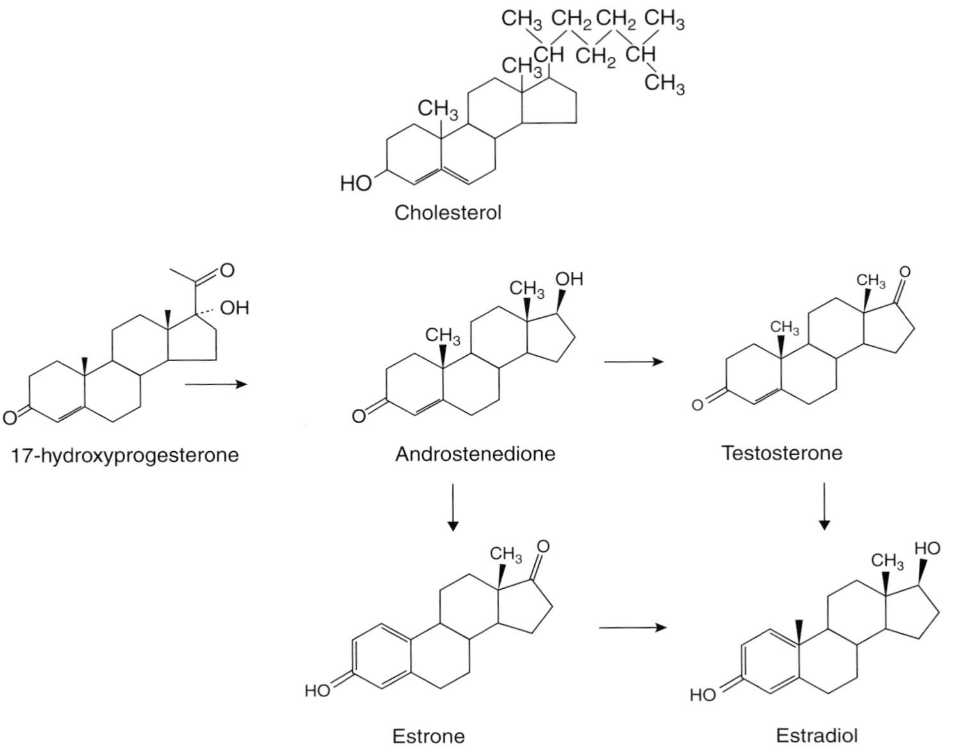

Figure 3.16 **Cholesterol and derivatives.** Cholesterol is a four-membered steroid ring attached to a hydrocarbon chain (top). Aromatase converts androstenedione, testosterone and 16-hydroxy-testosterone into oestrogen, 17-estradiol and 17-,16-estriol (lower).

(hormone responsive elements or HREs) in the regulatory regions of genes. Inactive receptors are usually bound to proteins that repress their activity as transcription factors. The conformational change caused by ligand binding releases inhibitory proteins and permits co-activator proteins to bind to activate transcription. For some receptors, however, hormone binding creates a complex that represses transcription. The cell-specific expression of receptors and co-regulatory proteins determines the transcriptional profile of gene expression in response to individual hormones.

There are four major categories of steroid hormone receptors:

Type I (hormone binds to receptor in the cytosol causing dissociation of heat shock proteins and translocation to the nucleus. The HREs for type I receptors are two half-sites separated by a variable number of bases, the second site being an inverted repeat of the first). Examples: receptors for oestrogen, glucocorticoid, progesterone and testosterone (Fig. 3.17).

Type II (in the absence of hormone remain bound to DNA as heterodimers, usually with retinoid X receptor alpha (RXRα)). Hormone binding promotes exchange of co-repressor for co-activator proteins). Examples: receptors for vitamin D, retinoic acid and thyroid hormone (Fig. 3.18).

Steroid hormones

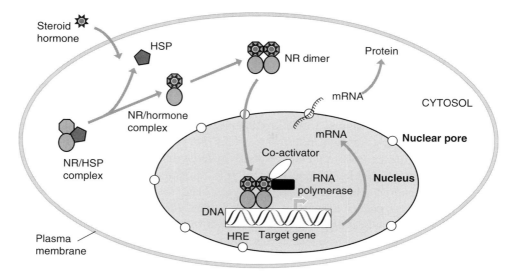

Figure 3.17 Type I steroid hormone receptor mechanism. Ligand (hormone) binding promotes dissociation of heat shock proteins, dimerisation and translocation to the nucleus where the receptor (NR) binds to a hormone response element (HRE) to activate transcription. For the oestrogen receptor the displaced heat shock protein is HSP90 and histone acetyltransferase is a co-activator of transcription. The single gene encoding the progesterone receptor generates two main isoforms, A and B. There are two oestrogen receptor (ER) genes and the encoded receptors are denoted as ERα and ERβ. However, there are at least three ERα and five ERβ alternatively spliced isoforms. There are three oestrogenic hormones that activate ERs, 17-beta-estradiol, estrone and estriol.

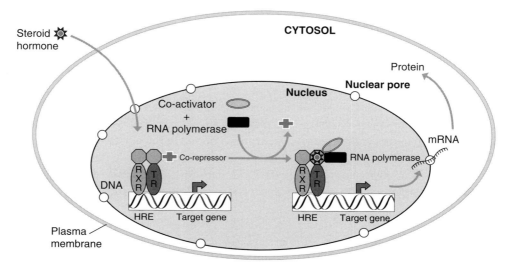

Figure 3.18 Type II steroid hormone receptor mechanism. The heterodimeric receptor, typically including RXRα, remains bound to its HRE even in the absence of ligand. Hormone binding promotes dissociation of co-repressor and recruitment of co-activator to activate transcription. For the thyroid hormone receptor the displaced co-repressor has histone deacetylase activity: the co-activator is histone acetyltransferase.

Type III (similar to type I but HRE is a direct rather than an inverted repeat). They are orphan receptors.

Type IV (receptors bind as monomers or dimers but HRE comprises a single half-site).

Steroid hormones are particularly important in cancer because oestrogens can stimulate the growth of breast and endometrial tumours and androgens may accelerate prostate tumours. Thus, for example, oestrogen receptors are over-expressed in 70% of breast cancers (referred to as ER-positive). Both ER and progesterone receptor status are used to predict response to endocrine therapy (i.e. treatment that modulates the effect of hormones). Thus ER^+/PR^+ tumours respond much better than do ER^+/PR^- tumours. Since 1980 tamoxifen has been the standard anti-oestrogen therapy for breast cancer because it can antagonise the action of oestrogen. Tamoxifen is a selective oestrogen receptor modulator (SERM). It is a pro-drug metabolised by cytochrome P450 to active forms that compete with oestrogen in binding to ERs. It is 'selective' because it shows tissue specificity: in the endometrium it acts as a partial agonist of the ER. In breast cells, however, ER/tamoxifen complexes bind to DNA and recruit co-repressors that inhibit transcription of genes activated by ER/oestrogen. Oestrogens are synthesised from androgens by the enzyme aromatase, and more recently aromatase inhibitors (anastrozole, letrozole) have been introduced as anti-cancer agents that work by lowering the level of oestrogen.

Conclusions

The critical feature of receptor tyrosine kinases (RTKs) is that when they are activated by their specific hormone ligands, their kinase activity is switched on to phosphorylate tyrosine amino acids in the cytosolic regions of the receptors themselves. This creates a 'scaffold' of phosphate-bearing tyrosines to which cellular proteins bind. These may be adaptors or enzymes and they initiate intracellular pathways that ultimately signal changes in the pattern of gene transcription in the nucleus to turn on the cell cycle that leads to proliferation. Highly conserved across all animal species is the mitogen-activated protein kinase (MAPK) pathway in which a molecular switch (the RAS protein) turns on a cascade of kinases leading to the phosphorylation (and hence activation) of protein transcription factors. One of the most important genes whose expression is essential for mammalian cell proliferation is MYC that acts as a 'master regulator' because it is a transcription factor that controls about 15% of all human genes. Although RTKs are the most prominent regulators of cell division, other types of receptor can be involved, notably G-protein-coupled receptors. Steroid hormones are a separate class of chemical messengers that cross the plasma membrane and, through specific intracellular receptors, interact directly with DNA to control gene transcription.

Key points

- Somatic mammalian cells can opt out of the cell cycle and enter a quiescent (G_o) phase.
- A wide variety of growth factors (or mitogens) can stimulate re-entry into the cell cycle.
- Many growth factors stimulate a range of intracellular responses but their actions are transduced by only a few types of receptor.
- A major trans-membrane signalling mechanism involves the activation of tyrosine kinase catalytic domains in the cytoplasmic region of growth factor receptors (RTKs).
- A major signalling pathway activated by such receptors involves RAS, RAF1 and the MAP kinase family and leads to activation by phosphorylation of transcription factors.
- Components of these pathways are frequently abnormal in cancers.

Future directions

- A major enigma is how cells channel a multiplicity of signal inputs to produce a discrete phenotypic response. Systems biology approaches are beginning to model responses but these are at a fairly embryonic stage. Of relevance to cancer is the question of the signalling mechanisms by which alternative pathways are up-regulated in response to specific RTK inhibitors.

Further reading: reviews

RECEPTOR TYROSINE KINASES (RTKS)

Schlessinger, J. (2000). Cell signalling by receptor tyrosine kinases. *Cell* **103**, 211–25.

ACTIVATING RAS AND MAPK

Buday, L. and Downward, J. (2008). Many faces of Ras activation. *Biochimica et Biophysica Acta-Reviews on Cancer* **1786**, 178–87.

SUSTAINED VERSUS TRANSIENT ACTIVATION OF RAS AND MAPK

McClean, M. N., Mody, A., Broach, J. R. and Ramanathan, S. (2007). Cross-talk and decision making in MAP kinase pathways. *Nature Genetics* **39**, 409–13.

McKay, M. M. and Morrison, D. K. (2007). Integrating signals from RTKs to ERK/MAPK. *Oncogene* **26**, 3113–21.

Pawson, T. (2004). Specificity in signal transduction: from phosphotyrosine-SH2 domain interactions to complex cellular systems. *Cell* **116**, 191–203.

MYC IS A CENTRAL REGULATOR OF CELL GROWTH AND PROLIFERATION

Wolfer, A. and Ramaswamy, S. (2011). MYC and metastasis. *Cancer Research* **71**, 2034–7.

CYTOKINE RECEPTORS

Wang, X. Q., Lupardus, P., LaPorte, S. L. and Garcia, K. C. (2009). Structural biology of shared cytokine receptors. *Annual Review of Immunology* **27**, 29–60.

G-PROTEIN-COUPLED RECEPTORS

Houslay, M. D. (2010). Underpinning compartmentalised cAMP signalling through targeted cAMP breakdown. *Trends in Biochemical Sciences* **35**, 91–100.

Palczewski, K. (2010). Oligomeric forms of G protein-coupled receptors (GPCRs). *Trends in Biochemical Sciences* **35**, 595–600.

LIGAND-GATED ION CHANNEL RECEPTORS

Dopico, A. M. and Lovinger, D. M. (2009). Acute alcohol action and desensitization of ligand-gated ion channels. *Pharmacological Reviews* **61**, 98–113.

STEROID HORMONES

Lin, S. X., Chen, J., Mazumdar, M. *et al.* (2010). Molecular therapy of breast cancer: progress and future directions. *Nature Reviews Endocrinology* **6**, 485–93.

4 | 'Cancer genes': mutations and cancer development

Cells are continuously exposed to DNA damaging events but due to the efficiency of the repair machinery only a small proportion of these are retained, that is, become somatically acquired mutations. For the development of most cancers the estimate is that between five and fifteen 'driver' mutations, that is, mutations in critical genes, are required. From whole genome sequencing we now know that these are typically part of a spectrum of tens of thousands of mutations acquired by tumour cell genomes. The two main classes of 'cancer genes' are oncogenes and tumour suppressors. In the former mutations confer gain-of-function and are dominant; in classical tumour suppressors gene function is lost. In addition, micro RNAs can also play important roles in cancer development through their capacity to regulate the expression of both oncogenes and tumour suppressor genes.

Collecting mutations

We saw in Chapter 2 that the list of things that can give us cancer has grown so long you might be inclined to wonder how it is that we don't all get cancer at a very early age. In fact most cancers develop very slowly, in part because we have evolved ways of repairing our DNA when it has been damaged so that only a small proportion of the lesions actually become 'fixed' in the genome. As we've seen, causes of DNA damage include high-energy electromagnetic radiation (γ rays, X-rays and UV radiation) and atomic particles emitted by radioactive atoms (α and β particles). This ionising radiation can damage DNA in two ways: (1) directly (e.g. by causing breaks that lead to chromosomal translocations and deletions); or (2) indirectly by interacting with water (radiolysis) to generate reactive oxygen species (ROS). The ROS formed are the hydroxyl radical, hydrogen peroxide and the superoxide radical.

$$H_2O \xrightarrow{\epsilon^-} \cdot OH \xrightarrow{\epsilon^-} H_2O_2 \xrightarrow{\epsilon^-} O_2 \cdot \xrightarrow{\epsilon^-} O_2$$

$$\underset{\text{Hydroxyl radical}}{} \quad \underset{\text{Hydrogen peroxide}}{} \quad \underset{\text{Superoxide radical}}{}$$

The hydroxyl (free) radical is one of the most reactive of all chemicals because it readily removes an electron from any molecule it encounters, converting that molecule in turn into a **radical** (usually called a free radical, a highly unstable, reactive molecule with an unpaired electron). This is a problem, given that we can't avoid external mutagens (radiation and chemical carcinogens). But it's worse than that because we make these very reactive free radicals as by-products of some normal cellular reactions.

We've also seen that some of our food contains cancer-causing chemicals (e.g. polycyclic aromatic hydrocarbons, aromatic amines, alkylating agents). Electrophilic carcinogens react with nucleophilic sites in the purine and pyrimidine rings of DNA, forming covalent links ('adducts'). These chemically modified bases in DNA can result in errors in repair and hence mutations. The main reason why smoking is such a powerful contributor to the cancer statistics discussed earlier is that tobacco smoke contains several potent carcinogens that form a wide range of DNA adducts.

We have three lines of defence. The first lies in what we eat. The reason why we are so often told to eat fresh fruit and vegetables is that they contain several antioxidants that scavenge ROS (e.g. vitamin C and α-tocopherol) and so confer a degree of protection. The second is that we have evolved a number of cellular enzymes (e.g. superoxide dismutases) that protect against free radicals. These provide a biochemical buffer by reacting with the dangerous by-products of normal metabolism and converting them into harmless substances.

Nevertheless, despite these protections, it is estimated that the genome in a normal adult human cell acquires ~20,000 DNA lesions (i.e. chemically damaging hits) per day. Many of these might result in mutations *if* the modified, damaged DNA was not repaired correctly. Thus, the third and most important protective mechanism is a variety of DNA repair processes that correct nearly all of the lesions formed. Of the 20,000 distinct DNA 'hits' per day, on average less than one of these remains in the DNA to be passed on when the genome is replicated. Mutations therefore accumulate at the rate of one every day or two to give ~10,000 mutations over a lifetime.

Genetic roulette

Thus about 0.0003% of the bases comprising the human genome are damaged beyond repair in the course of a normal lifetime. But the really interesting question is: what happens in a cancer cell? It's only in the last few years that we've been able to answer with other than a 'guesstimate' but now, thanks to the sequencing revolution, we know that in a typical human cancer cell the number of such mutations is ~10,000 although the range for different cancers is from 1,000 to 100,000. Note that these cancer mutations will be superimposed on the 'normal' mutational background. Recall that the human genome has three billion (3×10^9) base pairs with about 22,000 genes, the regions encoding proteins taking up <2% of the total DNA. Although the vast majority of mutations occur in inter-gene regions of the DNA, there are likely to be ~100 mutations that alter amino acids in each cell, i.e. <1% of the ~22,000 genes have acquired a mutation. Non-coding mutations can of course alter the behaviour of cells, but for the moment let's focus on the 100 coding mutations. Of the 100 mutated genes, only a small number actually *drive* the development of the cancer. With some exceptions it takes a long time for a critical set of mutations to accumulate in a single

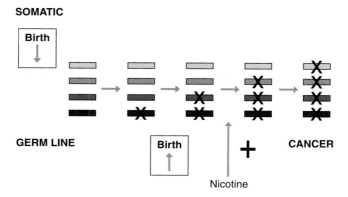

Figure 4.1 A mutational steeplechase leads to cancer. Of the tens of thousands of mutations that can accumulate in a cancer cell, some five to fifteen distinct 'driver' mutations are responsible for the development of the cancer. In hereditary cancers an individual is born with one such 'germ line' mutation, thereby conferring a high probability of cancer developing in the carrier. The rate at which mutations arise is increased by exposure to carcinogens, e.g. nicotine.

cell – that's why most forms of cancer are diseases of old age. Despite the sequence revolution, we still don't know for any cancer precisely what the critical number of mutations is but estimates range from five to fifteen distinct mutations (Fig. 4.1), although fewer may be required for some types, in particular leukaemias. Whatever the precise number, they make up a set of **'driver' mutations** sufficient to override the normal controls of cell proliferation. When their effects emerge at the earliest stage of tumour development the precursor cells are therefore monoclonal (derived from one single cell).

The five to fifteen distinct 'driver' mutations are thus a sub-group of the random mutations that become 'fixed' (i.e. remain) in the genome and the term 'driver' is used to distinguish mutated genes that specifically contribute to cancer development from the **'passengers'** that don't do much. A further distinction gives rise to the concept of 'restricted cancer genes' that are in effect 'drivers' for specific types of cancers, for example, the translocations that are critical for some leukaemias (e.g. *BCR-ABL1*, see below) or the *EWS-ETS* fusion in **Ewing's sarcoma**.

The cancer genomic landscape

Whole genome sequencing has already facilitated comparison of DNA sequences from hundreds of tumours with those of normal tissue from the same individuals. This permits the sequence comparison of essentially all genes to identify mutations that characterise specific tumours or sub-sets thereof.

Showing genes with mutations as a 'landscape' vividly illustrates their distribution across the genome. Imagine a separate dot for every gene (~20,000) on the map of Fig. 4.2. If there are 100 coding mutations, there should be 100 dots on the map.

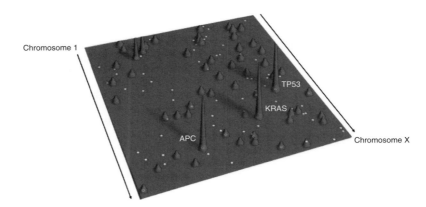

Figure 4.2 Two-dimensional map of genes mutated in a group of colon cancers. Chromosomes are arranged end to end with the short arm of chromosome 1 at the rear left corner. Of the ~100 mutations affecting proteins, five occur in most tumours ('mountains') and ~50 are present in about 5%. The mountains include *APC*, *KRAS* and *TP53* (aka *P53*) and two others (*PIK3CA* (chromosome 3) and *FBXW7* (chromosome 4)). The rest are scattered in other genes (not shown). (Wood *et al.*, 2007).

In Fig. 4.2 infrequent mutations (occurring in only one or two tumours) have been omitted. The rest fall into two groups: about 50 genes are mutated in about 5% of these tumours and these show up as 'hills' – i.e. they're fairly common. A small number (five) have arisen in almost all of these tumours – these are 'mountains' – that is, cancer drivers. Four of them – *APC*, *KRAS2*, *TP53* (aka *P53*) and *PIK3CA* (encoding the catalytic alpha subunit of phosphoinositide-3-kinase) – were anticipated as they were known to be frequently mutated in colon cancer. We will consider their function shortly and take up the commonly mutated gene, *PIK3CA*, in Chapter 6. The other gene affected, *FBXW7*, encodes a protein that causes degradation of an important regulator of cell proliferation, cyclin E: its inactivation leads to **genetic instability**.

Having introduced the expression 'cancer genes' we should note that it is really jargon: strictly there are no such things but it's a useful term if you define it to mean genes that, as a result of some change, have become abnormal in terms of the activity of the protein (or the RNA) they encode.

Mutations in 'cancer genes': oncogenes and tumour suppressor genes

Tumours result from subversion of the processes that control the growth, location and mortality of cells. This loss of normal control mechanisms arises from the acquisition of mutations in three broad categories of genes:

1. **Proto-oncogenes.** These encode components of signalling pathways that regulate proliferation. In their mutated form they can become **dominant** 'oncogenes'.

2. **Tumour suppressor genes.** These encode proteins whose loss of function leads to de-regulated control of cell cycle progression, protein degradation, cellular adhesion and motility. They generally exhibit recessive behaviour.
3. **DNA repair enzymes.** These proteins maintain genomic integrity and mutations causing loss of function therefore attenuate the repair processes we have discussed and promote genetic instability.

Mutations in these genes are also presumed to produce changes in cell surface protein expression, protein secretion and cell motility that contribute to metastasis, no mutations having been specifically associated with metastatic development. Genes shown to have a functional association with specific cancers now number some 500 oncogenes and about 100 tumour suppressor genes.

Oncogenes

Oncogenes were first identified in viruses capable of transforming cells in culture or inducing tumours in animals. The classic experiment of Peyton Rous in the early years of the twentieth century showed that cell-free filtrates from chicken tumours can give rise to **sarcomas** (tumours of **connective tissue**) when inoculated into normal chickens. The infectious agent was identified by electron microscopy in the 1950s as a virus – the Rous sarcoma virus – eventually revealed to be a **retrovirus**, i.e. having an RNA genome. Studies of temperature-sensitive mutants of the virus showed that sustained expression of a viral protein was required for transformation. The gene encoding this protein had been picked up by the virus from the DNA of its host during the normal life cycle of the retrovirus in which viral RNA is converted to DNA and inserted in the host genome. The cellular gene that had been captured by the virus was *Src*, which encodes a tyrosine kinase highly conserved throughout the animal kingdom. Many other oncogenes have now been identified both in retroviruses and by other means, including, most recently, whole genome sequencing. They have in common their derivation from normal genes (proto-oncogenes) that are highly conserved across all species, they function mainly in growth factor signalling pathways and they act in a dominant manner (e.g. a single allele mutation is activating). Mutant forms of the virus were also found that multiplied normally in infected cells but did not cause transformation. The non-transforming mutants had lost all or part of the gene that had been acquired from the host genome in the generation of the transforming virus (Fig. 4.3).

Although many retroviruses can cause tumours in animals, particularly the feline and bovine leukaemia viruses, they are rarely associated with human cancers. The RNA human immunodeficiency virus (HIV) destroys cells of the immune system so that victims become susceptible to infection. Infection by the Kaposi sarcoma associated herpes virus (KSHV or HHV8) causes the gradual development of Kaposi's sarcoma (which is not really a sarcoma because it starts in the lymphatic system). This is because KSHV encodes two proteins that act directly on cell signalling pathways to override normal controls of proliferation and survival. HIV is a member of the human T-cell

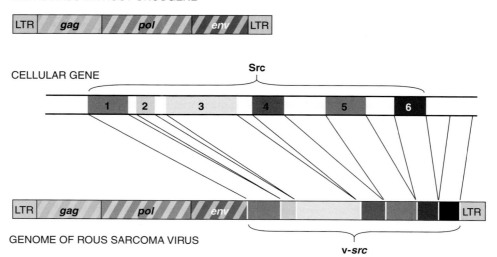

RETROVIRUS WITHOUT ONCOGENE

CELLULAR GENE

GENOME OF ROUS SARCOMA VIRUS

Figure 4.3 Acquisition of a cellular gene by a retrovirus. The three major structural and replicative retroviral genes are *gag* (encoding group-specific antigens: viral capsid proteins), *pol* (reverse transcriptase and integrase) and *env* (the viral envelope proteins). The six exons of Src are numbered. These are incorporated into the RSV genome together with a small deletion of exon six and additional 3′ sequence (black box).

lymphotropic virus (**HTLV**) family (it's HTLV-III). HTLV-I weakens the immune system and so increases the risk of opportunistic infection by bacteria, fungi, viruses or protozoa. The HTLV-I virus is thought to cause the rare adult T-cell leukaemia, a form of non-Hodgkin's **lymphoma**, but this cancer takes a very long time to develop and only a small fraction (1 in 1,500) of those infected with HTLV-I develop the disease. HTLV-II and HTLV-IV have not been specifically linked to any disease.

The first human oncogene

The key experiment showing that the human version of a gene that had been acquired by an oncogenic retrovirus can cause cancer, independent of any virus, came in 1983. In principle it couldn't have been simpler: extract DNA from a human tumour, fragment and transfect into cells in culture (Fig. 4.4). Pick out the cells that become transformed, inject them into mice, excise the tumour that develops and isolate the gene responsible from the tumour cells. If the transfection is into mouse cells, the cancer-promoting gene can be identified as human and thus derived from the original tumour. Through this kind of experiment Robert Weinberg and colleagues identified the first human oncogene, *RAS* – so named because a retroviral form causes *rat* sarcomas. We now know that there are three closely related human *RAS* genes (*NRAS,*

What turns a proto-oncogene into an oncogene?

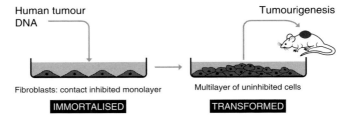

Figure 4.4 **Transformation of an immortal cell line by DNA from a human tumour.** Primary cultures of normal cells (freshly dispersed from tissues) have a limited lifetime before they stop proliferating and become **senescent** or die. However, it is possible to select from such cultures cells that have become immortal, that is acquired mutations that permit them to proliferate indefinitely (one characteristic of tumour cells). These cells still grow as a monolayer in which proliferation spontaneously stops when the cells cover the dish (**contact inhibition**). Transfection of immortal cells with human tumour DNA can cause **transformation** (loss of contact inhibition) and give rise to cells that form tumours when injected into mice. Human DNA can be detected in the transformed mouse cells by using a probe for the *Alu* family of repetitive sequences present every few thousand base pairs in the human genome. A second round of transfection distributes the human DNA among many recipient mouse cells: only a few *Alu* **sequences** remain adjacent to the oncogene and DNA from cells in separate foci can be shown to contain this sequence in a **Southern blot**. The oncogene fragment can be isolated and cloned by making a genomic DNA library from these cells and screening with the *Alu* probe. This method led to the discovery of *RAS*, the first human oncogene to be identified.

HRAS1 and *KRAS2*), each encoding a GTPase implicated in control of cell proliferation. Mutations that convert them from proto-oncogenes to oncogenes occur in about 20% of human tumours (see below).

What turns a proto-oncogene into an oncogene?

In the development of human cancers proto-oncogenes can be converted into oncogenes as a consequence of mutations (point mutations or deletions/insertions), chromosomal rearrangement (which includes gain of entire chromosomes (**aneuploidy**) and chromosome translocation) or gene amplification (Fig. 4.5). Amplification and chromosome translocation can give rise to elevated cellular concentrations of the normal gene product; chromosome translocation can also cause the expression of new proteins created by fusion of coding sequences from separate genes.

Figure 4.5 represents the basic ways in which mutations can affect cancer genes. The smallest is a single base change, something that will often have no effect at all but, if it happens to make a critical change in the amino acid sequence or in a transcriptional control element, may have dramatic consequences. More extensive changes to proteins can include the deletion of entire domains. Errors made by the chromosome replication machinery can multiply the number of copies of an otherwise normal gene – gene amplification. Conversely, complete loss of a gene (hence no protein product can be

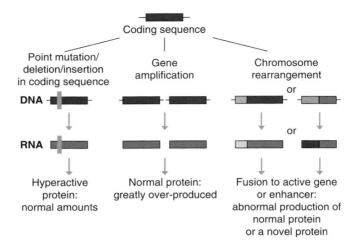

Figure 4.5 DNA mutations. The smallest mutation is a change of one base to another (which can change the activity of the protein) or the loss or insertion of a single base. Genes or large segments of chromosomes may be duplicated (amplified) many times (which can give elevated cellular concentrations of the normal gene product) or entire chromosomes may be duplicated or lost (aneuploidy). Segments of DNA may be shifted from one chromosome to another (chromosome translocation), which can generate new proteins through the fusion of coding sequences from separate genes or place genes under new control sequences.

made – so there is complete loss of function) is also a common feature of cancers and provides one means for the inactivation of tumour suppressor genes (see below). Finally, DNA shuffling in the form of chromosome translocations can produce a novel protein (a chimera) or a normal protein that comes under different control (i.e. it is made in the wrong place or at the wrong time or in inappropriate amounts).

A single base change: RAS

The normal RAS proteins are molecular switches that relay signals from cell surface receptors into the cell (see Chapter 3). Its capacity to act as a switch derives from its binding to the small molecule GTP (a nucleotide: deoxyguanine 5′-triphosphate). GTP binding stimulated by activation of receptors switches RAS, changing its conformation to open a binding site for the next protein (RAF) in the signalling pathway from the receptors to the nucleus (Fig. 4.6).

However, RAS is also an enzyme. It catalyses the breakdown of GTP by removing one of its phosphate groups, to give GDP (i.e. it hydrolyses guanosine *tri*phosphate to guanosine *di*phosphate). This simple change, the conversion from RAS-GTP to RAS-GDP, switches off RAS. That is, when GDP is bound RAS adopts a conformation in which the site of interaction with the next protein (RAF) in the signal cascade is masked. The most common mutation in RAS is a single base change that converts the amino acid at the 12th codon from glycine to valine. Glycine 12 plays a critical role in permitting bound GTP to be hydrolysed to GDP. Any other amino acid in this

What turns a proto-oncogene into an oncogene?

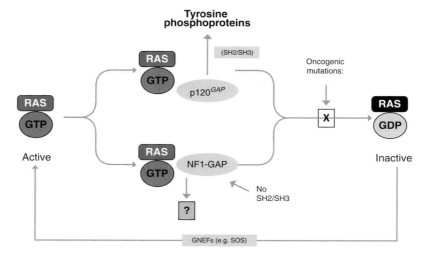

Figure 4.6 Regulation of RAS activity by GTPase activating proteins. In Chapter 3 we noted that RAS is activated by the dissociation of GDP, which is achieved by SOS. It is switched off by the conversion from RAS-GTP to RAS-GDP. GNEFs: guanine exchange factors.

position inhibits the reaction and RAS remains locked in the transition state. RAS has become 'hyper-activated' in the sense that, although its method of signalling is unaltered, it now signals constitutively: the molecular switch is permanently 'on'. Other activating point mutations can occur in RAS (Ala59 and Gln61) but all inhibit GTP hydrolysis either by diminishing GTPase activity or (for Ala59) modulating the rate of GTP-GDP exchange.

Other coding mutations: the epidermal growth factor receptor (EGFR)

The major cell surface receptors that signal via RAS are the receptor tyrosine kinases (RTKs: Chapter 3), of which the first to be identified was that for epidermal growth factor (EGF). The EGFR typifies the RTK family, having a large, single, trans-membrane polypeptide chain (~1,250 amino acids long). Epidermal growth factor signalling regulates proliferation, differentiation and survival. In common with RAS and SRC, the EGFR has been transduced by a retrovirus with an associated mutation that gives growth factor-independent kinase activity (Fig. 4.7). It is perhaps not surprising, therefore, that the EGFR is quite frequently abnormal in human cancers. In breast and ovarian tumours and in a form of **lung cancer** a frequent mutation deletes a large, extracellular segment including the ligand-binding domain. This corresponds to the viral mutation and, rather than switching off signals from the receptor, the opposite occurs: unhindered by the outer domains, the cytosolic (and trans-membrane) regions are drawn to each other and switch on the internal signalling pathway irreversibly, without the requirement for a growth factor. Cells carrying this mutant form of the EGFR signal regardless of growth factor levels: they've become autonomous and taken a big step towards becoming a tumour cell. Other mutations have been

'Cancer genes': mutations and cancer development

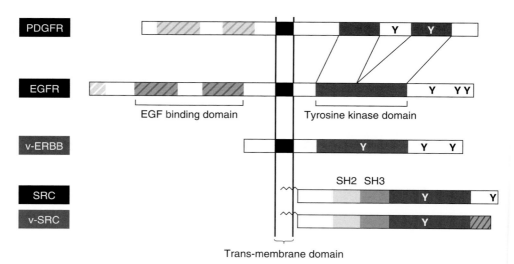

Figure 4.7 **Structures of tyrosine kinase proteins and retroviral oncoproteins.** Gene fusion of PDGFRA and VEGFR2 and intragenic deletions in PDFGRA occur in some gliomas. The product of the v-*erbB* gene, derived from the *Erbb1/EGFR* gene, lacks the extracellular epidermal growth factor binding domain and a tyrosine *trans*-phosphorylation site in the truncated C-terminus by comparison with the normal EGFR. Major deletions of the extracellular domain of EGFR occur in a variety of human cancers. In cellular SRC phosphorylation of a C-terminal tyrosine normally suppresses kinase activity. In v-SRC this region is deleted and the kinase is constitutively activated: some point mutations also occur in v-SRC. The corresponding mutation occurs in a sub-set of human colon cancers.

identified in the tyrosine kinase domain that constitutively activate the enzyme and thus are functionally equivalent to the extracellular domain truncation (Fig. 4.8). In addition, a variety of exon combinations may be deleted in diverse cancers and the EGFR may also be amplified or over-expressed (Fig. 4.9).

Other members of the RTK family behave aberrantly in a variety of cancers. Thus, for example, up-regulated expression of members of the PDGF family and their receptors has been reported and signalling through both forms of the receptor (PDGFRA and PDGFRB) is associated with invasive breast cancer. Mutation and amplification of *PDGFRA* occurs in some brain tumours and gastrointestinal stromal tumours frequently have activating mutations in either *KIT* or *PDGFRA*. Members of the fibroblast growth factor (FGF) family are important regulators of development and angiogenesis, and activating germ line mutations cause a number of serious conditions. Somatic mutations in FGFR2 and FGFR3 are associated with several types of cancer including breast and colon (Fig. 4.10).

Gene amplification: MYC

The duplication of entire genes occurs frequently in human cancers and was first discovered in the most common solid cancer in children – **neuroblastoma**. From a

What turns a proto-oncogene into an oncogene?

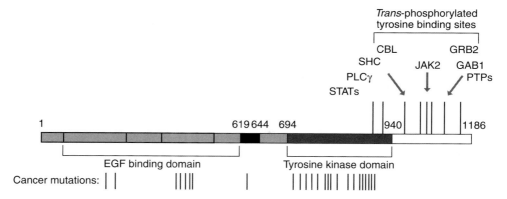

Figure 4.8 The EGFR: frequent sites of point mutations.

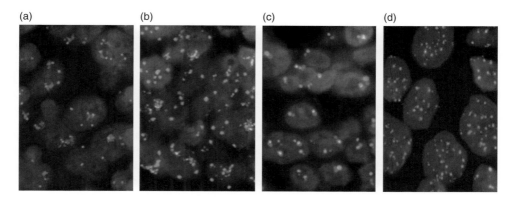

Figure 4.9 EGFR gene amplification in non-small cell lung carcinoma. The EGFR gene copy number was determined by fluorescence *in situ* hybridisation (FISH) assay using two fluorescent probes one targeting the centromere of the chromosome (green), the other the EGFR gene (red). The pictures show four patterns of EGFR gene amplification. (a) Large EGFR gene clusters; (b) co-localised clusters of EGFR and centromere signals; (c) large and bright EGFR signal, larger than the centromere signals in tumour cells; (d) high frequency of balanced EGFR and centromere signals (Varella-Garcia, 2006). (See plate section for colour version of this figure.)

search for novel proto-oncogenes there emerged *NMYC*, which strikingly may be amplified up to 1,000-fold in advanced cases. *NMYC* is a close relative of MYC, a transcription factor that is essential for cell proliferation. *MYC* was the second onco-gene to be discovered after *SRC* and was identified through the insight of Michael Bishop in analysing an **avian myelocytomatosis virus (AMV)**, so called because it causes tumours in **myeloid cells**. The virus also induces carcinomas, the most prevalent of human cancers, so Bishop reasoned that its genome might bear something at least as exciting as *Src*. It was a brilliant piece of educated guesswork because the gene in question turned out to be *Myc* (named from *my*elocytomatosis). It transpired, as we

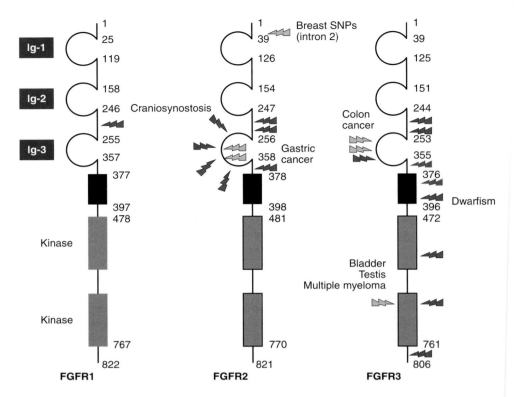

Figure 4.10 **Structural similarity and disease mutations within the fibroblast growth factor receptor (FGFR) family.** There are four closely related members of the FGFR family and their amino acid sequences are highly conserved both within the family and throughout evolution. Alternative splicing of exons encoding the extracellular domains of FGFR1, 2 and 3 results in multiple isoforms with unique ligand binding specificities and tissue localisations. A number of mutations in FGFR1, 2 and 3 cause **autosomal** dominant disorders of skeletal and cranial development (dark arrows). FGFR2 was first identified as an amplified gene in human gastric cancer and somatic mutations have been identified in FGFR2 and FGFR3 in human cancers. The gastric cancer FGFR2 mutations are identical to two of the germ line activating mutations responsible for craniosynostosis syndromes, suggesting that a similar mechanism of ligand-independent constitutive activation of the FGFR2 product is responsible for both diseases.

RAS and FGFR3 mutations are mutually exclusive, which may reflect the fact that both RAS and FGFR3 activate the MAPK pathway.

Two variations in the sequence of the second intron of FGFR2 alter transcription factor binding affinity (OCT1/RUNX2 and C/EBPβ): they cause an increase in FGFR2 expression, thereby increasing the risk of developing breast cancer.

noted in Chapter 3, that expression of *MYC* is essential for mammalian cell proliferation, and it is the oncogene most frequently associated with the development of human cancers. In most types of cancer cell the amount of MYC protein is increased relative to normal cells – in some colon cancer cells there are over 100,000 molecules of MYC compared to fewer than 1,000 in normal cells (Table 4.1). In most cases this is

Table 4.1 Representative concentrations of MYC protein, molecules per cell.

Quiescent cells	<800
Serum-stimulated fibroblasts	5,000
Lymphoblastoid cells (Bloom's syndrome)	48,000
Burkitt's lymphoma	100,000

Abnormal MYC expression occurs in most tumours, frequently as a result of amplification. **Bloom's syndrome** is a rare condition that pre-disposes to a range of cancers, arising from mutations in a DNA helicase that causes chromosomal damage. MYC is over-expressed in **Burkitt's lymphoma** as a result of a translocation (see below).

due to over-expression of the gene (synthesis of too much protein) rather than amplification (duplication of the gene itself), although in some tumour sub-sets MYC amplification is very prevalent (e.g. in breast tumours with *BRCA1* mutations). Thus mutations in MYC protein are not necessary to render MYC oncogenic although when they do occur they may enhance pathogenicity.

Gene amplification occurs as a result of errors during DNA replication but the mechanism is less important than the fact that it increases the number of copies of the normal gene. The number of duplications may range from two copies to the 1,000 sometimes found in neuroblastomas. The upshot is that more protein is made than would be produced in a normal cell. For *MYC* in particular it may be readily appreciated that if you have too much of a protein that promotes cell proliferation you might have a very good cancer driver.

The duplication of relatively long stretches of DNA within the human genome occurs quite frequently. About 5% of the human genome is made up of what's called segmental duplications – that is, segments longer than 1,000 bases that are virtually identical in sequence to stretches found elsewhere. Segmental duplications make up more than half the Y (male) chromosome. So gene amplification in cancer is only a more extreme version of something that has been a normal part of evolution, thought to have been favoured for the increase in genetic potential that it conferred. The prevalence and significance of this duplication is perhaps best illustrated by the fact that, from whole genome sequencing, we now know that, on average, about 80 genes differ in copy number between the genomes of any two individuals – and that difference in gene dosage (reflected by the amount of proteins made) makes a significant contribution to individual variation.

Chimeric protein: BCR-ABL1, PML-RARA and ETV6-RUNX1

Chromosome translocations are particularly common in cancers arising in bone marrow cells – leukaemias and lymphomas. Specific translocations are associated with particular types of leukaemia and the introduction of chimeric genes into mice or zebra fish has shown that they are sufficient to trigger development of the disease, even though additional mutations subsequently accumulate. The best known chromosomal

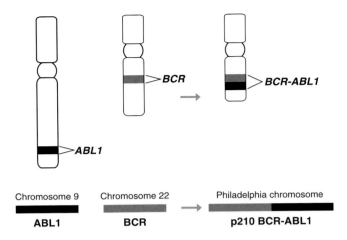

Figure 4.11 The Philadelphia chromosome. (Top) The 9:22 reciprocal translocation places *ABL1* within the *BCR* gene on chromosome 22: the product of the gene is p210 BCR-ABL1 (bottom). Patients with chronic neutrophilic leukaemia who have this translocation express another variant, p230 BCR-ABL1.

translocation was identified in 1960 by Nowell and Hungerford in a case of chronic myeloid leukaemia (CML). Janet Rowley subsequently showed that it involved a fragment from the tip of chromosome 9 displacing a large part of chromosome 22, producing a shortened, hybrid chromosome – the Philadelphia chromosome (Fig. 4.11). The effect of this fusion is that the chromosome 22 gene *BCR* is placed directly in front of the *ABL1* gene that is carried on the fragment of chromosome 9. Because the fusion is 'in frame' the transcription and translation machinery now produce a chimeric protein (BCR-ABL1) that is the driving force for leukaemia development. The main reason for this is that ABL1 encodes a tyrosine kinase enzyme whose activity is increased in the chimeric form. Tyrosine kinases can act as powerful proliferation signals and ABL1 appears to be particularly potent in this respect when expressed in bone marrow cells. It is notable that, in a similar manner to the *SRC* and *RAS* oncogenes, there is a mouse virus that has picked up a mutated version of the *ABL1* gene. This virus causes leukaemia in infected animals and the tyrosine kinase activity of the viral ABL1 protein is similar to that of chimeric BCR-ABL1 in humans. The Philadelphia chromosome is present in over 90% of CML cases and also occurs in 10% to 20% of patients with acute lymphoblastic leukaemias (ALL).

Acute promyelocytic leukaemia (APL), a sub-type of acute myelogenous leukaemia (AML), is characterised by the capacity of all-*trans* retinoic acid (ATRA) to induce remission. ATRA is the acid form of vitamin A and is used to treat acne (it's also called tretinoin). AML stems from chromosomal translocations involving the retinoic acid receptor alpha (*RARA*) gene and, most commonly, the *promyelocytic leukaemia* (*PML*) gene (Fig. 4.12). The PML protein is a transcription factor that functions as a tumour suppressor, regulating cell proliferation, survival and **senescence**. In part its tumour suppressor activity derives from interaction (of isoform IV: there are seven known isoforms) with the tumour suppressors p53 and RB1. The fusion protein PML-RARA

What turns a proto-oncogene into an oncogene?

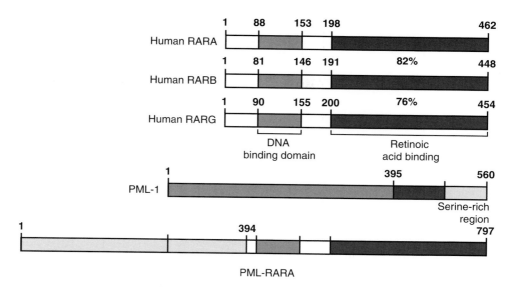

Figure 4.12 **Structures of (the retinoic acid receptor family) RARA, RARB and RARG, and PML and PML-RARA.** Alternative splicing of the normal PML gene generates multiple isoforms with differing C-termini. Recurrent chromosomal translocations involving the RAR locus are the hallmark of acute promyelocytic leukaemia (APL). RARA undergoes reciprocal translocation with PML in >95% of cases of APL but it can also fuse to various other partners leading to the expression of APL-specific fusion proteins with identical RAR moieties. Multiple isoforms of PML-RARA co-exist due to (i) variation in the chromosome 15 break point within three PML breakpoint cluster regions; (ii) alternative splicing of PML; and (iii) alternative use of two RARA polyadenylation sites.

essentially suppresses this suppressor function, so that PML-RARA acts as the transforming protein of APL. PML-RARA proteins form DNA binding complexes with members of the RXR family of retinoic acid receptors. Thus PML-RARA is a hormone-dependent transcription factor that acts as a dominant negative oncoprotein to inhibit myeloid differentiation and promote proliferation. ATRA is an RAR agonist that blocks the dominant negative effect of PML-RARA, permitting undifferentiated leukaemic blast cells to differentiate into neutrophils. The mechanism of ATRA action is a (rare) example of a therapeutic strategy in which tumour cells are induced to differentiate and hence cease to proliferate: they have exited the cell cycle and entered the G_0 quiescent phase or 'post-mitotic' state. In combination with other drugs ATRA gives remission rates of over 90%.

In the 1970s arsenic trioxide (As_2O_3) was also shown to give high remission rates for APL and both retinoic acid and arsenic are used for the clinical treatment of this disease (Fig. 4.13). Although they interact with different regions of PML-RARA, they both cause the protein to undergo attachment of **ubiquitin** that initiates its degradation by the **proteasome**.

When leukaemias either fail to respond to chemotherapy or patients relapse after a period of remission, the cause appears to lie in a very small population of 'leukaemia

(a)   (b)   (c)

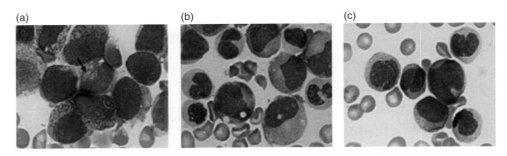

Figure 4.13 Acute promyelocytic leukaemia (APL): induction of differentiation in bone marrow blast cells. (a) No treatment: showing immature promyelocytes with characteristic granularity and Auer rods, elongated needles found in the cytosol of leukaemic blasts (see arrow). (b) Cells from patients treated for three weeks with retinoic acid. (c) Cells from patients treated for three weeks with As_2O_3. In (b) and (c) the granulocytes have started to differentiate and have large areas of clear cytoplasm. Nuclear remodelling is also observed, ranging from a simple indentation to the polylobular nuclei of terminally differentiated granulocytes (Zhu *et al.*, 2002). (See plate section for colour version of this figure.)

initiating cells' (LIC). Leukaemia initiating cells resemble normal hematopoietic stem cells in that they are generally quiescent. However, they are pluripotent and retain the capacity to replicate, and provide perhaps the strongest evidence for the cancer 'stem cell hypothesis' to be discussed in the next chapter. Because they are generally not proliferating they tend to be resistant to elimination by drugs. Thus, for example, although imatinib (see Chapter 6) is effective against chronic myeloid leukaemia (CML), LICs are likely to survive chemotherapy to cause relapse. PML is expressed at high levels in LICs from a mouse model of CML and also in blast cells from CML patients. It transpires that PML plays a critical role in maintaining LICs in a quiescent state. When PML is inhibited by As_2O_3, LICs re-initiate proliferation and thus become susceptible to anti-leukaemic therapy.

The leukaemias have been particularly revealing in terms of cancer development because individual cells carrying 'driver' translocations, e.g. *BCR-ABL1*, can be identified. **Fluorescence *in situ* hybridisation** (FISH) uses labelled probes to detect specific translocations, permitting identification of the leukaemic cells in a blood sample. DNA from these cells can then be sequenced to determine the complete mutational pattern in individual tumour cells. Perhaps the most informative evidence relating to stem cells and cancer, and indeed to the general way in which tumours develop, has come from childhood ALL.

Three quarters of children with ALL have a translocation that generates a chimeric protein from the *ETV6* and *RUNX1* genes (Fig. 4.14). However, each child is unique in the location of the breakpoints giving rise to the chimeric gene, revealing that the break has occurred in one cell in the fetus that has then multiplied. Although the incidence of *ETV6-RUNX1* fusions is ~1%, only about 0.01% of children get leukaemia – clearly indicating that the fusion is a 'promoter' event following which further disruptions are required (deletion of *ETV6* is common as a second event to *ETV6-RUNX1* fusion). This is consistent with the disease having a protracted latency

What turns a proto-oncogene into an oncogene?

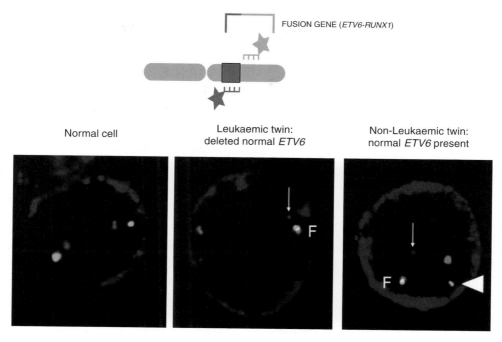

Figure 4.14 The *ETV6–RUNX1* translocation in childhood ALL. Fluorescent probes identifying the two genes show them on distinct chromosomes in a normal cell (left) and juxtaposed in cells from the leukaemic twin (centre) and healthy twin (right). *ETV6* is green, *RUNX1* red and the fusion gene *ETV6-RUNX1* is green-red (F). The weak red signals (arrows) are the remnants of the *RUNX1* locus left by the translocation. The green signal (open arrowhead) is the intact second *ETV6* allele. CD19 (blue) is a surface antigen expressed by these leukaemia cells. (Images contributed by Mel Greaves (Institute of Cancer Research) from Hong *et al.* (2008).) (See plate section for colour version of this figure.)

(that is, it may appear at any time up to about 14 years of age) and that transgenic mice with *ETV6-RUNX1* or *RUNX1-RUNX1TI* don't get leukaemias.

Analysis of the mutational pattern in individual cells reveals that they differ – in other words that the tumour is made up of different clones, albeit that have all been driven by the same initiating event (*ETV6-RUNX1* fusion). These 'genetic snapshots' reveal extraordinary diversity in mutational patterns: for example, a clone may acquire an extra copy of a gene and then subsequently lose it (Fig. 4.15). This picture of genomic plasticity can only be visualised by studying single tumour cells, to which the leukaemias particularly lend themselves. The approach has revealed what Mel Greaves has described as the '*clonal architecture*' of tumours whereby individual clones are present in different proportions, each having its own capacity for self-renewal. In other words, cancer development is not linear but branching and that there is no such thing as a 'leukaemia stem cell' – merely diverse populations of cancer propagating cells.

These conclusions are consistent with several examples of monozygous twins, where both carry the same initiating mutation but only one has developed ALL (i.e. they are discordant). The normal twin has the initiating fusion gene (*ETV6-RUNX1*) but has

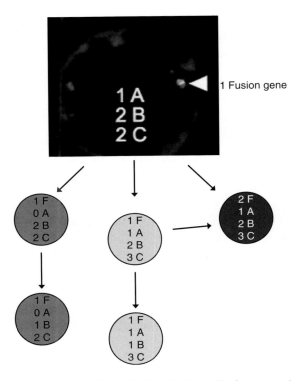

Figure 4.15 Differing proportions of genetically distinct cells from a patient with leukaemia caused by the *ETV6-RUNX1* translocation. The numbers refer to the copy numbers of the fusion gene *ETV6-RUNX1* (F), *ETV6* (A), *PAX5* (B) and *RUNX1* (C). Image contributed by Mel Greaves (Institute of Cancer Research); figure redrawn from Greaves, 2009. (See plate section for colour version of this figure.)

none of the copy number changes in the afflicted child. As happens in the game of genetic roulette, the unaffected twin has failed to acquire the mutation(s) necessary to render the 'driver' effect fully tumourigenic.

It is technically difficult to apply this method of single cell analysis to solid tumours but FISH has shown that, within a malignant brain tumour, combinations of three different RTKs (*EGFR*, *MET*, *PDGFRA*) may be amplified in patterns that vary between individual cells in the same tumour. For example, *EGFR* and *MET* amplification in the same cell or intermingled sub-populations of cells with mutually exclusive *EGFR* or *MET* amplification. Despite the heterogeneity of the tumour, subclones with specific RTK amplification(s) shared an early mutational event (e.g. *CDKN2A* deletion and/or *TP53* (aka *P53*) mutation), indicating their descent from a common precursor.

The alternative approach of 'single nucleus sequencing' utilises DNA amplification followed by next-generation sequencing and is sufficiently sensitive to assess copy number alterations in the genomes of single cells. Application of this method to cells from different regions of a primary breast tumour revealed that growth of a single clone had led to the release of one cell that seeded a metastasis and that, in terms of copy number alterations, the metastasis had not subsequently undergone significant evolution.

What turns a proto-oncogene into an oncogene?

Broadly speaking, molecular analysis is beginning to confirm that tumours are indeed heterogeneous, that is they are comprised of multiple clones. The independent modulation of mutational profiles and the consequent diverse proliferation rates may permit individual clones to emerge as dominant. In other words, as we might have suspected all along, cancers are a form of dynamic Darwinism.

Normal protein, abnormal control: MYC

Burkitt's lymphoma is characterised by chromosomal translocations involving the *MYC* gene but, in contrast to BCR-ABL1 in CML, these re-arrangements do not affect the encoded protein. So normal MYC is produced but its control is changed. The culprits in Burkitt's lymphoma are regions of immunoglobulin genes that find themselves juxtaposed to *MYC*. Sometimes translocated MYC may also pick up mutations that have some effect on its function or on the half-life of its mRNA, but the main feature of this type of mutation is that normal *MYC* finds itself regulated by a DNA sequence that usually controls a completely different gene. Examples of chromosome translocations associated with specific cancers are given in Table 4.2.

Complete loss of a gene

The converse of duplication is deletion – that is complete loss of a segment of DNA that includes a gene or genes. At first glance it may seem improbable that the loss of a gene, and hence the protein encoded, might promote cancer. However, we have seen several examples of how biological systems involve equilibria between forces pushing in opposite directions. Nowhere is this more true than in the cell cycle, which is driven forwards (cell proliferation) by the action of cyclin-dependent kinases but is also subject to negative regulation – that is, there are proteins that can hold up progression through the cycle (Box 3.1; Chapter 3).

Another major priority is that division should not occur before DNA has replicated and that this should not happen if the DNA is damaged in any way. Accordingly there are proteins that, in effect, arrest the cell cycle until the correct signals are received and until any DNA damage has been repaired. These proteins act as brakes and it's easy to see that if they are lost the cell will lose a regulator of normal division and become cancer-prone. The genes that encode such proteins are called tumour suppressor genes and they comprise the second major group of 'cancer genes'. These evolved to control normal cell growth, so they're not really 'tumour suppressor' genes at all: a more appropriate name might be 'growth suppressor' genes. The critical point, however, is that key members of this family are often disabled in cancers – commonly by loss of the entire gene or even a complete chromosome, although sometimes less drastic mutations also impair their function (e.g. a single base change in the promoter of the retinoblastoma gene, *RB1*, can block its transcription). Glioblastomas, the most common and aggressive type of human brain tumour, have very often lost one complete copy of chromosome 10 and with it at least two tumour suppressor genes, *PTEN* and *ANXA7*. In addition to losing chromosome 10,

Table 4.2 Examples of chromosomal translocations.

Gene (chromosome)	Translocation	Cancer
ABL1 (9q34.1) BCR (22q11)	(9;22) (q34;q11)	Chronic myeloid leukaemia
ABL1 (9q34.1) *ETV6/RUNX1* (12p13)	(9;12) (q34;p13)	Acute lymphoblastic leukaemia
BCL2 (18q21.3) IgH (14q32)	(14;18) (q32;q21)	Non-Hodgkin's lymphoma
CCND3 (6p21) IgH (14q32)	(11;14) (q13;q32)	Non-Hodgkin's lymphoma
ERG (21q22.3) *TLS/FUS* (16p11)	(16;21) (p11;q22)	Chronic myeloid leukaemia
INK4A (9p21-p22) TCRα (14q11)	(9;14) (p21–p22;q11)	Acute lymphoblastic leukaemia
MYC (8q24) IgH (14q32) Igλ (22q11) Igκ (2p12)	(8;14) (q24;q32) (8;22) (q24;q11) (2;8) (p12;q24)	Burkitt's lymphoma
PDGFRB (5q33) *ETV6/RUNX1* (12p13)	(5;12) (q33;p13)	Chronic myeloid leukaemia
RET (10q11.2) Protein kinase A regulatory sub-unit α (17q23)	(10;17) (q11.2;q23)	Papillary thyroid carcinoma
TRK (1q23-1q24) *TPM3* (1q31)	inv1(q23;q31)	Colon carcinoma
PDGFB (22q13) *COL1A1* (17q22)	(17;22) (q22;q13)	Dermatofibrosarcoma protuberans Giant-cell fibroblastoma (GCF)

glioblastomas also frequently have extra copies of part of chromosome 7: the amplified DNA includes the *EGFR* gene that, as we've seen, is a powerful oncogene when over-expressed or mutated.

Tumour suppressor genes

The notion that there might be such things as tumour suppressor genes came originally from fusing a tumour cell with a normal cell in vitro and showing that the resulting hybrids were generally non-tumourigenic. That was a slightly surprising result at the

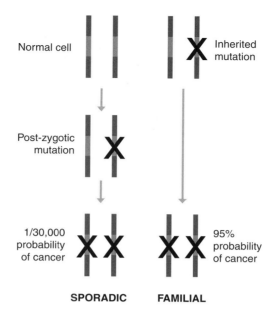

Normal cell

Inherited
mutation

Post-zygotic
mutation

1/30,000
probability
of cancer

95%
probability
of cancer

SPORADIC   FAMILIAL

Figure 4.16 **Retinoblastoma development.** The development of familial and sporadic forms of retinoblastoma, as originally hypothesised by Knudson.

time but we now have a ready explanation: the normal cell provides a gene or genes that have been lost from the tumour cell ('complementation'). Two such genes that can switch a cell from normality into cancer mode simply by ceasing to function are the retinoblastoma gene (*RB1*) and *P53*, both key cell cycle regulators.

The suggestion that some familial cancers may be caused by loss of function of both copies of a gene was first made by Robert DeMars in 1969, his idea being that an individual might be born with an inherited mutation in one copy of the gene (i.e. one allele) but be perfectly normal and that cancer would only appear if a subsequent (somatic) mutation knocked out the other copy (so the individual then became homozygous for the mutant cancer-causing gene). Alfred Knudson (1971) put his idea on a firm basis mainly by thinking about **retinoblastoma**, a rare, inherited childhood disease (incidence 1 in 20,000) in which tumours develop in the eye – specifically in photoreceptor cells in the retina. The cancer comes in two forms, sporadic and familial. In sporadic retinoblastoma there is no family history and just one tumour develops in one eye. In the inherited form tumours arise in both eyes (Fig. 4.16). Knudson suggested that the incidence and growth of retinoblastomas might indeed be explained if both copies of a gene had to be inactivated and that the two forms might arise if those inheriting the disease were born with one defective gene and went on to lose the other (by somatic mutation), whereas in the sporadic form individuals were born genetically normal but acquired somatic mutations in both genes within one cell.

The requirement for two genetic events gave rise to the term 'two-hit model' and it was a remarkably perceptive bit of thinking, based as it was on no molecular evidence whatsoever. It has turned out to be absolutely correct for the retinoblastoma gene although it was a very long journey from Knudson's hypothesis to the identification and cloning of *RB1*. It emerged that both copies of the gene are indeed defective in all retinoblastomas but the real importance of *RB1* in cancer is that it is also knocked out in a substantial proportion of many other tumours: it is lost, for example, in 20 to 30% of lung, breast and pancreatic carcinomas. *RB1* is therefore the classical model for a tumour suppressor gene in that *both* paternal and maternal copies of the gene must be inactivated for the tumour to develop (this behaviour is called 'recessive'). Oncogenes by contrast are 'dominant', because mutation of just one allele *is* sufficient for an effect to be seen. It's perhaps worth noting a possible confusion of terms here in that, although the *RB1* gene behaves in a recessive manner, within families germinal mutations in *RB1* are inherited in an autosomal dominant manner, that is, one copy of the altered gene increases the risk of cancer.

What does the retinoblastoma protein do?

Cell division proceeds in a series of steps, the order of which is rigorously regulated. The most important step before one cell divides into two is the replication of its DNA. The retinoblastoma protein is critical for this because it controls the cell cycle clock, permitting DNA replication only when the cells are big enough (i.e. they've passed through the G1 growth phase) and when growth signals are telling the cell to divide (Box 3.1, Chapter 3 and Fig. 4.17). RB1 functions by regulating transcription of a specific group of genes. RB1 is really a sort of master regulator because it works by associating with a set of conventional transcription factors comprising the E2F family.

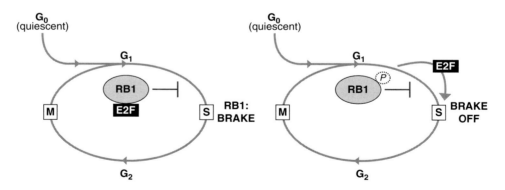

Figure 4.17 **The retinoblastoma protein controls the cell cycle clock.** RB1 has a 'binding pocket' that interacts with E2F proteins to inhibit cell cycle progression from G_1 to S phase. When RB1 is phosphorylated E2F is activated (as a *trans-activating* transcription factor – '*trans*' meaning that E2F binds to part of the molecule (DNA) containing the gene it controls, as opposed to an indirect (intermolecular) *cis* effect) and cells progress to S phase. The binding pocket also interacts with the E7 oncoprotein of the DNA human papillomavirus-16, part of the mechanism by which this virus subverts normal cell cycle control and promotes cancer.

RB1 binds to E2Fs, forming a complex that *prevents* transcription of a set of genes. When RB1 releases its hold on the E2Fs they become free to turn on this set. Some of these genes encode proteins required for DNA replication as well as regulators of cell cycle progression (e.g. MYC). The enzymes that actually make RNA from DNA (the RNA polymerases) are also controlled by the RB1-E2F complex. All of that means that not only does RB1 determine whether a cell can replicate its DNA and continue through the cell cycle but it also determines whether the cell can work at all – because the RNA polymerases in effect make all the components of a living cell.

RB1 loss is a significant step towards cancer because it de-regulates the transcriptional power of the E2Fs. Abnormal duplication of DNA ensues and the stage is set for the accumulation of mutations. RB1 is a target of the cyclin-dependent kinases that control cell cycle progression. Under normal circumstances these control the switch from transcriptional repression to transcriptional activation by phosphorylating RB1, which releases the E2Fs (summarised in Box 3.1, Chapter 3).

Tumour protein 53 (TP53 aka p53)

RB1 has an even more celebrated counterpart that emerged some years before the retinoblastoma gene and protein were isolated. TP53 (usually called p53) was first detected in a lysate of cells that had been infected with a DNA tumour virus and it was the first tumour suppressor gene to be cloned. What it did remained a mystery until the 1990s when two experiments gave very clear indications of its function. The first was the generation of a transgenic mouse by Allan Bradley's group in Houston that had both *P53* alleles knocked out – that is, the mice were unable to make any p53 at all. To general surprise, this didn't seem to bother the mice at all. They developed and grew into normal adult animals and they could even breed quite happily, although the females were a bit less efficient at it than their wild-type counterparts. Fortunately Bradley's group persisted by continuing to observe the transgenic mice, and they were rewarded over the next six months as it gradually emerged that these mice were extremely prone to cancer. So much so that about three quarters of the mice had developed **neoplasms** of one sort or another.

The second experiment that highlighted the role of p53 was carried out in Dundee by David Lane's group who directed a mild burst of sunlight (UV irradiation) onto their own forearms and then measured the amount of p53 protein in the skin cells. The result was amazing: before irradiation p53 was almost undetectable (as it is in most normal cells) but after a short exposure to sunlight the amount of p53 had dramatically increased.

Thus was p53 revealed. It isn't necessary for normal cells to grow and divide, indeed it's so unnecessary that normal cells hardly bother to make it. But subject a cell to any kind of stress that damages DNA – even mild UV radiation is very stressful in this respect – and p53 is made in large amounts. This explains why the knock-out mice were fine but it also explains why they were so susceptible to cancers: as we've noted, all animals are exposed to a wide range of DNA damaging assaults and p53 has evolved as part of a major mechanism for protecting cells against such assaults.

p53, cell cycle arrest, apoptosis and MYC

p53 is a transcription factor that is activated by DNA damage to switch on the expression of various genes (Figs. 4.18 and 4.19). These protect cells against DNA damage by stopping the cell cycle and division until the damaged DNA has been repaired, or by killing the cell, the precise effect depending on the specific genes activated by p53 (Fig. 4.20). The cell machinery that detects DNA damage is exquisitely sensitive: it can pick up just one break in the two metres of double-stranded DNA in each cell and as a result turn on the *P53* gene. One target of p53 is a gene that controls cell cycle arrest, *WAF1* (*w*ild-type p53-*a*ctivated *f*ragment *1*), by encoding a protein that blocks the action of several of the kinases that drive the cycle – so it's a cyclin-dependent kinase inhibitor (CKI). This buys time for DNA repair systems to rectify DNA damage, whether the mutations have been caused by UV light or by other mutagens.

Simply giving the cell more time may not be enough, however. When damage occurs the best anti-cancer tactic is to destroy the cell and a second critical role of p53 is as an activator of **apoptosis**, the controlled programme by which cells commit suicide. Among its targets are three genes, *BAX*, *PUMA* and *NOXA*, that stimulate apoptosis by permeabilising the outer mitochondrial membrane. This releases pro-apoptotic proteins, in particular the respiratory chain component cytochrome c, which activates the cell death programme. p53 has switched on the most effective cancer protection you could have – the elimination of a cell that is likely to turn into a tumour precursor.

We saw earlier that MYC is essential for cell proliferation and that it is often de-regulated in cancer, leading to hyper-proliferation that, in turn, can increase genetic instability and promote tumourigenesis. By way of contrast, MYC can also exert tumour suppressor effects that result in either apoptosis or senescence. This remarkable duality reflects the role of MYC in mediating stress responses that, in normal cells, lead to cell death. One action of MYC that induces apoptosis is the suppression of *WAF1* transcription: this prevents cell cycle arrest and switches p53 to the activation of apoptotic mediators (BAX, PUMA and NOXA). In addition, MYC also inhibits expression of anti-apoptotic genes in the BCL2 family (Chapter 6).

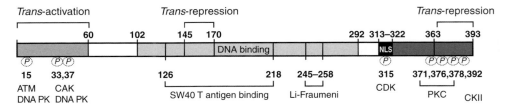

Figure 4.18 Features of p53. *Trans*-activation occurs via the N-terminal domain and there are two *trans*-repression domains. Major phosphorylation sites that regulate the activity of p53 include targets for the ataxia telangiectasia protein (ATM), DNA-activated protein kinase (DNA PK), cyclin-activated kinase complex (CAK), cyclin-dependent kinases (CDK), the nuclear localisation signal (NLS), protein kinase C (PKC) and casein kinase II (CKII).

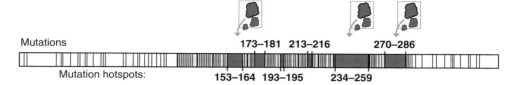

Figure 4.19 **Mutation sites in p53.** p53 is only 393 amino acids long but participates in numerous cellular functions. It is principally a transcription factor, being either an activator or a repressor depending on the target gene and the cellular environment. Vertical bars indicate mutations: black boxes dense regions thereof. Hot spots are shown as filled boxes, including the region in which germ line mutations mainly occur in the rare, autosomal dominant Li–Fraumeni syndrome: 50% of the carriers develop diverse cancers by 30 years of age, compared with 1% in the normal population. p53 (and its relatives p63 and p76) has several different promoters, and multiple isoforms are produced as a result of alternative splicing. The resultant complexity in p53 function is indicated by the fact that tumours in which a wild-type allele is still present often express abnormal levels of isoforms. The symbol indicates mutations specifically caused by tobacco smoke.

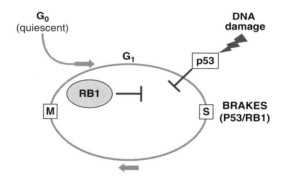

Figure 4.20 **p53 in cell cycle control.** DNA damage raises cellular p53 levels: *trans*-activation by wild-type (but not mutant) p53 can then cause cell cycle arrest, promote DNA repair or drive apoptosis. *Trans*-activation by wild-type p53 is prevented by mutant p53 proteins or by some viral proteins, e.g. human papillomavirus-16 E6.

Thus in normal cells low levels of MYC expression are required for growth and division whereas abnormally high concentrations confer protection to stress by initiating cell death. However, if cells lose a critical component of the tumour suppressor network, particularly p53, elevated MYC activity becomes a powerful tumourigenesis driver.

p53 and cancer

The critical role of p53 led David Lane to describe it as 'guardian of the genome'. As cancers almost always involve disruption of the genome it is unsurprising that *P53* is one of the most frequently mutated of all cancer genes. *P53* is on chromosome 17 in a region that is often deleted in human cancers but, in addition to complete loss of the

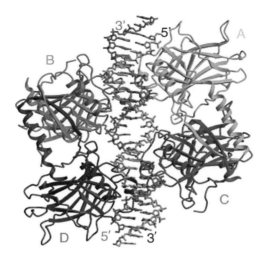

Figure 4.21 **The protein p53 is a transcription factor, binding to DNA.** The figure shows the three-dimensional structure of two p53 dimers bound to a p53-response element. The four p53 molecules (green, yellow, red and blue ribbons) are labelled A to D: in essence they form a clamp around DNA (cyan). The individual p53 molecules bind to grooves in DNA (blue): amino acids in mutational hot spots can directly contact DNA. (From Emamzadah *et al.*, 2011.) (See plate section for colour version of this figure.)

gene, over 6,000 *P53* mutations have been detected. All told, over 70% of human cancers have a mutation that affects the activity of p53. Mutations are usually somatic (not inherited) and occur with high frequency in all types of lung cancer, in over 60% of breast tumours and in ~40% of brain tumours (**astrocytomas**), frequently in combination with the activation of oncogenes. *P53* provides a good example of how a specific effect of a carcinogen can promote cancer. Studies of the X-ray crystallographic structure of p53 complexed with DNA reveal that there are several mutational hotspots at amino acids that directly contact DNA and are therefore critical for its function as a transcription factor (Fig. 4.21). Guanine residues in these codons react with a specific component of cigarette smoke (benzo[*a*]pyrene), which makes this polycyclic aromatic hydrocarbon a potent mutagen and carcinogen.

Although most *P53* mutations are somatic, some unfortunate individuals inherit one defective allele, as can happen in retinoblastoma with *RB1*. This gives rise to the rare, autosomal dominant disease Li-Fraumeni syndrome: 50% of the carriers develop diverse cancers by 30 years of age, compared with 1% in the normal population.

p53 is not entirely a tumour suppressor

The interaction of p53 with DNA requires four molecules – a tetramer. A mutation in one allele that affects the function of the encoded protein could therefore produce a p53 tetramer containing mutant and normal (from the unaffected allele) forms that might behave abnormally. When this occurs it can exert 'dominant' effects more severe than the loss of an allele. Thus, although *P53* is considered to be a member of the tumour suppressor family, it is more complex than *RB1*. Such complications are not confined to p53: several tumour suppressors can acquire mutations in one allele that change the function of the protein, resulting in a phenotype even with one normal allele. So it isn't as simple as just knocking out both copies of the gene. For p53 the situation is further complicated by the fact that the gene contains an alternative promoter and can generate multiple splice variants with tissue-dependent expression (Fig. 4.18) – these are differentially expressed in tumours compared with normal tissue. Some of the variant forms of p53 produced by alternative splicing actually behave like oncoproteins rather than tumour suppressors – they actively promote abnormal proliferation.

One way in which mutant p53 can modulate cell phenotype is by interacting with a close relative, p63, which is also a transcription factor. This can mis-target p63 to a set of genes that promote invasion. *P53* is thus remarkable in that, although the wild-type gene is a potent tumour suppressor, mutations (of just a single amino acid) can generate proteins that are selected for and increase invasive and metastatic potential.

DNA repair

The role of p53 is central to the capacity of the cell to repair almost all of the thousands of chemical hits that DNA receives every day from radiation, carcinogens, metabolic by-products and various other assaults. The fact that mutations caused by many different agents are repaired so effectively that fewer than one per day remains fixed in the genome suggests that a number of processes are involved. In fact several hundred genes are devoted to various aspects of DNA repair and these too represent an important class of tumour suppressor (Table 4.3). Mutations that impair their function will permit the replication of potential cancer cells and, for obvious reasons, mutations in these genes are said to cause genetic instability.

The autosomal recessive disease xeroderma pigmentosum arises from a genetic defect in the system for repairing DNA damaged by UV light. Sufferers are prone to

Table 4.3 DNA repair genes and sources of genomic instability in cancers.

Gene family function	Examples
DNA repair:	
Mismatch repair	*MSH2 (MutS), MSH3, MSH6, MLH1 (MutL), PMS2*
Homologous recombination	*BRCA1, BRCA2*
Non-homologous end joining	*NBS*
Other	*ATM, MGMT*
DNA replication	Flap endonuclease *(FEN1)*
Chromosome segregation	*APC, MAD, BUB*
Cell cycle checkpoints	*RB1, P53*
DNA methylation	DNA methyltransferase (DNMT)
Tumour suppressors	*ARF, INK4A, BRCA1*

skin cancers and need to avoid sunlight. Their problem is not with p53, however, which is activated and stops cells dividing. The time bought is of limited use, however, because DNA repair is defective, the fault most commonly being in the nucleotide excision repair system. In addition to xeroderma pigmentosum, a number of conditions that predispose individuals to cancer are due to DNA repair defects including Werner syndrome, Bloom's syndrome and Fanconi anaemia. Each occurs because of mutations in proteins that signal or enact the repair process (Table 4.3). **Ataxia telangiectasia**, a rare, inherited neurodegenerative disease that carries an increased risk of cancers, particularly **leukaemia** and **lymphoma**, is also often included in this category. It arises from mutations in the ATM protein, a kinase that detects double-strand breaks in DNA and then activates p53: ATM is the reason why p53 is so sensitive to DNA damage. The *BRCA2* gene, mutations in which confer a high risk of breast cancer and also a risk of ovarian cancer, encodes a protein that has a role in DNA repair and hence an inherited BRCA2 defect will contribute to genetic instability.

Hereditary cancers

The mutations in DNA repair genes that increase the susceptibility of carriers to cancer are usually inherited. However, as we've seen, the broader class of tumour suppressor genes can also carry mutations from generation to generation and, although the cancers that develop from these mutations only represent about 10% of all cancers, they have been remarkably informative about what happens as tumours evolve. This is mainly due to the fact that families carrying mutant genes have been intensively studied, including members who have not shown any symptoms, and this has helped to build a model of the stages of tumour development.

Hereditary breast cancer

Breast cancer is perhaps the best known of the cancers that 'run in families', that is, show 'familial aggregation'. It is about twice as common in first-degree relatives of women with the disease than it is in the general population. About 5% of all female breast cancers (men can get the disease too but very rarely – about 1% of all breast cancers) arise from inherited mutations. In the 1990s the genes *BRCA1* and *BRCA2* were identified and studies of large populations showed that inherited mutations in either gene confers a lifetime risk of the disease of over 50%, compared with an average risk of breast cancer for women of ~10%. Since then variants in a number of other genes (*CHEK2*, *ATM*, *BRIP1* and *PALB2*) have been shown to increase the risk by about two-fold. Like *BRCA1* and *BRCA2*, these are all tumour suppressor genes and they all play roles in DNA repair. Nevertheless, taken together, they only account for about 25% of the risk factors, which means that in three quarters of the cases of familial breast cancer the genetic cause, or causes, are unknown. This rather depressing situation is not for want of effort: it is simply that studies trying to link mutations in specific genes to inherited breast cancer have failed to come up with anything. It seems very probable, therefore, that rather than there being more *BRCA*-type genes waiting to be discovered, inherited susceptibility to breast cancer usually results from the combined effect of lots of genetic variants (**single-nucleotide polymorphisms**), each making but a small contribution to a polygenic disease.

Hereditary colon cancer

Cancers of the bowel have perhaps been the most informative as far as the pattern of mutations required for tumour development. This is particularly true of familial adenomatous polyposis (FAP): this hereditary syndrome is characterised by the presence of hundreds of adenomatous lesions (polyps) in the colon at an early age, any one of which can, over time, convert into a malignant **adenocarcinoma**. It is relatively easy to examine the lining of the bowel (by endoscopy) and thus screen individuals from susceptible families and, if pre-malignant polyps are detected, the condition can usually be treated by surgery. Only if one of the polyps develops beyond the adenoma stage, penetrating the underlying **basement membrane** and thus allowing tumour cells to pass into the circulation, does the condition becomes malignant and hence life-threatening (Fig. 4.22).

In the majority of FAP cases the predisposition arises from a defective *APC* (adenomatosis polyposis coli) gene. Cells from transgenic mice show very beautifully that the APC protein plays a part in the separation of duplicated chromosomes during mitosis and that APC mutations disrupt this process. Damaged APC causes chromosomes to be torn apart in an unregulated fashion as cells divide. This destabilises the genomic integrity of a cell, giving rise to the genetic instability characteristic of cancers. For

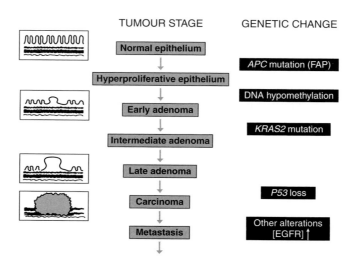

Figure 4.22 Progression of familial adenomatous polyposis. Schematic of the development of malignant adenocarcinoma initiated by a defective *APC* gene. The subsequent sequence of major genetic abnormalities occurs frequently but not invariably.

obvious reasons this form of genetic instability is referred to as chromosomal instability (CIN). Mutations in APC therefore prime the cell to acquire yet more mutations and, from studies of families with such mutations, it has emerged that very frequently the oncogene *KRAS2* is mutated and subsequently *P53* function is lost – a kind of mutational signature. Other mutations occur but we have already seen why these two genes are often involved in cancer and they are a very good example of mutations in different signalling pathways having convergent effects in driving the development of cancer (Chapter 6).

DNA repeats itself

Human DNA contains a large number of short stretches of repeated sequence – so much so that these make up about half of our genome. These sequences are important in cancer because some are the cause of another type of genetic instability.

There are two main categories of repeat sequences: simple-sequence DNA (or satellite DNA) and interspersed repeats. In satellite DNA the length of the repeat unit is from one to 500 base pairs. Because these repeated blocks have an individual frequency of nucleotides (A, T, C and G) they have a slightly different density to bulk DNA so that, when they are separated on a density gradient, they show up as 'satellite' bands. There are two types: **microsatellites** (repeated sequences of up to 6 base pairs) and **minisatellites** (between 14 and 100 bases long). Microsatellites may have arisen during DNA replication if the daughter strand 'slips' on the template strand so that the same sequence is copied twice.

Microsatellite instability (MIN) is a second form of genetic instability. It is caused by mutations in genes that carry out a specific type of DNA repair – necessary when the machinery that replicates DNA makes a mistake and an incorrect base is incorporated in newly synthesised DNA. That machinery is pretty efficient – it makes such a mistake only about once every 100,000 bases it joins together. When it does so, however, there is a brief window in which the proteins of the mis-match repair system can detect an error (before the newly synthesised DNA strand becomes methylated), cut out the offending base and replace it with a correctly matched one. This gives an overall error rate of one in 10^9 bases. Defects in DNA repair proteins produce a 'mutator phenotype' that shows up particularly as errors in microsatellites (these normally have a very low mutation rate). The question of whether a mutator phenotype is essential for cancer cells to evolve or whether the normal mutation rate will suffice is controversial but it is one into which, fortunately, we need not be drawn here.

Polymorphisms and cancer

In terms of DNA sequence human beings are ~99.9% identical but the 0.1% shortfall means that about one in every 1,000 bases differs between individuals. Many of these 'single nucleotide polymorphisms' (SNPs pronounced 'snips') have no effect but, overall, their effects are what distinguishes us from one another and some subtly modulate our susceptibility not only to cancers but also to many other diseases (Fig. 4.23). These minor variants may explain why some individuals with healthy lifestyles (non-smoking, good diet, low alcohol intake, etc.) develop cancer, while others with unhealthy lifestyles do not.

Because SNPs are randomly distributed most do not occur in coding regions, so that the effects they exert are subtle, for example, slightly altering the rate at which an mRNA is made. It seems likely that the cumulative effect of many such minor variants accounts for the missing cause of hereditary breast cancer, mentioned earlier, and indeed, several SNPs associated with a two-fold increase in breast cancer risk have been identified (these affect the genes *TGFB1*, *FGFR2*, *TOX3*, *MAP3K1* and *LSP1*). The study of monozygous (genetically identical) twins is a powerful way of determining the overall genetic contribution to disease susceptibility. Monozygotic twins show a strong correlation in their risk of developing breast cancer, indicating that a significant proportion, perhaps as high as 50%, arise in a genetically susceptible minority of women.

Modulating gene expression without mutation: epigenetic changes

So far in thinking about the causes of cancer we have only discussed changes in the sequence of DNA, that is, mutations. Mutations can be critical if they change the structure of a protein or the way in which genes are expressed and so impact on cell behaviour – the phenotype. However, gene expression can also be changed without alterations in the underlying sequence of DNA. What's more, such changes can become fixed in the genome, just as if they were mutations, and be passed between generations. These are epigenetic changes, perhaps the best known of which is X-inactivation in female mammals where one of the two copies of the X chromosome is silenced in every cell. X chromosomes are inactivated by packaging them into a compressed form of

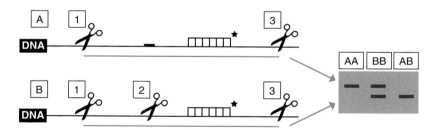

Figure 4.23 Single nucleotide polymorphism (SNP). A SNP is a difference in one base in an otherwise identical stretch of DNA sequence. Shown here in DNA sequences A (upper) and B (lower). B carries restriction site 2 (in addition to 1 and 3) that is absent in A because of a SNP. These could be two alleles in one individual's DNA or DNA from two different people. The star represents a label attached to a probe that recognises a specific short sequence of DNA (an oligonucleotide probe). When the DNA fragments are separated in a gel (right) they can be detected by the labelled probe.

DNA (heterochromatin) that suppresses gene expression. Tortoiseshell cats show this effect because each X chromosome carries a fur colouration gene (black and orange): inactivation of one chromosome produces the fur colour of the other.

In cancer, however, there are two other major routes to epigenetic effects. The first is through modification of proteins that associate with DNA and help it to fold up in the nucleus. The proteins are histones and the combination of DNA and histones makes up chromatin – so chromosomes are really chromatin. Because the structure of chromatin is critical for the transcription of genes, so is that of the histones and, like other proteins, they can be covalently modified after translation by the addition of, for example, phosphate, methyl or acetyl groups to their constituent amino acids.

The second route is by addition of methyl groups to cytosine bases next to guanine bases in DNA. Concentrations of such 'CpG' motifs occur in the promoter regions at the start of about half of human genes – CpG islands of between 0.5 and 4 kb in length. In normal cells CpG islands are usually unmethylated but hypermethylation, leading to transcriptional silencing of specific genes against a background of general hypomethylation, occurs widely in human cancers. The pattern of genomic DNA methylation correlates with transcription in that transcriptionally active chromatin domains are hypomethylated and inactive regions are hypermethylated. A number of carcinogens can cause epigenetic effects and hence promote cancer.

Micro RNAs (miRNAs)

Relatively recently a new class of regulators of mammalian cell function have emerged in the form of non-protein-coding genes. The RNAs these genes encode are short – just 18 to 24 bases in length, hence they are 'micro RNAs' – and their sequences are complementary to sequences in *bona fide* messenger RNA (that is they can form pairs with bases in the mRNA sequence in just the same way that DNA base-pairs in the double helix). When micro RNAs bind to their target mRNA the effect is to stop the message being processed by the ribosome – in other words to inhibit protein synthesis (Fig. 4.24). Sometimes it's just temporarily blocked and sometimes the association leads to the mRNA being broken down – a very effective way of preventing protein translation from that message.

There are at least 800 human miRNAs and they are involved in the regulation of cellular differentiation, proliferation and apoptosis. It will come as no surprise therefore to learn that miRNAs are important in cancer: the pattern of expression of many of them changes in tumours by comparison with the corresponding normal tissue. Quite often the level of miRNAs is lower in tumours, which suggests that when switched on they act as a protection against tumour development. Even so, the patterns of expression vary markedly between different types of tumours – for example, leukaemias can be distinguished from solid tumours by their miRNA profiles and, for some tumours, sub-sets of miRNAs may actually be switched on as part of tumour development. In addition miRNA networks regulate tumour metastasis and the tumour cell microenvironment.

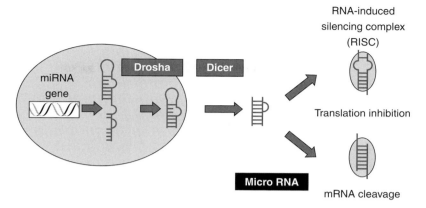

Figure 4.24 **Micro RNA synthesis.** Pri-miRNA is processed to pre-miRNA in the nucleus. The proteins Dicer and Drosha generate miRNA, one strand of which is incorporated into the RNA-induced silencing complex (RISC). miRNA directs RISC to degrade mRNA or inhibit translation. Dicer and Drosha act as tumour suppressors in some cancers, high levels are associated with increased survival whereas deletion of one allele of Dicer reduces survival (Kumar *et al.*, 2009; Merritt *et al.*, 2008).

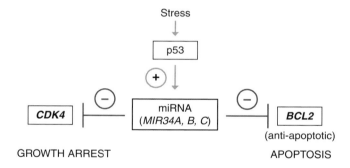

Figure 4.25 **Acting as a transcription factor, p53 switches on a micro RNA (*MIR34*) which has two mRNA targets (*CDK4* and *BCL2*).** Blocking the synthesis of CDK4 causes cell cycle arrest; BCL2 is an anti-apoptotic protein so that preventing its synthesis permits apoptosis.

Armed with our knowledge of p53 and how it can drive cells to suicide, we can look at one illustrative example of a miRNA in action. The gene is *MIR34* and it is one of the many targets of p53: when p53 attaches to a specific binding site in the promoter the effect is to 'turn on' the gene – in other words, more *MIR34* is made (Fig. 4.25). *MIR34* has two specific mRNA targets: *CDK4* and *BCL2*. The destruction of these two mRNAs of course reduces the amount of CDK4 and BCL2 protein. CDK4 is one of the important kinases that drives the cell cycle, whereas BCL2 is a powerful inhibitor of the suicide pathway (it binds to BAX – see Chapter 6) and blocks cell death. Reducing the levels of these proteins delivers a double anti-cancer blow: proliferation is blocked and apoptosis is activated. *MIR34* is thus one of the ways in which p53, when induced by DNA damage, can ensure that a cell does not become a tumour precursor.

Conclusions

Cancer is driven by a mutational steeplechase in which DNA damage that fails to be repaired becomes fixed in the genome. Somewhere between five and fifteen critical mutations – so called 'drivers' – are thought to be necessary for most types of cancer to develop fully. Two major classes of 'cancer gene' contribute drivers – oncogenes and tumour suppressors. Oncogenes exert dominant effects – that is, the mutation that converts a 'proto-oncogene' into an oncogene confers gain-of-function. Tumour suppressor genes lose function as a result of mutation. Mutations in DNA span the entire range from a single base change through loss of large regions of DNA to complete chromosomal deletion and may also include shuffling of segments giving rise to new genes (chromosomal translocation). Conversely, regions of DNA or entire chromosomes may be duplicated giving rise to 'gene amplification'. Some mutations have no effect on function. However, even a single base change, if it affects a critical amino acid can produce a 'driver', as it does in activating RAS in about 20% of human cancers. Gene or chromosome loss that involves a tumour suppressor removes one of the brakes on cell proliferation. Two major tumour suppressors that are frequently inactivated in human cancers are *P53* and the retinoblastoma gene *RB1*. The juxta-position of normally distant regions of the genome can generate novel proteins if the gene fusion event maintains the coding sequence 'in-frame'. Chimeric proteins thus produced are a distinctive feature of leukaemias but they also occur in solid tumours. More subtle effects on cancer development may arise from polymorphisms that may, for example, slightly modulate the efficiency of gene transcription. Micro RNAs regulate protein synthesis from mRNAs and can exert either oncogenic or tumour suppressive effects.

Key points

- Two major classes of genes are involved in the development of cancers: oncogenes and tumour suppressor genes.
- Oncogenes are normal genes ('proto-oncogenes') that are either mutated and/or abnormally expressed, either in amount or in location.
- Oncoproteins generally function in the pathways that regulate cell proliferation and apoptosis.
- Tumour suppressor genes encode proteins that normally function to restrain cell growth or to control cell location.
- Loss of function of tumour suppressors contributes to the development of cancers, and mutations in tumour suppressor genes combined with the activation of some oncogenes are associated with most if not all cancers.
- Micro RNAs are a third category of cancer genes. Different micro RNAs can, in effect, act as oncogenes or tumour suppressors.

Future directions

- The power of whole genome sequencing is revealing the entire panoply of mutations in individual tumours. An initial challenge is to extract from the enormous amounts of data which of the accumulated mutations are major forces in tumourigenesis, i.e. are 'drivers'. This is an important challenge because these offer the most attractive therapeutic targets.

- A major difficulty in delineating links between polymorphisms and diseases is the requirement for large numbers of cases and controls to be compared to provide the statistical power necessary to reveal weak associations. However, on average, only one polymorphism in one thousand will be in coding sequences. For the remainder devising assays to determine their functional effects is a substantial challenge.

- Micro RNAs are involved in complex regulatory networks that are only slowly being dissected. As these are resolved they provide a new field for therapeutic intervention.

Further reading: reviews

THE CANCER GENOMIC LANDSCAPE

Bell, D. W. (2010). Our changing view of the genomic landscape of cancer. *Journal of Pathology* **220**, 231–43.

MUTATIONS IN 'CANCER GENES': ONCOGENES AND TUMOUR SUPPRESSOR GENES

Hesketh, R. (1997). *The Oncogene and Tumour Suppressor Gene Facts Book*, 2nd edn. Academic Press.

ONCOGENES

Croce, C. M. (2008). Molecular origins of cancer: oncogenes and cancer. *New England Journal of Medicine* **358**, 502–11.

THE FIRST HUMAN ONCOGENE

Parada, L. F., Tabin, C. J., Shih, C. and Weinberg, R. A. (1982). Human EJ bladder carcinoma oncogene is homologue of Harvey sarcoma virus *ras* gene. *Nature* **297** (5866), 474–8.

WHAT TURNS A PROTO-ONCOGENE INTO AN ONCOGENE?

Meyer, K. B., Maia, A. T., O'Reilly, M. *et al.* (2008) Allele-specific up-regulation of *FGFR2* increases susceptibility to breast cancer. *PLoS Biology* **6**(5): e108 doi:10.1371/journal.pbio.0060108.

TUMOUR SUPPRESSOR GENES

Burkhart, D. L. and Sage, J. (2008). Cellular mechanisms of tumour suppression by the retinoblastoma gene. *Nature Reviews Cancer* **8**, 671–82.

TUMOUR PROTEIN 53

Junttila, M. R. and Evan, G. I. (2009). p53 – a Jack of all trades but master of none. *Nature Reviews Cancer* **9**, 821–9.

DNA REPAIR

Pallis, A. G. and Karamouzis, M. V. (2010). DNA repair pathways and their implication in cancer treatment. *Cancer and Metastasis Reviews* **29**, 677–84.

HEREDITARY CANCERS

Lin, J., Gan, C. M., Zhang, X. *et al.* (2007). A multidimensional analysis of genes mutated in breast and colorectal cancers. *Genome Research* **17**, 1304–18.

POLYMORPHISMS AND CANCER

Dong, L. M., Potter, J. D., White, E. *et al.* (2008). Genetic susceptibility to cancer: the role of polymorphisms in candidate genes. *Journal of the American Medical Association* **299**, 2423–36.

MODULATING GENE EXPRESSION WITHOUT MUTATION: EPIGENETIC CHANGES

Jones, P. A. and Baylin, S. B. (2002). The fundamental role of epigenetic events in cancer. *Nature Reviews Genetics* **3**, 415–28.

MICRO RNAS (miRNAs)

Meltzer, P. S. (2005). Cancer genomics: small RNAs with big impacts. *Nature* **435**, 745–6.

LEUKAEMIA

Greaves, M. (2009). Darwin and evolutionary tales in leukemia. The Ham–Wasserman Lecture. *Hematology* **2009**, 3–12.

THE RETINOBLASTOMA (RB1) GENE

Ianari, A., Natale, T., Calo, E. *et al.* (2009). Proapoptotic function of the retinoblastoma tumour suppressor protein. *Cancer Cell* **15**, 184–94.

5 | What is a tumour?

Most aspects of normal cellular behaviour are subverted in the development of tumours. These changes include a switch to aberrant signalling in both pro- and anti-proliferative pathways, the acquisition of the capacity to avoid cell death and to replicate indefinitely, and perturbation of the normal metabolic profile. In addition, dynamic interactions between tumours and normal cells in their environment can progressively co-opt inflammatory and immune responses so that they support rather than inhibit tumour growth and can recruit host endothelium to provide a blood supply. The defining feature of malignant tumours is the ability of cells to migrate through adjacent tissue and eventually to colonise distant sites. This process of metastasis is poorly understood at the molecular level. It remains essentially untreatable and is the major cause of cancer death.

Introduction

Cancer is a group of diseases characterised by abnormal cell growth through which cells may acquire the potential to disperse (metastasise) from the site of origin (primary tumour) to other sites in the body (secondary tumours). The word tumour comes from the Latin '*tumor*' referring to the swelling that occurs as a consequence of these abnormal growths and is now used interchangeably with 'neoplasm', meaning new or abnormal cell growth. This definition of neoplasm leads to a major division of cancers into **malignant** and **benign**. The terms 'tumour' and 'cancer' have also come to be used synonymously but a distinction might be made in that metastatic cancer occurs because a tumour has acquired the capacity to invade its surroundings, the first step in spreading to secondary sites. This involves the destruction of other cells, critically some that make up the vessels of the circulatory (blood and lymphatic) systems. Once the tumour cell can get into the circulation it can be carried to other locations: it has become malignant. The implication, of course, is that there are tumours that are not malignant.

Benign tumours

Because the term neoplasm simply means new, abnormal growth it refers to both malignant tumours and **benign tumours**. This second category lacks the distinguishing features of malignancy – that is, the cells can't invade surrounding tissues and therefore they do not metastasise. Benign tumours are usually encapsulated, that is, surrounded by a membrane that restrains their invasive capacity. They can arise in any tissue, their cells generally resemble those of the tissue of origin, in contrast to malignant cells, and for the most part they are pretty harmless. Despite being slow growing, they can reach a considerable size (as big as a grapefruit) and if they compress other tissues (e.g. blood vessels or the brain) can have serious effects that require surgical treatment. In addition, some benign tumours can have harmful, indirect effects if they occur in **endocrine** tissues and result in abnormal levels of hormone production, for example, adenomas in the thyroid, adrenocortex or pituitary glands. This effect suggests that the neoplasms are of cell types that normally make up the tissues in which they arise. This is often a feature of benign tumours to the extent that usually they do not give rise to any symptoms and are only detected by chance. The general name for such tumours is hamartoma, the counterpart of which is chor-istoma when comprised of normal cells growing in an abnormal location. Benign tumours and the tissues with which they are associated are summarised in Table 5.1.

Most of us probably try quite hard not to think about cancer, at least in a personal sense. Even so, almost all of us will have given it a moment's thought when contem-plating the various birthmarks, moles and warts that adorn most human bodies. The thought will have come as two questions: are these cancers and what should I do about them? To which the answers are 'no' and 'nothing' – almost always. In a strictly technical sense they are indeed neoplasms because they are an abnormal growth of skin but the best thing to do is simply regard them as a blemish. Some birthmarks, for example strawberry marks that generally occur on the face, gradually disappear of their own accord. For marks that do not and are felt to be disfiguring it may be possible to reduce their prominence by laser treatment. But there is just one word of warning – hidden in the 'almost always' above. The medical fraternity refer to birthmarks as naevi, the most common naevus being composed of melanocytes, melanin-producing cells in the bottom layer of the epidermis. Thus moles are benign tumours formed of clusters of pigmented skin cells. The only real problem with moles is that just occa-sionally one of them may turn nasty and develop into a fully malignant tumour but there are two reasons why even this should not keep you awake at night. Firstly, that event usually needs some help from you – something that large numbers of us provide by lying in the sun without any protection. Secondly, because these are skin growths you should notice any change in their behaviour. Which is why you are advised to use sun creams and to take action if any of your moles change appearance – get bigger, blacker or itchier or start bleeding.

Warts have much in common with birthmarks and moles but there is one big difference: warts are caused by viral infection. For that reason we aren't born with

Benign tumours

Table 5.1 Benign tumours.

Benign tumour	Cell type	Comment
Fibroma	Fibrous or connective tissue	Also called fibroid tumours or fibroids. Uterine fibroids (also known as uterine leiomyoma, myoma, fibromyoma, leiofibromyoma, fibroleiomyoma and fibroma) are the most common benign tumours in females
Chondroma	Cartilage-forming (chondrocytes)	Chondrosarcomas are cancers of the cartilage
Choristoma	Tissue at an abnormal site	Myelolipoma: choristoma of the adrenal gland
Hamartoma	Most commonly lung; also occur in heart (cardiac rhabdomyoma from myocytes), hypothalamus, kidney, spleen and other vascular organs	Cowden syndrome, characterised by multiple hamartomas, is associated with increased risk of breast and thyroid cancers
Lipoma	Fat (lipocytes)	Present in ~1% of population: several forms (angiolipoleiomyoma, chondroid lipomas, etc.)
Papilloma	Epithelial (skin, alimentary tract or bladder)	Malignant forms are carcinomas[a]
Phaeochromocytoma	Adrenal medulla (inner region of adrenal gland)	Abnormal release of adrenaline and noradrenaline

Benign tumours with malignant potential	Cell type	Malignant derivative
Adenoma	Epithelial (glandular tissue); colon polyps are adenomas with a tubular structure	Adenocarcinoma[a]
Astrocytoma	Astrocytes (brain)	Astrocytoma: may progress to anaplastic astrocytoma. Usually does not metastasise beyond central nervous system
Meningioma	Meninges (membranes around brain and spinal cord)	Sarcoma
Mesothelioma	Epithelioid and spindle (lining the coelomic cavities, usually the pleura or peritoneum)	Mesothelioma

Table 5.1 (*cont.*)

Benign tumour	Cell type	Comment
Naevus	Melanocytes (pigmented skin cells)	Melanoma. BRAF mutations in most naevi indicate they are pre-malignant
Schwannoma	Schwann (cells that produce the myelin sheath surrounding and insulating peripheral nerve cells). Acoustic neuroma (more correctly vestibular schwannoma because it occurs in the vestibular portion of the eighth cranial nerve) is probably the best known schwannoma, affecting about 1 in 100,000 people worldwide	Neurofibrosarcoma: ~1% of schwannomas become malignant. They develop in a small proportion of individuals (~5%) who have the inherited condition neurofibromatosis (NF). There are two sorts of NF, 1 and 2, depending on whether the inherited mutation is in neurofibromin (NF1) or merlin (NF2)
Teratoma	Normal derivatives of (usually) all three germ layers (e.g. may contain cells with characteristics of brain, liver, lung, etc. and even organs, e.g. hair, teeth, bone, etc.)	Usually benign but malignant forms known. Thought to be congenital

[a] Benign – until basement membrane breached by **epithelial cell** tumour.

them and indeed they are rare in babies. Nevertheless, most of us get them at some point, often before we are twenty. A recent survey of children in the UK revealed that almost all of them had warts of some description. Fortunately, almost all warts are harmless and disappear of their own accord – though some take years to do so. Warts are rough lumps of skin that commonly arise on the hands or feet or in the anogenital area. Palmar warts occur on the palm of the hand, plantar warts, otherwise known as verrucas (*verruca plantaris*), on the soles of the feet. Warts are contagious because of their viral cause, the virus in question being human papillomavirus (HPV). There are more than 100 types of HPV, different types causing different wart variants. HPV produces warts because it causes cells in the top layer of skin (epidermis) to make excessive amounts of the protein keratin. Although it is not possible to get rid of HPV once infected, most warts can be treated either chemically or by either freezing (cryosurgery), burning (cauterisation) or by laser treatment. We commented on the impact of the relatively small number of oncogenic HPVs in Chapter 1 and we will return to their mechanism and the development of vaccines in Chapters 6 and 7.

The benign–malignant boundary

It will have become obvious that, although there are some benign tumours that never become malignant and some that very rarely do, the boundary between the two is not very clear. Can we clarify things by looking at the underlying molecular biology? If cancers arise because of mutations in genes, maybe benign tumours are mutation free and simply arise because of a local imbalance in growth factors. There's much less information available for benign lesions compared with tumours but a quick glance dashes any hopes of a simple answer.

Some benign tumours never become malignant yet carry mutations in genes that in other tissues are associated with malignancy. On the other hand there are tissues in which mutations occur in both benign and malignant tumour forms, albeit sometimes with quite different frequency. Table 5.2 contrasts mutation patterns in a number of benign and malignant tumours.

We're stuck with a typical cancer problem. The distinction between benign and malignant tumours is crucial: one of them can kill you. Unfortunately, even the power of modern molecular biology cannot identify what it is that converts a relatively harmless form of abnormal growth into the fatal variety.

Table 5.2 Mutation patterns in benign and malignant tumours.

Patterns of gene mutations in benign and malignant tumours

Seborrheic keratoses and epidermal naevi: benign skin tumours that never become malignant but have somatic mutations in *FGFR3* and/or *PIK3CA*. Mutations in these genes are associated with carcinoma in other tissues, e.g. colon, cervix, bladder, that do not have mutations in benign tumours.

Gastrointestinal stromal tumours: frequently have activating mutations in *KIT* in malignant tumours (62%) but much less often (16%) in benign tumours.

Pheochromocytomas (in the chromaffin cells of the adrenal medulla): 90% benign. Those that become malignant metastasise to bone, lung, liver or lymph nodes. Loss of chromosome 17p occurs frequently in these tumours but neither that nor mutations in *P53* appear to associate with malignancy.

Adrenocortical tumours: mutations of β-catenin occur in about one-third of both benign and malignant forms. However, *P53* mutations are mainly associated with adrenocortical carcinomas whereas mutations in *PRKAR1A* (that activate cyclic AMP) occur in benign, secreting tumours.

Breast: *P53* mutations common in human cancers and in the breast can occur in normal or benign tissue.

Melanoma: specific mutation ($BRAF^{V600E}$) frequent (~66%). However, the same mutation is detected in congenital and anogenital naevi, indicating that it doesn't simply arise from sun-bathing. What's more, although the mutation is present in most malignant melanoma cells, it's also present in most naevi but they never progress beyond the benign stage, so this mutation is not an overwhelming force in driving tumour development.

Types of malignant (non-benign) cancer

Cancers can be divided into two broad classes: solid and liquid tumours. Most human and animal cancers are solid tumours. Approximately 85% are **carcinomas** (strictly 'carcinomata') – malignant tumours of epithelial cells – which includes malignant tumours of glandular epithelial tissue that are termed **adenocarcinomas**. Carcinoma *in situ* is a pre-malignant change in which cells proliferate abnormally (hyperplasia) within their normal location. The epithelium (particularly of the cervix or skin) shows many malignant changes but does not invade the underlying tissue. Carcinoma *in situ* is a feature of many cancers. For example, ductal carcinoma *in situ* (DCIS) is one of the two forms of mammary ductal carcinoma, the most common type of breast cancer, characterised by hyperplasia in the wall of the milk ducts. It carries a risk of developing into the invasive ductal carcinoma (IDC) form in which the cells have become malignant.

Most other cancers are **sarcomas**, tumours of tissue derived from the mesenchymal layer (connective tissue, bone, cartilage, muscle, fat, blood vessels) that are often highly malignant. Sarcomas are therefore relatively rare and they are not thought to have a pre-malignant (*in situ*) phase.

The remaining ~3% of human cancers are **leukaemias** and **lymphomas** that arise from the abnormal proliferation of white blood cells or, very rarely, of red blood cell precursors. White cells, together with red cells (erythrocytes), make up the population of cells that continuously circulates in the bloodstream. Because leukaemias and lymphomas arise in the blood and the tissues where blood cells develop they are collectively named haematological neoplasms. The word leukaemia comes from the Greek words for white and blood (leukos λευκός and aima αἷμα) and refers to cancers arising from abnormal proliferation of white blood cells in the bone marrow. These cells are sometimes called **leucocytes** but, as that term covers all white blood cells including **lymphocytes**, it's apt to be a bit confusing. Lymphomas are cancers of lymphocytes, the cells of the adaptive **immune system** that fall into the two main classes of T cells and B cells. The two major divisions of this cancer are Hodgkin's disease (**Hodgkin's lymphoma**) and **non-Hodgkin's lymphomas**.

What makes a cancer cell a cancer cell?

Making a simple statement about what distinguishes a normal cell from a tumour cell is straightforward – it's mutations. Even when the initiating factor is infection by viruses that are not directly mutagenic, the effects of their oncogenic proteins are equivalent to mutations that occur in genes that are critical regulators of proliferation and apoptosis. Similarly, mutations also arise from the effects of chronic bacterial infection. Furthermore, specific mutations or groups of mutations are often associated with some cancers. This was first established for some hereditary colon cancers and it is now possible, for example, to classify acute myelogenous leukaemia and other cancers into sub-groups reflecting their behaviour on the basis of patterns of mutations and to predict whether metastases will develop from primary breast tumours (Chapter 7). Revealing mutational signatures may permit differentiation between cancers that are clinically indistinguishable, provide information about how tumours develop and, of course, assist in the rational design of drugs to target abnormal gene products (Chapter 8).

The requirement to accumulate a specific hand of mutations to drive tumourigenesis is, of course, the reason why most cancers don't appear until late in life. If this idea of cancers selecting groups of mutations, so to speak, and using them to their advantage sounds familiar it's because it closely parallels the concept of Darwinian evolution. That is, the development of a tumour cell may be compared with the evolution of a species: the driving force for both comes from the acquisition of specific mutations and the selection of those helping survival and proliferation, until the cancer cell gradually acquires a growth advantage over normal cells. The cancer always requires its host to survive (it's a parasite) but eventually it is capable of diverting nutrients from the host to support its own uncontrolled expansion. Thus the cancer patient may continue to eat normally but nevertheless lose weight – the observation that led Warburg to call it the 'wasting disease' some 80 years ago.

Cancer cell characteristics

The extraordinarily detailed molecular picture of cancer that is now emerging derives from a massive archive of cellular information that has given us a reasonably clear picture of the phenotypic changes associated with tumour cells and their neighbours as cancers evolve (Fig. 5.1). These may be classified as follows:

1. They have a reduced dependence on external signals for growth.
2. They ignore external signals telling them not to grow.

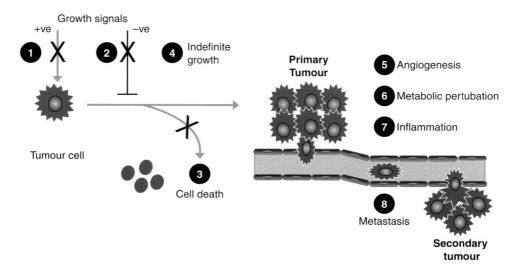

Figure 5.1 Eight main features of cancers.

What is a tumour?

3. They avoid suicide, a powerful anti-cancer strategy when cellular DNA is damaged.
4. They can grow indefinitely.
5. They can induce the formation of their own blood supply.
6. Their metabolism is perturbed relative to that of normal cells.
7. They may promote inflammation and activation of an immune response.
8. They can spread from their primary site to other places.

Let us now look at each of these characteristics in a little more detail.

Cell growth is largely independent of normal signals

One definition of a multicellular organism is an aggregate of cells that communicate with each other so that each knows how to behave to permit the entire organism to work. One critical and recurrent question for an individual cell is whether to proliferate, that is to grow and divide into two daughter cells, or whether to remain in a quiescent, non-proliferating state. All normal cells rely on cues from the rest of the animal to make this decision. These cues come in the form of molecules that are either directly presented to the target cell by adjacent cells, or that diffuse either from nearby cells or from further afield (via the bloodstream, for example). Both positive ('divide') and negative ('don't') signals contribute to the outcome (Fig. 5.2).

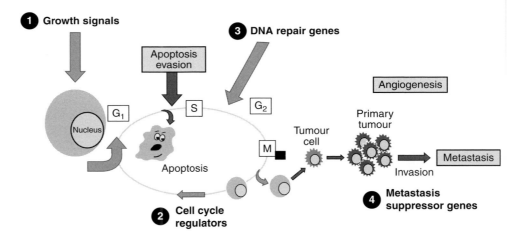

Figure 5.2 **Genetic signals in cancer development.** There are three broad functional categories in which mutations can arise: (1) and (2) pathways driving cell proliferation and controlling cell cycle progression and apoptosis; and (3) DNA repair genes acting to minimise the accumulation of genomic defects: mutations in these genes can thus increase genetic instability. A distinct category of regulators, micro RNAs, includes both oncogenes and tumour suppressor genes (Chapter 4). In addition to those driving proliferation, some messengers deliver anti-growth signals, notably **transforming growth factor β** (TGFβ). Metastasis is regulated by modulation of the expression of **metastasis suppressor genes** (4) that encode proteins normally involved in the suppression of tissue invasion. 'Angiogenesis' refers to the development of new blood vessels from existing vasculature: a requirement for tumour development beyond a diameter of ∼1 mm.

A key feature of a cancer cell is that it has to a substantial degree become independent of these molecular cues. One of the main ways cancer cells achieve this independence is by changing their internal wiring so that they can function as if they were being told to divide when they are not. In addition, aberrant expression of cell-surface receptors may render the cell hyper-responsive to growth factors, and we discussed examples of mutations that give rise to such effects in the previous chapter. As ever in cancer, it's a little bit more complicated than that because, rather than completely switching off contact with the outside world, tumour cells can persuade their neighbours to change sides, so to speak; that is, normal cells adjacent to the tumour start to make factors that promote growth of the tumour itself. At the same time, as part of their re-wiring, tumour cells may even start to make their own growth-promoting factors and we'll return to these points shortly in the tumour microenvironment.

So a critical point about cancer cells is that they change the way in which they sense the world around them and they also change the way their environment sees them. By so doing they make themselves able to grow in a way that is largely independent of that world.

Resistance to inhibitory growth signals

Animal cells are continuously responding to the question of whether they should embark on the process of division or remain quiescent. As we've seen, the positive signals for proliferation come in the form of growth factors, the pathways that they activate ultimately communicating with the machinery of the cell cycle. In a counter-balancing act, some **cytokines** deliver anti-growth signals, the most notable of these being the transforming growth factor beta (TGFβ) family that inhibit the growth of most normal types of cell. The TGFβs signal in essentially the same way as other protein hormones but in normal cells their effect is to inhibit cell division, the most critical target in this context being transcriptional repression of *MYC*. The anti-proliferative effect of TGFβ can thus combine with the restrictive effects of members of the tumour suppressor family in limiting cell cycle progression. Prominent among the latter are RB1 and p53 and these, as we have seen, are frequently inactivated in cancers. The inhibitory effects of TGFβ are also frequently ablated in tumour cells in much the same way that growth factor signalling can become constitutively activated: by the acquisition of mutations in pathway components. For TGFβ the outcome is that either its signals are no longer transmitted to the cell or intracellular signalling is inhibited or subverted. However, the role of TGFβ is more complex than merely being a negative regulator of cell proliferation. While this cytokine certainly inhibits cell cycle progression and indeed can promote suicide in many types of normal cell, in a variety of tumours these responses are lost but the signalling function of TGFβ is switched to promote an invasive and metastatic phenotype. TGFβ can contribute to a malignant phenotype by activating synthesis of metalloproteinases and it also can induce an angiogenic response in the **endothelium**. Thus, signalling pathway mutations appear to convert TGFβ from tumour suppressor to tumour promoter (Chapter 6).

Resistance to cell death

We have seen that normal cells carry on a balancing act with regard to proliferation and that the resulting homeostasis is a feature of all multicellular organisms. There's nothing particularly startling about this concept, although the scale and range of the associated activity is breathtaking. Just to stay as we are, human beings make one million new cells every second. After development is completed we make almost no new neurons, skeletal muscle cells or heart muscle cells (cardiac myocytes) and red blood cells, once released from the bone marrow, cannot divide and they survive in the circulation only for about 120 days. The inability of myocytes to proliferate is one reason why heart attacks are so serious – any resultant loss of heart cells cannot be replaced. On the other hand, epithelial cells that line organs such as the skin, lung and intestine, proliferate rapidly throughout life. The lining of our blood vessels is the endothelium: it's made up of a specialised sort of epithelial cell and, although endothelial cells don't usually proliferate, they switch on rapidly if we injure ourselves, as part of the process that repairs our blood vessels. This turns out to be very important in the development of solid tumours and we shall return to it shortly.

So some cells are dividing very rapidly while others have completely lost the capacity to divide at all and some, like circulating B lymphocytes, are just waiting for the right cue to switch on division. But as we saw in Chapter 4, mammalian cells have an inbuilt capacity for committing suicide, for example as a reaction to DNA damage, by apoptosis (Fig. 5.3) and it has emerged that this suicide programme is inhibited in most, perhaps all, cancer cells.

There are two major ways in which individual cells can be killed: necrosis or apoptosis. In necrosis (from the Greek for dead: νεκρόσ) some parts of the cell are broken down but in essence it simply bursts and releases its contents to the surrounding tissues. The chemicals released by the cell may then cause inflammation at the site. Necrosis can be caused by infection or injury and it occurs, for example, in myocardial infarction (a heart attack) when the blood supply to the heart is interrupted. For the same underlying reason it can also be important in cancer: that is, the central regions of solid tumours may have a poor blood supply, as we shall discuss in detail later, and when this happens tumour cells can be killed through lack of blood-borne nutrients. When necrosis blocks the blood supply to the limbs it can lead to gangrene – the body tissues themselves start to decay: the affected region turns black because of the iron released from haemoglobin after blood cells have lysed.

Apoptosis was first observed by John Kerr, Andrew Wyllie and Alastair Currie in 1972 who showed that it is involved in cell turnover in many healthy adult tissues and also in the elimination of cells during normal embryonic development. Cells undergoing apoptosis break up into fragments that are either shed and carried away by the circulation or taken up by other cells and rapidly degraded. In other words the evidence is rapidly removed, in contrast to necrosis. One benefit of this is that cells that are no longer required are removed without causing inflammation, a potential cancer-promoting event that we'll come back to later. The idea that there is a suicidal

Figure 5.3 Apoptosis or programmed cell death. A HeLa cell undergoing apoptosis (centre) surrounded by healthy cells visualised by scanning electron microscopy (by Thomas Deerinck and Mark Ellisman, NCMIR, UCSD). Typically cells undergoing apoptosis begin to shrink as the cytoskeleton is broken down by **caspases**, chromatin condenses and the nuclear envelope breaks down as DNA fragments. Blebs form in the cell membrane and the cell dissociates into vesicles that are phagocytosed by **macrophages**. **Phosphatidylserine**, normally located only on the cytosolic face of the plasma membrane, translocates to the outer surface where it can be detected by the binding of annexin V.

component built in to the normal development of animals might seem counter-intuitive so we should remind ourselves that many animal structures attain their final form by selective degradation of specific parts. Nowadays many parents-to-be will know this if they have ultrasound scans of their unborn baby, which reveal how our fingers develop from what is really a web until the destruction of cells removes the tissue between the digits.

Kerr and colleagues found that apoptosis could always be detected in untreated malignant neoplasms. A number of earlier studies had noted that a component of cell loss/disappearance occurred in growing tumours, the rate of which could approach 50% of the proliferation rate of cells in the tumour. It might come as an even bigger surprise to discover that tumours eliminate some of themselves than to find this going on during normal development. This may, therefore, be a useful point to remind ourselves that normal tissues reflect a balance between proliferation and cell loss, and that tumours merely represent a slight perturbation of the normal state (Fig. 5.4). We described in Chapter 4 how apoptosis is switched on by the tumour suppressor protein p53 in response to DNA damage, leading to the destruction of mitochondria and the dissolution of the entire cell.

There are two major **apoptosis** pathways: the **intrinsic** (mitochondria-associated) and **extrinsic** pathways. The intrinsic pathway represents a response to internal stress signals (e.g. DNA damage or oncogene expression); whereas the extrinsic pathway is

What is a tumour?

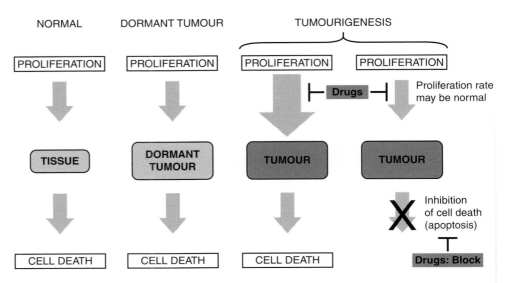

Figure 5.4 **Tissue homeostasis: the balance between cell proliferation and cell death.** Tissues are normally regulated by a balance between cell proliferation and cell death (apoptosis). In tumours an increase in proliferation and/or an inhibition of apoptosis cause abnormal growth. Specific labelling of cells that are proliferating shows that only a minority (usually <10%) of tumour cells are in the cell cycle, that is, proliferating at any one time. This shows that even within tumours there are substantial restrictions on the proliferative cell cycle – a fact that can readily be deduced by noting that a single cell replicating every 24 hours would form a tumour of about 2 cm diameter in four weeks. If proliferation and apoptosis are in equilibrium or both are switched off, the tumour is effectively dormant. Tumour growth thus depends on the balance between the net rate of proliferation and the rate of apoptosis. The pathways that signal abnormal proliferation or apoptosis provide potential targets for drug therapy. Apoptosis can be detected by labelling the broken ends of DNA and it occurs in most tumours.

externally activated by hormones or cytokines. However, although arising from diverse signals, these pathways converge to activate proteases that are normally latent (caspases 8 and 9) leading to mitochondrial membrane permeabilisation and the progressive proteolytic destruction of the cell (Fig. 5.5 and Chapter 6). The caspases are a family of proteases that degrade proteins – mainly each other – at aspartic acid (Asp) residues.

In a normal, healthy cell the protein BCL2, located on the outer membranes of mitochondria, protects the cells from death by inhibiting apoptosis. As part of the stress response the cytosolic protein BAX moves to the mitochondria and inhibits BCL2. Recall that *BAX*, *PUMA* and *NOXA* are genes switched on by p53 as part of the DNA damage response (Chapter 4). This interaction between BAX and BCL2 has the effect of making the mitochondrial membrane permeable and releasing cytochrome c. Once released cytochrome c associates with APAF1 (apoptotic protease activating factor 1) to form the 'apoptosome', a multi-protein complex that activates a caspase cascade from caspase 9 to the major 'executioner caspases', caspase 3 and caspase 7.

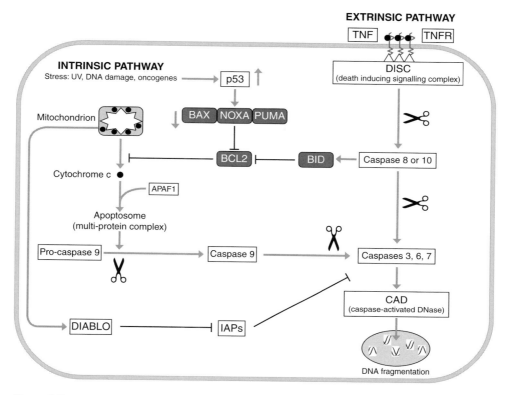

Figure 5.5 Apoptosis: intrinsic and extrinsic pathways. The intrinsic pathway is a response to internal cell stress, for example radiation, DNA damage by drugs or the expression of oncogenes. The extrinsic pathway responds to death ligands acting on their receptors (e.g. Fas or tumour necrosis factor receptor 1). IAPs refer to the inhibitor of apoptosis protein family that includes survivin. DIABLO, released from mitochondria, inhibits IAPs. Scissors indicate pro-caspase activation by cleavage.

The result is that caspase-activated DNase (CAD) is released to digest nuclear DNA. Cancer cells typically have an elevated internal pH (~7.5) with respect to normal cells (~7.2) and, as cytochrome c-mediated caspase activity declines with increasing pH, this has the effect of conferring a degree of resistance to apoptosis. The fragments of DNA generated by caspase action can be separated on a gel as a 'ladder' – the characteristic signature of apoptosis.

The extrinsic pathway is essentially the response system to an extracellular suicide note. It is activated by death ligands binding to their receptors that are members of the tumour necrosis factor receptor (TNFR) superfamily (Chapter 3). Following receptor activation, protein complexes form that lead to the activation of caspases 8 and 10 and the eventual phagocytosis of the cell.

The two pathways can cross-react through the action of caspase 8 on BID so that it can then bind to BCL2, inhibiting the anti-apoptotic action of BCL2. A family of inhibitor of apoptosis proteins (IAPs) bind to caspase 3 to prevent its activation – the most notable being survivin (encoded by *BIRC5*), which is commonly expressed in tumours and associated with resistance to chemotherapeutic drugs. Other proteins released from damaged mitochondria counteract IAPs (e.g. DIABLO), so that the overall apoptotic response reflects the balance between multiple factors that act in the two pathways.

It seems probable that self-destruction by apoptosis is the default option in multi-cellular animals and that the apoptotic pathways are normally repressed by survival signals, for example, from hormones and nutrients, that sustain the expression of anti-apoptotic factors. We'll come back to the apoptotic signalling pathways in more detail in Chapter 6 (Figs. 6.5 and 6.6).

Apoptosis is one way in which cells can respond to stress, in particular the stress of damaged DNA. Another type of stress response is a kind of cellular re-cycling referred to as **autophagy**, activated particularly as a response to nutrient deficiency. Enzymes break down organelles, including mitochondria, and use their constituent molecules in a re-cycling that, in effect, directs available energy into essential cellular processes. Autophagy probably occurs in every cell and is responsible for the degradation of most of the cell mass, but it may also permit cells to survive conditions in which they would otherwise perish. Tumour cells can use autophagy to ameliorate the effects of **hypoxia** (inadequate oxygen diffusion to the cell) and metabolic perturbation and they can also increase autophagic activity as a way of promoting resistance to anti-cancer agents.

Unlimited replicative capacity

If you remove a piece of tissue from an animal, break it up into single cells and put those cells into a suitable medium they will grow and duplicate themselves quite happily. They'll keep reproducing for somewhere between 20 and 60 doublings of the cell population – and then stop. They have then entered the state of senescence or replicative senescence – cellular old age. The cells are not dead, although they will eventually die, but they can exist for long time in culture in this state of suspended animation: they have barely detectable metabolic activity and require almost nothing in the way of nutrients.

One reason for cells becoming senescent is that they have lost DNA from the ends of their chromosomes – recall that our DNA is split up into 23 pairs of chromosomes, so that means there are 92 ($23 \times 2 \times 2$) free ends of double-stranded DNA within the nucleus. These are a problem for the machinery that replicates most of our DNA because it can't deal with the ends. Every time DNA is duplicated, therefore, which has to happen whenever a cell reproduces itself, the chromosomes are shortened. This might be serious at a very early stage if the DNA lost were important (e.g. encoded a protein). Presumably for this reason chromosomes are 'capped' at each end by repeated sequences of DNA – **telomeres**. Because telomeres are made up of a repetitive sequence that doesn't code for protein, loss of these sequences does not affect the viability of individual cells. Nevertheless, telomere loss would be incompatible with survival of the species if it happened in all cells and for that reason germ line cells and some stem cells

Unlimited replicative capacity

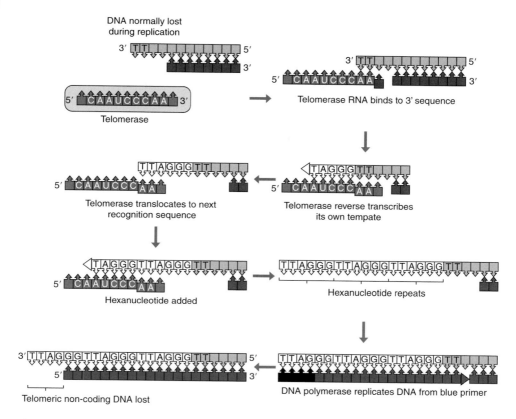

Figure 5.6 Telomerase. The enzyme is a complex of RNA, reverse transcriptase and other proteins. The reverse transcriptase creates single-stranded DNA using the telomerase RNA as a template.

express an enzyme, **telomerase**, that can replicate the ends of chromosomes (Fig. 5.6). Telomerase has its own RNA template, which, in effect, allows chromosomes to be temporarily extended to solve the problem. In all other types of cell telomerase is almost undetectable.

In about 90% of human primary tumours examined, however, there is substantial telomerase expression (in benign tumours the figure is ~25%). This is a consequence of the mutations that make a tumour cell and it enables cancer cells to maintain the length of their telomeres and thus escape mechanisms that restrict the doublings of normal cells to a finite number. They can grow indefinitely. Tumour cells that do not express significant telomerase use another mechanism to get round the problem – alternative telomere lengthening (ALT) – a process of recombination between the ends (telomeres) of different chromosomes. It will be evident, however, that in the early stages of cancer development telomerase activity may be inadequate to compensate for chromosome erosion. Cells with critically shortened telomeres will then enter a growth 'crisis' that generally results in death. Cells that survive this crisis and emerge as a proliferating clone have usually lost p53, the key monitor of genomic integrity. The unprotected ends of chromosomes have undergone end-to-end fusions with the production of dicentric chromosomes and the initiation of the chromosomal breakage–fusion–bridge cycle. This results in unequal chromosomal segregation during mitosis that can give rise to gene amplification and deletion. Thus, for these cells, inadequate telomerase activity may promote malignant progression.

Despite the fact that telomerase is largely suppressed in somatic cells, low levels of activity can be detected in most cells. This suggests that, yet again, nature works by balancing forces – in this case chromosome length. On the one hand you might wish your chromosomes to stay the same length so that you don't grow old but, on the other, we know that very active telomerase is a major contributor to cancer development. Amazingly, we may be able to adjust our telomerase activity, and hence the rate at which our chromosomes disappear, by changing our lifestyle. In other words, telomerase may turn out to be another biochemical marker for stress. Individuals who endure stress over prolonged periods (e.g. people who care for patients with severe disabilities or are HIV positive but without overt symptoms) have lower telomerase activity and shorter telomeres than their less stressed brethren.

In addition to its action in telomere maintenance, other functions of the telomerase catalytic subunit are emerging including effects on chromatin structure, mitochondrial RNA processing and modulation of the WNT signalling pathway through its capacity to interact with β-catenin (Chapter 6). The convergence of the reverse transcriptase activity (TERT) of telomerase and the WNT pathway is required for stem cell proliferation, a finding that has implications for tumour development as yet unresolved. In some human tumour cell lines at least TERT has a second mitochondrial activity in being able to influence cellular redox status, thereby limiting levels of **reactive oxygen species (ROS)**. ROS generation can occur, for example, as part of the p53-mediated DNA damage response leading to BAX translocation to mitochondria and the release of pro-apoptotic proteins. This effect on ROS endows TERT with yet another tumourigenic activity – the capacity to confer survival on cells that would otherwise be eliminated.

Induction of angiogenesis

J. B. S. Haldane, described by Peter Medawar as 'the cleverest man I ever met', observed in his essay *On Being the Right Size* that sheer size very often defines what bodily equipment an animal must have. Thus: 'Insects, being so small, do not have oxygen-carrying bloodstreams. What little oxygen their cells require can be absorbed by simple diffusion of air through their bodies. But being larger means an animal must take on complicated oxygen pumping and distributing systems to reach all the cells.'

Cellular nutrients, along with oxygen, are delivered in mammals by the bloodstream. Oxygen diffuses from circulating erythrocytes to cells that must be within about ten cells distance (~100 μm) of a blood vessel to avoid hypoxia. Like normal cells, cancer cells require a supply of nutrients and oxygen if they are to grow. They are, nonetheless, a pathological growth that the body isn't set up to accommodate and won't

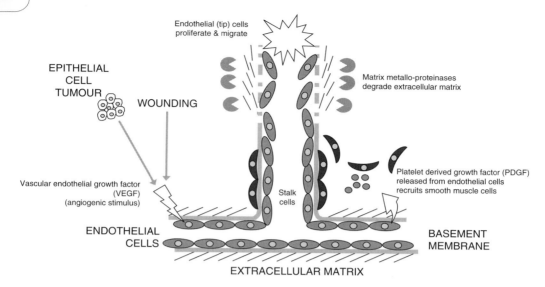

Figure 5.7 Angiogenesis. The initial stimulus is shown as the release of **vascular endothelial growth factor** (VEGF) by tumour cells. This promotes the activation of matrix metalloproteinases (MMPs) that degrade the **basement membrane** and the **extracellular matrix** into which proliferating endothelial tip cells migrate. As endothelial cells proliferate and migrate out of the capillary wall, they release platelet derived growth factor BB (PDGF-BB), a potent chemoattractant for smooth muscle cells that adhere to the endothelial tube, inhibiting proliferation and migration and stabilising the neo-vasculature. VEGFA also activates Notch signalling (Chapter 6) to inhibit proliferation of stalk endothelial cells.

provide with a blood supply in the way it would a tissue during normal development. This represents a major defence against cancer: even though a 'micro-tumour' – a small cluster of cancer cells – may have established itself, it cannot grow much beyond about one millimetre in diameter unless it can acquire a blood supply. Cancer cells achieve this by releasing chemical signals that stimulate the formation of new blood vessels – the process of angiogenesis (Fig. 5.7).

In switching on angiogenesis tumour cells are not doing anything novel as far as the organism is concerned. This process in which new capillaries sprout from pre-existing vessels is crucial for embryonic development although is almost absent in adult tissues. Transient, regulated angiogenesis does occur, however, as a repair process in adult tissues during the female reproductive cycle and also in wound healing. Much less usefully, angiogenesis occurs in over 70 diseases, including heart and vascular disease, rheumatoid arthritis, Crohn's disease, psoriasis, endometriosis and proliferative retinopathy (a common consequence of diabetes). The critical feature of angiogenesis is that signals activate the proliferation of the endothelial cells that line the vessels of the circulatory system. Because this lining is just one cell thick it is quite fragile and therefore it can easily be damaged, thereby providing a route from the circulation to the surrounding tissues.

What is a tumour?

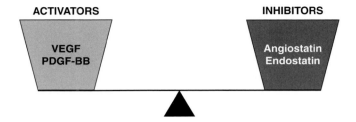

Figure 5.8 Angiogenic balance. The angiogenic switch is activated by an imbalance between activators (including VEGF and PDGF-BB) and inhibitors (e.g. angiostatin and endostatin).

Tumour vascularisation

Tumour cells, if they are to grow significantly, thus need to establish an 'angiogenic phenotype' – i.e. acquire their own vascular system. This is thought to be a critical step in tumour development that must happen first in the primary tumour and then subsequently when cells from the primary have spread to secondary sites, if those metastases are to expand. Angiogenesis is therefore an essential supporting component of most solid tumours. There is also a range of pre-malignant conditions (that includes, for example, some *in situ* carcinomas) within which both elevated levels of **vascular endothelial growth factor (VEGF)** and of microvessel density have been detected, indicating that, at least in some tumours, angiogenesis may be activated at a very early stage of development. Regardless of when it is initiated, the formation of a vascular network within a tumour creates a transport system waiting, so to speak, for metastatic cells to launch themselves once they are able to do so.

Angiogenic regulators

Like so many other facets of living systems, angiogenesis is a delicate balance: too much or not enough at any moment might be fatal. It will come as no surprise, therefore, to find that it is a multi-step process and that the equilibrium is under continuous regulation by both positive (pro-angiogenic) and negative (anti-angiogenic) factors (Fig. 5.8). Vascular endothelial growth factor is the most potent angiogenic factor known and when tumours induce a local excess of VEGF angiogenesis is initiated. VEGF binds to specific receptors on the surface of endothelial cells and one of the responses is the release of specific proteases that cut collagen, a major component of the basement membrane that surrounds blood vessels. Once this border has been breached endothelial cells can proliferate and 'sprout' towards the micro-tumour. Sprouting blood vessels then infiltrate the micro-tumour, providing the oxygen and nutrients that permit it to grow beyond the avascular limit. The clinical importance of tumour vascularisation is illustrated by the correlation between high levels of expression of the pro-angiogenic factor VEGF (that promotes dense formation of tumour micro-vessels) and poor prognosis in a variety of human tumour types.

The first anti-angiogenic protein to be identified was angiostatin, so named because it stops the growth of endothelial cells in culture and can block the growth of human tumour cells when they are inoculated into mice. Over 20 angiogenesis inhibitors are now known, some of which become activated by enzymatic cleavage of larger molecules that have a quite different function. Thus angiostatin is the centre region of plasminogen (from which plasmin, that breaks down the fibrin clots that protect wounds, is also made). Endostatin has similar properties to angiostatin and it is derived from a form of collagen. The gene that encodes this collagen, and hence endostatin, is on chromosome 21. Chromosome 21 is perhaps familiar because people with Down's syndrome have three copies (trisomy 21) instead of the normal two. Strikingly, Down's syndrome individuals have about half the normal lifetime risk of developing most cancers, although children have a 20- to 30-fold increased risk of leukaemia, and if they are diabetic never develop diabetic retinopathy. Endostatin may contribute to this protection although another chromosome 21 gene encodes a negative regulator of VEGF and the extra copy exerts an anti-angiogenic effect.

Discovery of angiostatin in a mouse model of metastasis

Angiostatin was isolated in the 1990s by Judah Folkman and his colleagues at the Boston Children's Hospital. They inoculated mice with cells from a tumour that metastasises to the lungs: over a few weeks the cells grew to form primary skin tumours about 1 cm in diameter at which point they were surgically removed (Fig. 5.9). At this stage micro-metastases could be detected throughout the lungs but none had developed beyond the critical size of about 1 to 2 mm or created their own blood supply – they were dormant. However, once the primary tumour was removed, the dormant tumours developed rapidly so that, after a further three weeks, the lungs doubled in weight due to the metastases. This suggested that the primary tumour itself had been producing an anti-angiogenic factor that prevented blood vessel development in the dormant tumours. If a protein spreads throughout the body it's pretty likely that some will be excreted. Accordingly Folkman and colleagues collected urine from their mice and analysed the proteins, eventually isolating angiostatin. Purified angiostatin given daily to mice after removal of a primary tumour completely prevented the development of micro-metastases. Angiostatin is active against primary tumours established in mice from inoculated human tumour cells and it also inhibits the proliferation of endothelial cells in culture (Fig. 5.10). As so often happens in cancer, **clinical trials** of angiostatin and endostatin were disappointing and they have been superseded by a recently developed angiogenesis inhibitor, bevacizumab (Avastin®),

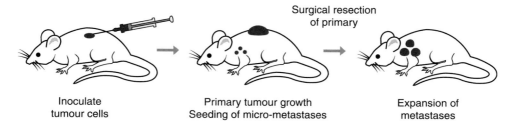

Figure 5.9 Mouse metastasis model.

(a)   (b)

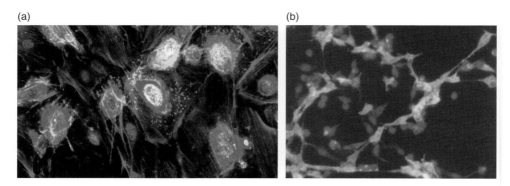

Figure 5.10 Endothelial cells in culture: effect of reduced oxygen. (a) Shows human endothelial cells grown under normoxic conditions (21% oxygen). The cells have been stained with three fluorescent markers (antibodies detecting actin (green) and the endothelial cell-specific von Willibrand factor (red) and the nuclear stain DAPI). (b) The same cells after nine hours under hypoxic conditions (oxygen <2%). These cells have started to form tubular structures as they would in vivo when initiating the growth of new blood vessels. (Photographs by Emily Hayes.) (See plate section for colour version of this figure.)

a monoclonal antibody against VEGFA, which has been used with some success in treating metastatic colon cancer and glioblastoma (see Chapter 7 for a discussion of antibody therapy).

Vascular mimicry by tumour cells

In an astonishing illustration of their adaptability, some types of tumour cells are able to differentiate into endothelial-like cells, thereby forming a neovasculature that is not host derived. This behaviour has been detected in glioblastoma, a particularly aggressive type of brain tumour, and confirmed both by immunohistochemical analysis of tissue sections and by showing that the tumour-derived endothelial cells share their mutational profile with the parent tumour. Thus their neovasculature is derived from tumour cells rather than, for example, from a fusion with host endothelial cells. This differential flexibility extends to the generation of smooth muscle cells and when these tumours are transplanted into mice their growth is supported by the expansion of the human, tumour-derived vascular tissue.

Abnormal metabolism

Tumour blood flow and the flexible cancer cell

Tumours are abnormal growths and it is not surprising that the blood vessels they create are also pretty weird. Usually they don't have any discernible pattern: sometimes they just stop like a kind of cul-de-sac, sometimes blood flows into them from both ends generating a form of traffic chaos and, in general, they are leaky and tortuous and have unstable blood flow patterns compared with normal tissues.

This picture of chaotic structure and flow suggests that some parts of a tumour may get less oxygen than others and measurements show that not only is this true but that cells within tumours often survive on much less oxygen (lower partial pressure, pO_2) than normal cells in adjacent tissue. In fact the centre of a growing tumour often becomes so hypoxic that the cells die, forming a necrotic core, while the outer regions of the tumour continue to grow. The instability of tumour vasculature suggests that the degree of hypoxia varies not only between different regions of a tumour but also with time. There is considerable evidence that what has been termed 'cycling hypoxia' can occur both relatively rapidly, due to fluctuation in blood flow through tumour micro-vessels, and more slowly, probably as a result of vascular re-modelling. In the face of this behaviour cancer cells have shown great versatility in the adaptations they have evolved to help their survival. These centre around the metabolic pathways that convert glucose into the principal energy currency of the cell, ATP.

There are two main stages: glycolysis and oxidative phosphorylation (Box 5.1). Glycolysis converts glucose to pyruvate: this pathway is an anaerobic reaction (it doesn't need oxygen and it parallels what happens when yeast makes alcohol, except that yeast converts pyruvate to ethanol rather than lactic acid). The most important thing, however, is that glycolysis generates two molecules of ATP for every glucose that is turned into pyruvate, the final step in the pathway being catalysed by pyruvate kinase. The pathways of the tricarboxylic acid cycle (TCA cycle) and oxidative phosphorylation take the pyruvate generated by glycolysis and turn it into water and carbon dioxide. This process requires oxygen (it is aerobic), it takes place inside a specialised organelle – the mitochondrion – and the energy released is converted into about 30 ATP molecules per glucose molecule consumed.

Since cancer cells are growing and multiplying, thus using much greater amounts of ATP, one might suppose that they would use the highly efficient aerobic pathway whenever possible, but this is not the case.

The first person to spot that there was something odd about metabolism in cancer cells was the biochemist Otto Warburg and he eventually demonstrated that they obtain most of their energy from glucose using the glycolytic pathway, rather than the second, oxygen-using stage. Typically, the glycolytic flux in rapidly growing tumour cells is a hundred-fold greater than in the normal cells from which the tumour originated. This conversion of glucose into lactate, even in the presence of abundant oxygen, is known as the Warburg effect or aerobic glycolysis. Presumably it reflects the fact that tumour cells have adapted to being fed by the disorganised blood supply they create, which gives rise to regions of low oxygen pressure. One consequence of this switch is that tumour cells mirror what happens in normal cells that are dividing in that carbon skeletons are diverted into the pentose phosphate pathway for macromolecular synthesis. Warburg suggested, incorrectly, that cancers actually occurred because mitochondria developed faults and it was disrupted metabolism that drove tumour formation. We now know that it is mutational changes in sets of genes that act as 'drivers' and metabolic perturbation is just one of the consequences. A critical contributor in this context is hyperactive MYC, which

Box 5.1 The major metabolic pathways involved in cancer

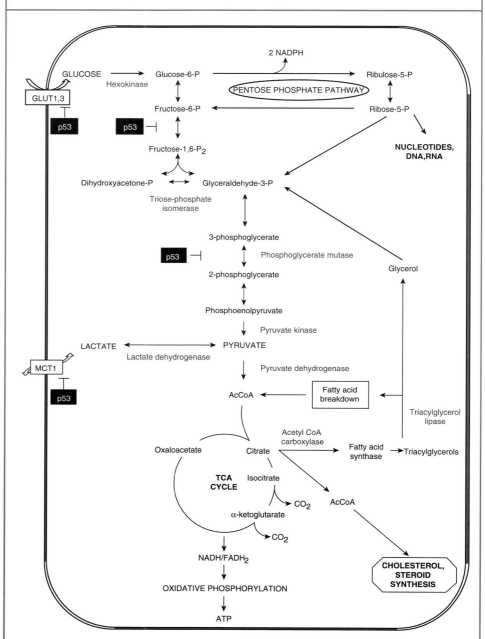

The breakdown of dietary glucose to convert its chemical energy into phosphate bonds in ATP takes place in three steps:

Box 5.1 cont'd

1. Glycolysis (conversion of glucose to pyruvate)
2. The tricarboxylic acid cycle (TCA cycle: releases NADH and $FADH_2$), and
3. Oxidative phosphorylation.

Overall about 30 molecules of ATP are formed when glucose is completely oxidised to CO_2. Rather than being converted to citrate, pyruvate can be converted to lactate (in muscle cells and red blood cells). Lactate thus produced is exported via protein channels in the cell membrane and diffuses in the blood to the liver where it can be converted to glucose. The pentose phosphate pathway is an alternative route to glycolysis for glucose that generates NADPH and pentoses (5-carbon sugars). These are used in fatty acid and nucleotide synthesis, respectively, so, although glucose is broken down, the pathway is essentially anabolic as it produces precursors for DNA and RNA synthesis. Enzymes are shown in grey. Four points at which p53 can act as a negative regulator of glycolysis are shown.

regulates many glycolytic genes in its transcription control portfolio and thereby contributes to the Warburg effect.

Warburg's discovery, reported in the 1920s, languished largely ignored in the literature for some 70 years until technical advances began to reveal the extraordinary behaviour that lies behind his observation. Notwithstanding, there were some significant observations in the intervening period. In the 1940s and 1950s it was shown that the lethal effect on normal cells of radiation used to treat cancers can largely be blocked if the tissue has previously been exposed to low levels of oxygen (anoxia). Conversely, tumour cells exposed to high levels of oxygen become sensitive to radiation treatment. In an ingenious experiment in which the two halves of a tumour were irradiated separately, one half when the patient was breathing normal air, the other when he was inhaling hyperbaric oxygen (three times atmospheric oxygen levels), the effect of radiation was shown to be much greater on the half irradiated under high oxygen conditions. To understand why this is we must first answer a critical question: how do cells detect changes in oxygen levels.

How do cells sense oxygen levels?

Various cellular responses, including angiogenesis, are affected by oxygen levels. Indeed the transcription of many genes is activated when cells are exposed to low levels of oxygen, for example, *VEGFA*. In the 1990s a number of transcription factors – regulatory proteins that control gene expression – were found to respond to hypoxia and named therefore hypoxia-inducible factors (HIFs). It was subsequently

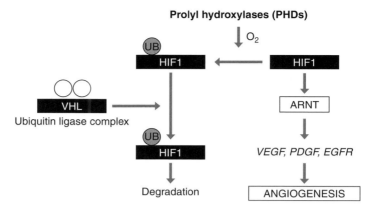

Figure 5.11 **Regulation of HIF activity.** In the presence of oxygen, hypoxia-inducible factor transcription factors (HIFs) are tagged with a hydroxyl group by prolyl hydroxylase (PHD) enzymes. This targets them for ubiquitin-mediated proteolysis via the VHL (von Hippel–Lindau) protein that forms a ubiquitin ligase complex with other proteins. Low oxygen levels increase the lifetime of HIF proteins that interact with ARNT to switch on transcription of *VEGF, PDGF* and the epidermal growth factor receptor (*EGFR*: see Chapter 3).

shown that the regulatory regions of oxygen-responsive genes contain specific sequence motifs to which HIFs bind – and hence control gene expression. Hypoxia-inducible factors are heterodimeric transcription factors composed of alpha subunits (*HIF1A*, HIF2A (for which the approved gene name is *EPAS1*) or *HIF3A*), and a beta subunit (the aryl hydrocarbon nuclear translocator (*ARNT*)). The levels of HIF1A and HIF2A are frequently up-regulated in tumour cells and, in addition to hypoxia, they can respond independently to activated oncoproteins (e.g. RAS). Among the genes they regulate are *VEGFA*, glucose transporters (*SLC2A1*/GLUT1 and *SLC2A5*/GLUT3), enzymes of the glycolytic pathway and regulators of apoptosis. HIFA proteins do not actually 'sense' oxygen but their very existence depends on proteins that do. Enzymes that use oxygen directly are oxygenases, and a group of these control the amount of HIF proteins in cells and consequently the cell's response to oxygen (Fig. 5.11). When normal levels of oxygen are present (normoxia) HIFA proteins are made but rapidly tagged with an –OH (hydroxyl) group: this is recognised by the cellular machinery responsible for breaking down proteins, so that normally HIFA proteins have a very short half-life. The enzyme that does the tagging is an oxygenase (prolyl hydroxylase) and it acts as a sensor of oxygen concentration in the cell. As cells become hypoxic the activity of the oxygenase declines, levels of HIFA proteins rise and their target genes are switched on.

In oxygenated cells the rapid degradation of HIFA proteins occurs via the **ubiquitin-proteasome pathway**, mediated by the von Hippel–Lindau (VHL) protein as part of a multimeric ubiquitin ligase complex. VHL is thus a critical regulator of cellular responses to hypoxia and is a 'tumour suppressor'. Loss of VHL function means that proteins that would normally be degraded by the proteasome continue to function, and some of these contribute to tumour development.

HIF gives tumour vascular chaos a helping hand

Mammals have at least three forms of the prolyl hydroxylase enzymes that tag HIFs and one of them, PHD2, has a dramatic effect on the structure of the endothelial cells that line blood vessels. In the disorganised tumour blood vessels endothelial cells are chaotically arranged with some cells floating in the lumen, leaving gaps in the lining. Surprisingly, reducing the levels of PHD2 (by knocking out one allele of the gene in mice) restores normal vessel structure within tumours. The tumour cells then find it much harder to invade their surroundings and metastasise. How might this happen? One member of the HIF protein family (HIF2A) is sensitive to PHD2 levels: the more PHD2 the less HIF2A. HIF2A in turn regulates expression of a soluble isoform of a receptor for VEGF (VEGFR1) and vascular endothelial cadherin (VE-cadherin), a protein that controls intercellular junctions. These two factors appear to be important in making normal vasculature in which the endothelial cells form a continuous, sealed lining. When PHD2 is made in hypoxic regions of tumours it reduces the amounts of these two orchestrators of normal endothelium, via its action on HIF2A, resulting in the chaotic form characteristic of cancer.

The return of Otto Warburg

One consequence of using glycolysis for ATP production in the Warburg effect is that tumour cells make a lot of lactic acid and release this into their surroundings. There are four human isoforms of the glycolytic enzyme that produces lactate (pyruvate kinase) and in many tumour cells PKM2 is the predominantly expressed form. PKM2 may be inhibited by signalling from growth factor receptors or via oxidation by ROS (the target being the sulfhydryl group of a cysteine amino acid). The levels of ROS are increased by oncogenic stress (e.g. signals from oncogenic RAS) and the consequence of suppression of PKM2 activity is accumulation in the level of its substrate, phosphoenolpyruvate. This in turn is a competitive inhibitor of the upstream glycolytic enzyme triose-phosphate isomerase (which catalyses the interconversion of dihydroxyacetone phosphate and D-glyceraldehyde 3-phosphate, Box 5.1). The result is the diversion of metabolic flux into the pentose phosphate pathway and the synthesis of nucleic acid precursors and NADPH. One role of NADPH is to minimise oxidative stress by providing reducing power for the generation of oxidised glutathione. By this means the level of ROS, a significant source of which is mitochondria, is reduced, a necessary measure if tumour cells are to survive. The overall effect is a re-direction of glycolysis to establish a new metabolic equilibrium – the Warburg effect – a measure that is costly in energetic terms but one that reflects the critical importance for tumour development of maintaining the redox balance.

Astonishingly, in human cancer cells PKM2 executes an additional, quite distinct function. When SRC is activated by signalling from the epidermal growth factor receptor, one of its phosphorylation targets is β-catenin (Chapter 6), which, in association with members of the LEF/TCF family, switches on transcription of genes that drive cell proliferation (*MYC* and *CCND1* (cyclin D1)). However, association of PKM2 with phosphorylated β-catenin is required for promoter binding, the effect of the complex being to dissociate **histone deacetylase** 3 (HDAC3), leading to histone H3 acetylation and transcription. The significance of this role is illustrated by the correlation of nuclear levels of PKM2 with the developmental stage of human glioblastomas and their prognosis.

The lactate released by hypoxic tumour cells can be taken up by cells in oxygenated regions of the same tumour and used as the fuel for oxidative phosphorylation, the most efficient way of making ATP. This is a remarkable symbiosis that has similarities to the way in which fast-twitch muscle fibres use glucose and produce lactate that is taken up by slow-twitch fibres that use oxidative phosphorylation.

As we noted, one of the targets of HIF1 is a gene that encodes a glucose transporter (GLUT1): hypoxic cells can thus take up glucose and they also express a protein (MCT4) that carries the end product of glycolysis, lactate, out of the cell. Aerobic tumour cells, in contrast, don't have GLUT1 or MCT4 but they *do* make a close relative, MCT1, that carries lactate from outside to in, thus making it available to fuel oxidative phosphorylation (Fig. 5.12).

p53 has emerged as a key negative regulator of glycolysis in that it can suppress the transcription of glucose transporters, inhibit the production of fructose-1,6-bisphosphate, promote the degradation of phosphoglycerate mutase and repress the expression of the MCT1 lactate carrier (Box 5.1). There are other points of regulation by p53 and mutant forms of the protein can actually promote glycolysis by, for example, stimulating transcription of hexokinase. Through the complexity of these multiple potential interactions, the essential concept to grasp is that the tumour suppressor mediates the balance between glycolysis and oxidative phosphorylation, and the switch to predominantly glycolytic ATP production occurs when normal p53 function is lost.

Because the levels of the proteins that determine the metabolic status of the cell are indirectly controlled by the amount of oxygen available, you might suppose that cells could switch this status if their oxygen supply changed, as it might well do in the chaotic environment of a growing tumour experiencing 'cycling hypoxia'. There is indeed evidence that this happens and that over a period of an hour or so a cell may flip from being a lactate user to an exporter.

The essential point to emerge therefore, is that solid tumours have both aerobic and anaerobic cells: cells showing the Warburg effect of aerobic glycolysis produce lactate that can be taken up by other cells for the synthesis of ATP.

This amazing flexibility of tumour cells that enables them to make the most of an uncertain situation offers possible targets for therapy. MCT1 is inhibited by α-cyano-4-hydroxycinnamate that has the effect of switching the fuel used by tumour cells from lactate to glucose. This drug has been used both in mouse tumour models and in human trials. It slows tumour growth because aerobic cells are prevented from taking up lactate, so they switch to using glucose, thereby depriving nearby hypoxic cells of their fuel. The hypoxic cells die and the aerobic cells become more sensitive to radiation.

In a further unpredictable corollary to the lactate story it has emerged that endothelial cells also express MCT1. The result of these cells taking up lactate appears to be the production of ROS leading to the activation signalling by NF-κB, cell migration and angiogenesis. These findings provide an exciting link between the Warburg effect, through which cancer cells can exploit hypoxic surroundings, and the promotion of a

(a)

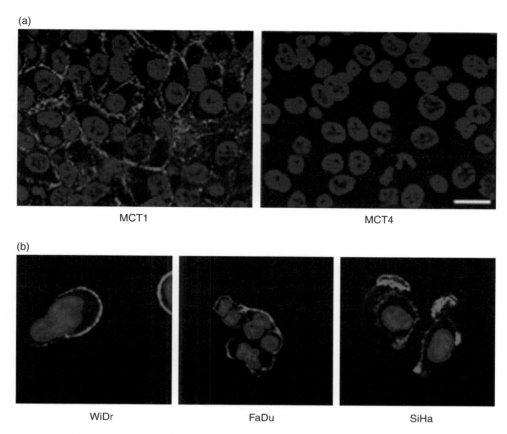

Figure 5.12 **Expression and activity of the lactate transporter MCT1.** (a) Fluorescent labelling of MCT1 (red) and MCT4 (red) and nuclei (blue) showing that oxidative tumour cells express high levels at the plasma membrane of the protein that carries lactate in (MCT1) and very little that carries it out (MCT4). Scale bar: 20 μm. (b) A variety of oxidative tumour cells express high levels of the protein that carries lactate in (MCT1) and very little that carries it out (MCT4). Fluorescent labelling of three cell lines (WiDr, FaDu and SiHa) shows the plasma membrane location of the lactate transporter MCT1 (red) and nuclei (blue). (c) MCT1 and hypoxia detected in cryoslices of a primary human lung cancer by confocal microscopy. MCT1 staining (green) and the red hypoxia marker hypoxia (2-[2-nitro-1H-imidazol-1-yl]-N-[2,2,3,3,3-pentafluoropropyl] acetamide [EF5] do not overlap. (d) Intravital microscopy shows that sustained MCT1 inhibition completely prevented angiogenesis of Lewis lung carcinoma tumours in mice. The MCT1 inhibitor was α-cyano-4-hydroxycinnamate (CHC). (Images kindly contributed by Dr Pierre Sonveaux, University of Louvain Medical School, reproduced by permission from Sonveaux et al., 2008, 2012.) (See plate section for colour version of this figure.)

key step in malignancy. The use of glucose as a fuel by hypoxic tumour cells has been exploited in the diagnostic method **positron emission tomography (PET)** in which the uptake of labelled glucose is followed to identify metastases (Chapter 7).

Disruption of energy metabolism in tumours can also occur through mutations in the cytosolic or mitochondrial forms of isocitrate dehydrogenase (IDH1/2,

(c)

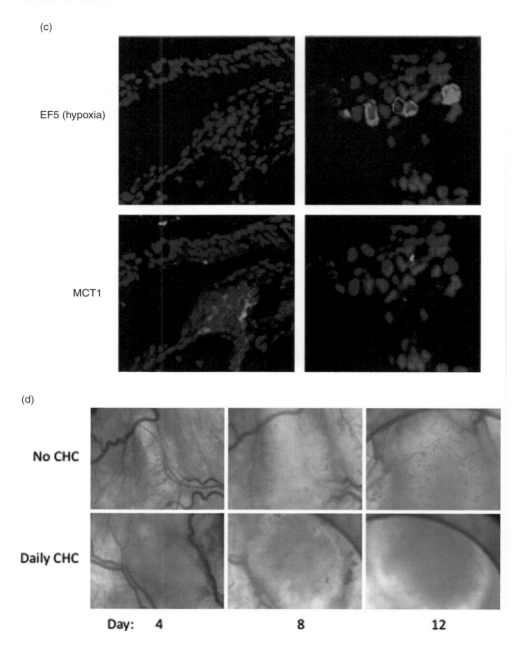

(d)

Figure 5.12 (*Cont.*).

respectively). The mutational effect is a gain-of-function so that enzymes normally catalysing the interconversion of isocitrate and α-ketoglutarate (αKG) now produce 2-hydroxyglutarate (2HG). IDH1 and IDH2 mutations occur in some brain tumours and leukaemias and one rationale for the emergence of these mutations is that 2HG may stabilise HIF1A through prolyl hydroxylase inhibition. However, the more aggressive grades of gliomas in which these mutations have been detected do not show pronounced angiogenesis and acute myeloid leukaemias (AMLs) carrying IDH mutations lack strong HIF expression. A more probable alternative is that mutant forms of IDH promote malignancy by affecting DNA methylation. The DNA demethylase enzyme, TET2, is regulated by αKG: IDH mutants, generating 2HG, inhibit its activity. Mutations in TET2 also occur in AML but they are mutually exclusive with IDH1/2 mutations and this sub-group of leukaemias has a specific hypermethylation signature.

Dormant tumours

Microscopic foci of tumour cells that do not expand were first detected early in the twentieth century. More recently autopsies of road traffic accident victims came up with the rather perturbing finding that many adults have accumulated a substantial number of microscopic colonies of cancer cells (also known as *in situ* tumours). These contained $\sim 10^5$ cells, occurred in a variety of organs and tissues and would have been undetectable had not accidental deaths made these tissues available for pathological analysis. These micro-tumours were clearly dormant: their carriers died in accidents and they had shown no signs of cancer. Furthermore, knowing what we do about the time course of cancer development, we can be sure that most of them would not have gone on to manifest cancer for many more years or even decades. Clearly the dormant tumours had spontaneously ceased to grow, the presumption being that this was due to inability to switch on angiogenesis. Considerable evidence from mouse models of tumour dormancy now supports this conclusion. Thus, tumour cells that are not angiogenic and remain dormant when injected into mice initiate growth when angiogenic factors are supplied. Furthermore, human tumour cells (breast carcinoma, glioblastoma, osteosarcoma and liposarcoma) that remain dormant for prolonged periods in mice before switching to a rapidly growing phenotype show a similar change in their pattern of gene expression as they do so. Notably the switch involves down-regulation of the angiogenesis inhibitor thrombospondin and up-regulation of the PIK3 and EGFR signalling pathways (Chapters 4 and 6).

Consistent with these observations are the dozen or so cases of organ transplants where the donor had previously been treated for melanoma and the same cancer subsequently developed in the recipient. The elapsed time between the donor undergoing surgery for melanoma and organ transplantation ranged from 6 months to 16 years. In each case the graft had carried with it melanoma cells, despite the donor being free of any evidence of secondary disease and detectable metastases at the time of his death, and these developed into tumours in the recipients.

These observations in humans have a parallel in some long-standing experiments in rats in which injected tumour cells failed to develop into liver metastases unless the rats were subjected to surgery. Following up to three repeats of this trauma, all the animals developed tumours. Although not confirmed, it seems probable that the activation of angiogenesis in response to surgery had the side-effect of switching on the generation of blood vessels to supply dormant tumours. An alternative explanation for dormancy has emerged from recent experiments using transgenic mouse models. These suggested that, rather than inhibition of angiogenesis, it is the action of the immune system that suppresses growth of disseminated tumour cells. Moreover, these studies provide further evidence that tumour cells can dissociate from a primary tumour during the earliest stages of its development, long before the primary tumour can be detected. The mice (in which the human RET oncogene (Chapter 3) is expressed in melanocytes) develop spontaneous melanomas (in the eye) and cells derived from the primary tumour can be detected in distant organs (notably the lung) within three weeks of birth. Genomic sequencing confirmed that the very early metastases were indeed composed of cells from the primary tumour. Nevertheless, the development of these distant metastases into tumours is suppressed for prolonged periods – the median age is about one year for the lung tumours.

The suppression of growth in these early metastases is critically dependent on a subpopulation of lymphocytes ($CD8^+$ T cells, also called cytotoxic or killer T cells): when these are depleted metastatic growth is switched on. In other words, $CD8^+$ T cells that would normally be present in the circulation inhibit proliferation of tumour cells in the metastases.

The concept of tumour 'immunosurveillance', proposed some 40 years ago, holds that malignant cells are generally killed by the action of the immune system, a major reason why so few develop as metastases. These results show, however, that signals generated by lymphocytes (interferon-γ (IFNγ) or tumour necrosis factor-α (TNFα)) can suppress proliferation in disseminated tumour cells so that they remain in a dormant state for prolonged periods.

Inflammation and the immune system

We noted in Chapter 2 that chronic infection can promote the development of cancers and indeed some 20% of cancers worldwide are caused by bacteria, viruses and parasites. Thus various oncogenic viruses and infectious organisms such as *Helicobacter pylori* are classified by the World Health Organization as carcinogens. These can exert mutational effects, DNA tumour viruses by inactivating tumour suppressors and *H. pylori* by enhancing free radical production and hence the frequency of oncogenic mutations in the host. All have in common the fact that sustained infection invariably causes inflammation. Inflammation is the first manifestation of the response of the immune system to tissue damage by, for example, pathogens or irritation, that leads to removal of the injurious agent and damage repair. Given that neoplasms are abnormal and disrupting growths, it is perhaps surprising that a critical role for the inflammatory process in cancer has only recently been recognised. However, it is now clear that inflammatory factors modulate the development of many, perhaps all, solid tumours and that chronic inflammation is also associated with some lymphoid malignancies (e.g. gastric lymphoma, which originates in the stomach and is most commonly caused by *H. pylori*). Thus hepatocellular carcinoma always arises in the presence of hepatitis (inflammation of the liver), a generalisation that extends to chemically induced tumours in mice. One facet of the inflammatory response that has been demonstrated in *H. pylori*-infected mice is the recruitment of fibroblasts from the bone marrow that express tumour-promoting factors and help to form a specific niche for the growing cancer.

The tumour cell environment

Solid tumours are supported by a framework of connective tissue, the **stroma**, composed of a variety cells and intercellular material (the extracellular matrix). The predominant cells are fibroblasts (that secrete the extracellular matrix protein collagen), with endothelial cells, **myofibroblasts**, lymphocytes, **mast cells** and macrophages also present. This multi-component environment infiltrates tumours so that it closely interacts with the cancerous cells in a two-way communication, mediated by proteins synthesised by both tumour cells and normal cells.

In principle there are two ways in which tumours can activate an inflammatory response: (1) as a result of necrosis caused by an inadequate blood supply; and (2) by the tumour cells themselves secreting pro-inflammatory cytokines. Note that the first of these will also occur after therapy in which radiation or drugs have caused necrotic cell death and that surgery will also result in inflammation. The subsequent activation of the immune system will bring additional cells into the tumour environment, in particular macrophages and lymphocytes, a phenomenon first observed by Virchow in the nineteenth century (Fig. 5.13).

There is much evidence that the inflammatory and immune response can inhibit cancer and eliminate malignant cells. Thus, for example, T and B lymphocytes can recognise tumour cell antigens and this is associated with a favourable prognosis, while patients taking immunosuppressants have an increased risk of solid tumours and lymphomas. Mice that cannot produce mature lymphocytes (Rag2-null) develop spontaneous cancers. Nonetheless, despite the immuno-protection that can be conferred, the fact that tumours do develop indicates that this can be overcome and indeed that the tumour microenvironment can become immunosuppressive and tumour-promoting. This switch should not be altogether surprising because, although the first role of the inflammatory response is to remove the damaging agent, a second feature is to restore the tissue to its normal state. This requires increased cell division in what has been called compensatory proliferation that, of course, requires survival and growth signals to be generated as part of the inflammatory response, signals that also have the potential to drive tumour progression.

A central role in directing both the innate and adaptive immune responses is played by the NF-κB transcription factors, principally by promoting T-cell maturation and proliferation. NF-κB is readily activated by a number of inflammatory signals, for example, interleukin 1 (IL1) and tumour necrosis factor (TNFα: Fig. 5.14).

NF-κB is constitutively activated in many cancers, as is signal transducer and activator of transcription 3 (STAT3), and the two cooperate in the progression of, in particular, colon, pancreatic and liver cancers. Together they promote expression of anti-apoptotic proteins (BCL2 and BCL$_{XL}$) and STAT3 can also drive proliferation by activating *MYC* and several cyclin genes. Their activation is generally the result of cytokine signalling in the tumour microenvironment, as evidenced by TNFα-null mice developing fewer skin tumours than normal animals in response to application of the **tumour promoter** tetradecanoyl phorbol acetate (TPA) and administration of a chemical mutagen.

STAT3 signalling is important in colitis-associated cancer (CAC) through its activation by IL6, a cytokine secreted by both T cells and macrophages that can have

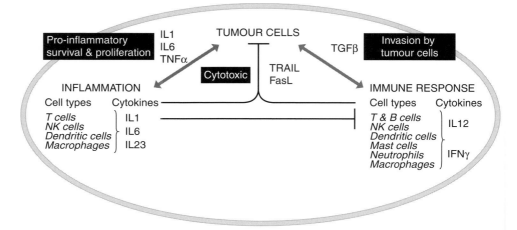

Figure 5.13 Balance between the immune system and inflammation in the tumour micro-environment. The scheme shows the major cytokines involved in regulating the balance between inflammatory and immune responses within solid tumours that eventually shifts from inhibition towards tumour promotion as malignancy develops. Interleukin 1 (IL1A and IL1B) are pro-inflammatory proteins synthesised by a variety of cell types, including activated macrophages and B lymphocytes. IL6 promotes B cell differentiation and can act in either a paracrine or autocrine manner to promote the growth of some types of tumour cell, although it can inhibit others. Both TNFα and IL6 are secreted by macrophages and are survival signals for metastatic cells. IL6 also acts with IL1 to activate T cells. Activated T cells and **natural killer (NK) cells** can kill tumour cells by the action of apoptosis effectors that include TRAIL (TNFSF10) and FasL (FASLG). There are two major categories of macrophages: M1 cells are activated by IFNγ and secrete high levels of proinflammatory cytokines (TNFα, IL1, IL6, IL12 and IL23). M2 macrophages are activated by several interleukins, including IL4. Most tumour-associated macrophages (TAMs) resemble M2 cells and contribute to tumour cell proliferation and invasion as well as promoting angiogenesis. Many types of cell secrete the multi-functional peptide TGFβ, which generally inhibits the proliferation of normal cells. However, it promotes invasion and metastasis of a variety of tumour cells and mutations in the TGFβ signalling pathway occur in diverse cancers (see Chapter 6).

pro-inflammatory or anti-inflammatory effects. Colitis-associated cancer is the most serious complication of inflammatory bowel disease and it increases the risk of colorectal cancer by 1% per year, so that this condition accounts for ~5% of all colorectal cancers. Inflammatory bowel disease affects the epithelial cell lining of the intestines and includes ulcerative colitis and Crohn's disease (not to be confused with irritable bowel syndrome). These are quite common conditions: 1 in 500 have ulcerative colitis and 1 in 1,000 Crohn's disease. Colitis-associated cancer can be reproduced in mice by treatment with two chemicals (azoxymethane and dextran sodium sulphate). This raises the levels of IL6, which is also present in abnormally large amounts in patients with colon cancer. IL6 acts via the transcriptional regulator STAT3 that, as we have seen, collaborates with NF-κB. When STAT3 is inactivated in mice both the size and number of colon tumours in the CAC model are greatly reduced. It appears

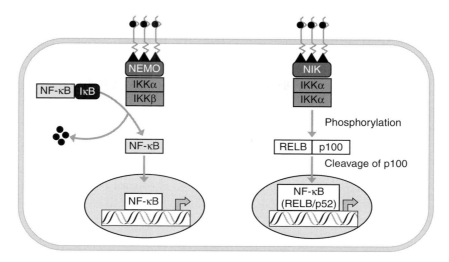

Figure 5.14 NF-κB **signalling.** Two pathways lead from receptors to activation of the NF-κB transcription factor ('canonical' and 'non-canonical'). In the first a complex of proteins (two kinases, IKKα and IKKβ, and a regulatory protein NEMO) is recruited to the activated receptor: this phosphorylates IκB causing it to be broken down, releasing NF-κB (to which is was attached) to enter the nucleus, associate with co-activator proteins and activate gene transcription. In the non-canonical pathway an IKKα complex is activated by NIK to release NF-κB. The NF-κB family has five members (p50, p52, REL, RELA and RELB) that act as homodimers or heterodimers.

that IL6/STAT3 is good at promoting tumour development because it both protects cells from apoptosis and provides a signal that drives cell proliferation.

In contrast to its pro-inflammatory effects, NF-κB can exert an anti-inflammatory action: its activation is controlled by the IκB kinase complex and deletion of this activator in myeloid cells inhibits both CAC and diethylnitrosamine-induced hepato-cellular carcinoma (HCC) in mice.

A major source of cytokines regulating the inflammatory and immune response are two main types of macrophage that are recruited to the vicinity of tumours (Fig. 5.13). M1 macrophages secrete mainly pro-inflammatory cytokines (e.g. TNFα, IL1, IL6), whereas the M2 type is considered to be anti-inflammatory, producing factors that promote cell growth (and hence tissue repair) and angiogenesis (e.g. IL10 and trans-forming growth factor β (TGFβ)). Tumour-associated macrophages (TAMs) constitute the bulk of the immune cells found in tumour microenvironments and their phenotype resembles that of M2 cells. TAMs, through the cytokines they release, activate the expression of proliferation, angiogenic and metastasis genes (*VEGF, COX2, EGFR* and *MMPs*).

The 'balance of power' in the tumour locale can be assessed from the cell populations and cytokine concentrations present and this is emerging as a powerful prognostic indicator for human cancers. David DeNardo and his colleagues have identified an 'immune signature' (CD68[high]/CD4[high]/CD8[low]) that indicates a high risk of metastasis and reduced survival for patients after breast cancer surgery. CD68 is a marker for TAMs, helper T cells are $CD4^+$ and cytotoxic T cells are $CD8^+$. Thus low levels of TAMs and helper T cells, together with high numbers of cytotoxic T cells ($CD68^{low}$/$CD4^{low}$/$CD8^{high}$), correlates with a good outcome after surgery. The relative proportions of TAMs and cytotoxic T cells also affect tumour response to anti-cancer drugs. Thus $CD68^{high}$/$CD8^{low}$ tumours respond relatively poorly to paclitaxel given before surgery ('neo**adjuvant**' chemotherapy, often prescribed for lymph node-positive patients). It seems probable from mouse models that drugs such as paclitaxel promote release of colony stimulating factor 1 (CSF1) from tumour cells that is an attractant for macrophages carrying the CSF1 receptor. TAMs can be depleted by blocking this interaction and a further notable finding is that, when combined with paclitaxel, this greatly reduces blood vessel density (by 50%) with associated tumour destruction. TAMs also secrete the cytokine CCL18 that promotes breast cancer cell metastasis.

Taken together these findings illustrate the importance of the tumour neighbourhood in progression and reveal both the prognostic value of a three-marker immune signature for breast cancer survival and a biomarker for neoadjuvant chemotherapy.

A more dramatic strategy by which cancer cells can co-opt normal cellular resources to promote metastatic conversion is by fusing with cells of the immune system. This concept is a century old but only recently has direct, in vivo evidence been provided by using fluorescent labelling to show that bone marrow-derived cells can indeed form hybrids with cancer cells during tumourigenesis. The transgenic mouse tumours were intestinal polyps, the epithelial cells of which fuse with macrophages. Perhaps the most unexpected facet of this extraordinary event is that, in addition to expressing a transcriptional profile that reflects both types of parent cell, fused cells also express a unique set of transcripts. The inference is that the hybrid **transcriptome** confers the migratory and invasive properties of macrophages upon cells of the tumour.

A further constituent of the microenvironment has been identified in the form of a sub-population of cells that defends human cancer cells from the immune system. These stromal cells are defined by the expression of fibroblast activation protein-alpha (FAP-α) and, although they comprise only about 2% of the tumour mass, their depletion permits rapid necrosis mediated by TNFα and IFNγ.

The overall effect of cytokine signalling on any tumour is therefore a balance between pro- and anti-inflammatory forces that are themselves unstable. That is, both macrophages and cytokines in the closely interwoven tumour microenvironment show extremely flexible behaviour and whether the overall inflammatory response is anti- or pro-tumourigenic appears to depend on the context – itself subject to change as tumours develop and, for example, respond to cycling hypoxia. This plasticity has given rise to the concept of 'immunoediting' by which a two-way communication modulates both the immune response and the development of the heterogeneous tumour. This suggests that the balance of this complex signalling will influence the rate at which tumours expand and indeed might produce sustained periods of stasis equivalent to the phenomenon of tumour dormancy.

In some mouse models the expression of HIF2A in macrophages contributes to their recruitment and hence to their role in driving tumour progression. In other types of tumour HIF2A appears to act as a suppressor. This paradoxical behaviour may result from the fact that responses to HIF2A are both tissue and concentration dependent, with high levels being tumourigenic. Reasonably effective drug treatments are available that reduce inflammation or suppress the immune system. However, the evidence discussed above that the immune system, at least in mice, inhibits the growth of micrometastases cautions against immunosuppression as a cancer therapy.

The uniqueness of malignancy

The seven features of tumours we've considered are characteristic of cancer cells but may also occur to some extent in benign tumours. Thus the size to which benign tumours can grow indicates that they acquire a degree of independence from growth factors, both positive and negative, escape apoptosis, continue to divide, have a degree of metabolic abnormality and possess angiogenic potential. Moreover, inflammation is also a common cause of benign tumours. Thus, although these properties are generally associated with tumour development, they do not permit the critical distinction to be made between benign and malignant growths. It is also worth noting that some malignancies do not manifest all of these seven characteristics. Thus, for example, pancreatic ductal adenocarcinomas have few blood vessels and yet can form aggressive tumours in an avascular environment. This may reflect a peculiarity of the pancreas in that tumours can be nourished without the requirement for extensive angiogenesis shown by most solid tumours. Whatever the explanation, these pancreatic tumours are an exception to the general notion that sustained angiogenesis occurs in all solid tumours. They also present a particular problem for systemic drug delivery because of their limited circulation.

The distinctions we have drawn thus far between malignant and benign, although very informative, have therefore been to some extent a matter of degree. The eighth property, however, absolutely defines malignant cancers: the capacity of cells to invade surrounding tissue and spread to other sites.

Metastasis and metastatic potential

Primary and malignant tumours

In a developing primary tumour the boundary is generally well defined and problems only arise if its size impacts on the physiological function of neighbouring organs or if substances that affect normal physiology are secreted (e.g. hormones such as insulin). At this stage tumours are usually treatable by surgery or other therapies. Local recurrence is often a consequence of incomplete removal (or killing) of all the tumour cells. The majority of primary tumours therefore do not kill: most cancer deaths (about 90%) arise from metastasis, the acquisition by some of the tumour cells of the capacity to invade surrounding tissue and spread to other sites in the body.

What is a tumour?

The steps that a cell has to go through to form a secondary metastasis are, in principle, simple: (1) migrate from the primary tumour; (2) burrow through the wall of a blood or lymphatic vessel (intravasation); (3) survive in the circulation until it sticks to a target site; (4) burrow through the wall of the vessel (extravasation); and (5) start to grow (proliferate) in its new location. These events do not have to occur in a continuous sequence and we've noted evidence that some tumour cells lie dormant in secondary sites for many years before their growth is re-activated. We should also record that some malignant tumours do not metastasise via blood or the lymphatics. Malignant gliomas arise from glial cells in the central nervous system and tend to infiltrate normal brain tissue rather than metastasise to distant sites in the body. Some grades of sarcomas, rare malignancies that develop in the bones or soft tissues, do not generally metastasise.

To take the first steps down the metastatic road a cell must start to release enzymes (proteases) that degrade the basement membrane defining the boundary of the tissue/organ in which they originated. It can then enter the blood or lymphatic system and circulate until it encounters a region of the vascular bed through which it can exit (extravasate) into normal tissue. In experimental mouse models of metastasis newly formed micro-metastases are often clustered around the blood vessel from which they left the circulation (perivascular cuffs) and they use that vessel as their initial life support system. In this environment micro-metastatic lesions may undergo 'colonisation' to form macroscopic tumours.

Early ideas about metastasis

This broad picture of the way in which cancers spread was recognised almost 200 years ago. While Laënnec was the first person to record that cancer could spread to secondary sites, the introduction of the term metastasis is credited to a French surgeon, Récamier, who succeeded him as a member of the Collége de France. The initial demonstration that the metastasis of tumours derived from epithelial cells is caused by malignant cells leaving the primary site and spreading through the body was provided by the German surgeon, Karl Thiersch. This was at odds with the notion proposed by the more famous Rudolf Virchow, father of modern pathology, that cancers spread within an organism through some sort of liquid medium that somehow changes connective tissue cells at a secondary site into metastases. It was, therefore, a while before the notion that cells migrate from a primary tumour to form a secondary became established but, once it did, the obvious question arose: when cells metastasise, how do they know where to stick? Or, as Stephen Paget more elegantly phrased it in a landmark paper of 1889: 'What is it that decides what organs shall suffer in a case of disseminated cancer?'

The simplest answer would be that it just depends on anatomy: that is, cells leave a tumour and then adhere to the first tissue to which the circulation carries them. But Paget had noticed that quite often this simply didn't happen and in his paper he described his own evidence and summarised the work of a number of other luminaries to show that 'the distribution of secondary growths was not a matter of chance'. Paget's speciality was breast cancer and in 735 fatal cases he had found 241 with liver secondaries, 70 lung metastases, 30 kidney metastases and 17 metastases to the spleen. He also noticed that secondary tumours from the breast occur with marked frequency in the bones. Paget gives full credit to the contributions of others, and particularly records Fuchs's prescient suggestion in 1882 that certain organs may be 'predisposed' for secondary cancer. All of which led Paget to a botanical analogy for tumour metastasis: 'When a plant goes to seed, its seeds are carried in all directions; but they can only live and grow if they fall on congenial soil'. From this, then, emerged the 'seed and soil' theory as at least a step to explaining metastasis.

The great strength of the seed and soil view is that it conveys an interplay between the tumour cell and normal cells, and that their actions collectively determine the outcome. As we shall see shortly, that is at the centre of our current picture of metastasis. Nevertheless, even at the beginning of the twenty-first century, the honest answer to the question 'What controls the spread of tumours?' is 'We don't know'. However, the facility with which genomes can now be sequenced is revealing the extent to which mutational evolution can continue in secondary tumours and hence that each is as unique as its primary precursor. Despite this continuous evolution, the presumption is that there are some common molecular strands to the dissemination process and these are gradually coming to light with the increasing sensitivity with which proteins and sub-sets of cells can be detected.

How do tumour cells become metastatic?

If we accept that the sites of metastases are not simply a consequence of anatomy, perhaps we should take a step backwards and ask how cells acquire metastatic capacity. Unfortunately, even that remains a pretty murky area but it seems probable that a sub-set of the mutations that are acquired *early* in tumourigenesis, giving a proliferative advantage, are also able to promote metastasis in later generations of tumour cells that have acquired further mutations. This idea is intuitively attractive because it implies that the primary tumour, or at least a sub-set of the cells therein, gradually develops the capacity to make the critical step – dispatching cells to other locations – that will ultimately be fatal for the host. This scenario has been confirmed in pancreatic cancers by the application of whole genome sequencing to reveal that the primary tumours contain a mixture of sub-clones each localised to distinct areas within the tumour. Distant metastases in different tissues (liver, lung, peritoneum) are derived from specific clones, each of which in turn developed from a single, parental, non-metastatic clone. Based on estimates of proliferation rates, at least ten years are required from the initiating mutation to the appearance of the parental, non-metastatic cell and a further minimum of five years for metastatic capacity to appear. On average patients live for two years after this event.

Notwithstanding the logicality of that model, as ever there are observations indicating that at least some cancers do it differently. Occasionally metastatic growths appear when no primary tumour can be detected. These are classified as 'cancers of unknown primary' and they are not that uncommon, falling in the top 10% of diagnoses. Ascertaining the primary tissue of origin for these tumours by conventional histology is often difficult, a problem that is being alleviated by the use of gene expression profiling to identify diagnostic patterns (Chapter 7). Whatever the molecular events that promote these tumours, they clearly do not require prolonged development of a primary. In mice with human breast cancer tumour cells implanted in a mammary gland, the human cells can be detected in bone marrow when the mammary glands show only increased growth of normal cells ('atypical ductal hyperplasia') – the earliest pathologically detectable stage of breast cancer. Similarly, in humans equivalent numbers of tumour cells have been detected in the bone marrow of patients regardless of the development stage that their primary tumour has reached. That is, metastasis had occurred to much the same extent in patients at the earliest stage as at each of the later stages.

These results indicate that at least some primary tumours are capable of releasing cells that relocate to metastatic niches during the earliest phases of tumour development. This suggests that critical steps required to produce a fully malignant tumour can occur in metastases, not just in the primary tumour. This evidence also shows that, for breast cancer at least, the poorer prognosis of patients with late stage primaries is not simply because they have exported greater numbers of cells to metastatic sites. One possibility is that the larger primaries may be releasing factors that promote growth of metastases.

Regardless of when, in the development of a primary tumour, cells become able to leave and form secondaries, it is now possible to look at the entire pattern of genes being expressed in a population of cells. Such analyses show that the patterns of gene expression in metastatic cells differ from those of their non-metastatic counterparts in the primary tumour. What appears to happen is that, in response to signals they receive from nearby cells or other environmental triggers, cells start to make proteins that enable them to detach from the primary tumour and initiate the process of intravasation. Given the huge range of genetic disruption that occurs in cancer cells, there seems no reason why this shouldn't happen very early in the life of some primaries, although one might suppose that the older the tumour the greater the chance that it will come up with a recipe for spreading.

The epithelial to mesenchymal transition

Cells change their appearance as they become metastatic, going through a process referred to as the epithelial to mesenchymal transition (EMT). This is mainly apparent as a reorganisation of the actin cytoskeleton: the scaffold of protein filaments that gives the cell shape and facilitates movement (Fig. 5.15). The EMT is a developmental programme and can be followed by the disappearance of certain proteins (e.g. E-cadherin (CDH1) and cytokeratin 18, which are markers for epithelial cells) and the up-regulation of others (e.g. vimentin and α-smooth muscle actin) as the cells start to look more like fibroblasts (more spindle like).

The EMT occurs during normal development and also in wound healing (Fig. 5.16). However, if you take breast epithelial cells that will not form tumours and induce the EMT (e.g. by expressing the transcription factors Twist or Snail) you can show that

(a)

(c)

(b)

(d)

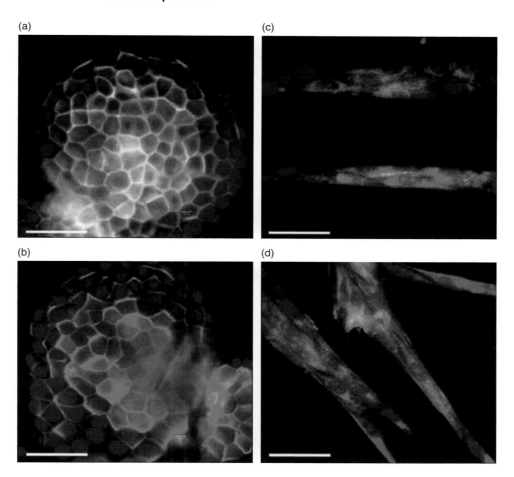

Figure 5.15 Images of the epithelial–mesenchymal transition. RAS-transformed mammary epithelial cells in collagen gel cultures form three-dimensional hollow, alveolar structures of polarised epithelial cells (a and b) expressing E-cadherin and a cortical actin ring but no mesenchymal markers (e.g. vimentin). Addition of TGFβ (6 days) induces these cells to undergo epithelial–mesenchymal transition (EMT), characterised by unorganised structures of fibroblas-toid cells expressing cytoplasmic actin and vimentin (c and d) but not E-cadherin. (Wiedemann, IMP) (See plate section for colour version of this figure.)

these changes are not the only ones to occur. Most of the mobile, mesenchymal-like cells that develop have a specific pattern of expression of two surface proteins, CD44 and CD24: they are CD44[high]/CD24[low]. This is precisely the pattern found on stem cells, cells that have the capacity for self-renewal and are pluripotent, that is, they can differentiate into any cell type. Because of their high capacity for self-renewal, stem cells are considered to be prime targets for the mutational assaults that produce tumours. As stem cells become progressively more genetically unstable they may promote changes in adjacent (stromal) cells that in turn develop into part of the expanding tumour.

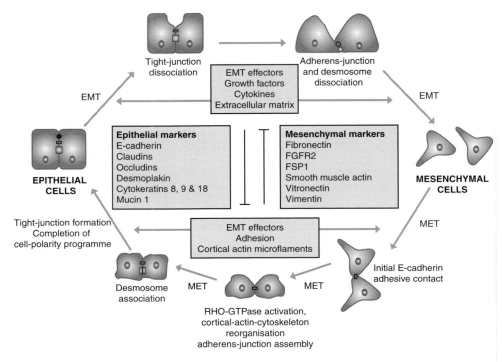

Figure 5.16 **The cycle of epithelial-cell plasticity.** The different stages during the EMT (epithelial–mesenchymal transition) and the reverse process MET (mesenchymal–epithelial transition) are regulated by effectors of EMT and MET that influence each other and modulate the tight junctions and the adherens junctions. The two states can be identified by characteristic markers, some of which are shown including FGFR2 (fibroblast-growth-factor receptor-2) and FSP1 (fibroblast-specific protein-1). (Adapted from Thiery and Sleeman, 2006.)

Stem cells

The theory that cancers arise from (mutations in) stem cells or germ cells is over 150 years old. It holds, in effect, that random mutations giving rise to a cell with the potential to develop into a tumour do not occur arbitrarily in terms of the affected cell (the 'stochastic model' of carcinogenesis in which any cell may be targeted) but arise specifically in precursor cells that retain some of the properties of stem cells. The main observations supporting this 'cancer stem cell hypothesis' are that (1) the 'initiating cells' in acute myelogenous leukaemia bear a phenotypic resemblance to normal hematopoietic stem cells (they are CD34$^+$/CD38$^-$); (2) major signalling pathways controlling stem cell self-renewal (WNT, Notch, Hedgehog, Chapter 6) can, when dysregulated, also promote tumourigenesis; and (3) when cells from a tumour are transferred between animals only a very small sub-population (0.1 to 0.0001%) forms new tumours, and the cells with that capacity carry surface proteins that are also found on stem cells. Populations defined as cancer stem cells have now been identified in cell lines derived from a range of cancers, notably brain, breast, colon and lung, on the basis of specific membrane proteins (e.g. CD133). In addition, cancer stem cells are held to resemble their normal counterparts by producing lower levels of ROS than do mature cells, thereby diminishing the extent of DNA damage they suffer. This may in turn contribute to their resistance to radiation and they may also show elevated drug resistance.

The question of whether stem cells (or at least cells that take on stem cell characteristics) are the critical components of developing tumours is of great importance when we come to consider the impact of drug treatment. Take a tumour that has reached the size of a 1 cm diameter sphere, which means about 10^9 cells: a drug treatment kills 99.9% of the tumour cells so that the tumour is undetectable unless you cut out the tissue to analyse it. Nonetheless, despite the efficiency of the drug, one million tumour cells remain: if only one of these has stem cell properties the tumour will recur. This is not, however, the only potential problem. When stem cell-like features are induced in breast epithelial cells yet another surprising change occurs: they become sensitive to a drug that has little effect on the precursor epithelial cells. The drug is salinomycin (which makes the cell membrane permeable to potassium) and it kills cancer stem cells but not other epithelial cells. This raises the problem that established cancer drugs, for example, paclitaxel, may be relatively ineffective against stem cells so that they produce the most undesirable of consequences: an enrichment of the stem cell pool from which tumours develop. This is a problem that has already been encountered in breast cancer patients who have been treated with conventional drugs (letrozole or docetaxel) where the residual cells carry stem cell marker proteins. The CD44^{high}/CD24^{low} cells that appear during the EMT have other properties that are characteristic of stem cells and this phenotype is also associated with human **basal-type breast cancers**, especially those arising from inherited mutations in the *BRCA1* gene frequently affected in breast cancer.

The implications for drug efficacy have focused attention on the cancer stem cell hypothesis. Nothwithstanding, it remains controversial and recent findings suggest that it might be less an illuminating model and more a confusing distraction. Most critically these have revealed that, far from being a fraction of one per cent, the proportion of tumour cells with tumourigenic potential may exceed 25%, depending on the precise experimental conditions. Moreover, some of the surface markers that have been proposed do not in fact discriminate between tumourigenic and non-tumourigenic cells. Most persuasively, transcription profiling has shown that embryonic stem cells and cancer cells share gene expression signatures. Moreover, the embryonic stem cell programme includes a distinct MYC-regulated module – recall that MYC is a 'master transcription factor'. Thus the coincidence in signatures and, by implication, other features of cancer cell 'stemness', arise from the central role of MYC and the fact that it is over-expressed in a high proportion of human cancers.

How do metastatic tumour cells know where to go?

By the time many patients present in the clinic with a primary tumour causing overt clinical symptoms there are already metastases in distant tissues. However, although some types of cancer cell metastasise preferentially to particular tissues (e.g. secondary

What is a tumour?

breast tumours often arise first in the lymph nodes and then in bone, liver and lung) how metastasising cells select their target sites is not at all clear.

For the best part of one hundred years Paget's aphorism of 'seed and soil' pretty well summed up our knowledge of how metastasis worked. Eventually, the use of radio-labelled tumour cells provided conclusive evidence for dissemination through the vasculature and for metastatic development at selective sites, with the inference that these are determined by interactions between tumour cells and their host. Despite such advances, we've had to wait until the twenty-first century for any further, significant insight into this process.

Malignant cells may retain sufficient phenotypic features to identify the tissue from which they originated, despite their frequently abnormal appearance, and they are often described as having a **de-differentiated**, embryonic phenotype. Metastasis per se does not require genetic changes. In Fig. 5.2 we alluded to the notion of 'metastasis suppressor genes' that restrain tissue invasion but do not affect primary tumour growth. These include, for example, *NM23* that indirectly restricts cell division to repress the spread of melanoma, breast and colon cancer cells. Conversely, evidence that activation of a sub-set of genes *promotes* the migration of tumour cells has given rise to the concept of '**metastatic signatures**'. Many of the components of these profiles of activity affect growth factor signalling – specifically the mitogen-activated protein kinase (MAPK) pathway (see Chapter 3). They may also include members of the interleukin family and we have already encountered IL4 and IL6 playing important roles in the tumour microenvironment. These cytokines are capable of stimulating the invasiveness of breast carcinoma cells and over-expression of another member of the family (IL11) has been associated specifically with colonisation in bone. IL11 is a regulator of bone cell proliferation and differentiation and can collaborate with another cytokine, osteopontin, which also regulates bone tissue homeostasis. Both of these genes may be further activated by TGFβ, abundantly present in bone matrix. These and related findings are beginning to reveal cooperative interactions that enhance metastatic activity. Even so, our current picture of the molecular basis of metastatic control remains cloudy.

A major breakthrough has come from mouse studies revealing that cells in primary tumours release proteins into the circulation and these, in effect, tag what will become landing points for metastasising cells. In other words these sites are deter-mined *before* any tumour cells actually set foot outside the confines of the primary tumour (Fig. 5.17). The process is more complex than merely tagging by a single type of protein released by the tumour. As well as sending a signal to the target landing strip, the tumour also releases proteins that act on cells in the bone marrow. This is the site of synthesis of circulating red cells and white cells, and the arrival of signals from the tumour causes the release of haematopoietic precursor cells (HPCs). These are a specific response to the signal from the tumour, carrying a marker in the form of a protein on their surface (in fact for one of the receptors to which VEGF binds, VEGFR1). These cells diffuse through the circulation until they reach the chosen location for metastasis, recognisable because the tumour signal tagged the landing site by making cells in that region produce a surface protein (fibronectin) to which

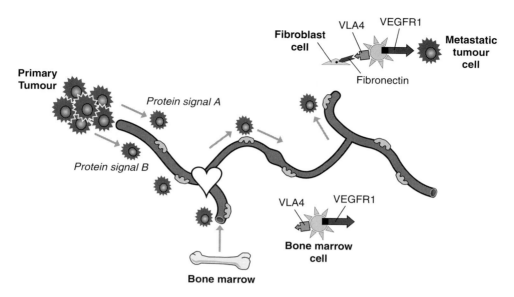

Figure 5.17 **Metastatic book–marking.** Integrin $\alpha_4\beta_1$ (also called VLA4) is a receptor for fibronectin.

the VEGFR1-bearing cells bind. They do this through a second cell surface protein, integrin $\alpha_4\beta_1$. In other words the target is recognised by HPCs because it provides a sticky patch for any cell with integrin $\alpha_4\beta_1$ on its surface. The presence of VEGFR1 on the attached HPCs then provides a marker for tumour cells to home in on, a process that David Lyden and his colleagues have named 'cellular bookmarking'. Matrix metalloproteinases released by HPCs are then able to reshape the microenvironment for the tumour cells to come.

This seems an extraordinarily elaborate mechanism for directing tumour cells to a target. How might it have come about, given that tumour cells cannot evolve in the sense of getting better at being metastatic: they just have to go with what they've got? We don't know but the most likely explanation is that they are taking advantage of natural defence mechanisms. We've commented several times that tumour cells are, in a sense, 'foreign'. That is, they are an abnormal growth and activate the immune system. Perhaps what is happening is that the signals emitted by the tumour cells are just a by-product of the genetic disruption that characterises cancer cells. Nevertheless, they may signal that there is 'damage somewhere in the body' to the cells of the bone marrow. That at least would explain why the bone marrow decides to release cells that are, in effect, a response to the tumour. The second question is more difficult: why should tumours release proteins that mark specific sites? Well, we know that different types of tumour metastasise to different places. In fact, if tumour cells are cultured in vitro and the medium in which they have been grown is injected into mice – no cells, just the medium containing any proteins the cells have released – metastatic sites are generated just as if a primary tumour had been sending out signals. This must reflect the fact that specific proteins released by tumours encounter binding partners (receptors) that happen to be present on, say, bone cells or liver cells but not on most other cell types.

Metastatic footnote

Metastasising cells are like an iceberg that breaks away from a large mass and is carried off by the ocean currents. Sooner or later the iceberg will melt and vanish: much the same happens to almost all metastatic cells: they get picked up and destroyed by scavenger cells in the circulation. But some metastasising cells are not eliminated: they stick to a target site somewhere on the lining of the circulatory systems and then manage to cross the vessel wall into the surrounding tissue. This is similar to what happens after an egg is fertilised: the cell starts dividing to form a clump of cells (a blastocyst), which implants in the endometrium and, eventually, when it's reached a suitable size, makes its entry into the world. But at least 25% of pregnancies end in miscarriages and this figure rises to 75% with increasing age. Despite metastasis being the most life-threatening facet of cancer, the odds against escaping tumour cells are much worse. Only about 2% of circulating tumour cells manage to extravasate so that they can form micro-metastases, and only 1% of these manage to persist and expand into tumours. Attrition within the circulation arises from shear stress leading to mechanical damage and by a form of apoptosis activated when the cells are no longer anchored to a substrate (called **anoikis**). Some tumour cells can acquire a degree of protection by coating themselves with platelets, but even so, with a success rate of less than 0.02%, the overall efficiency of metastasis is very low. Given the importance of stromal cells in promoting tumour development, it is plausible that they might also modulate the efficiency of metastasis. Direct evidence for such a role has come from mice that ubiquitously express **green fluorescent protein** (GFP) with implanted tumours of cells expressing a red fluorophore. While over 80% of the tumour cells collected from circulating blood were single cells, there were also tumour clumps that included host (green) cells (Fig. 5.18). The aggregates, in which fibroblasts were the predominant stromal cell, are more efficient than single tumour cells at establishing metastatic growths in the lung. The suggestion that, as Rakesh Jain and his colleagues put it, metastatic cells may 'bring their own soil' in the form of associated stroma, has been confirmed by examining metastases of human primary tumours occurring in the brain. Because fibroblasts are not part of normal brain tissue, their identification in such metastases indicates that they must have been transferred from the primary site with the tumour cells, subsequently establishing themselves and proliferating. If this emerges as a common feature of human cancers it would raise the possibility of targeting tumour clumps as a useful anti-metastasis strategy.

Do metastases metastasise?

There's another, perhaps rather obvious question, you might ask about metastasis. If primary tumours shed cells into the circulation and some of these eventually become secondary tumours in a new location, what's to stop cells from a metastasis doing the same thing in reverse? At least in mice, the answer is 'nothing and they do'. This has given rise to the idea of what Joan Massagué of the Memorial Sloan–Kettering Cancer Center has described as 'tumour self-seeding' and it can be visualised by labelling metastatic cells and inoculating them into mice.

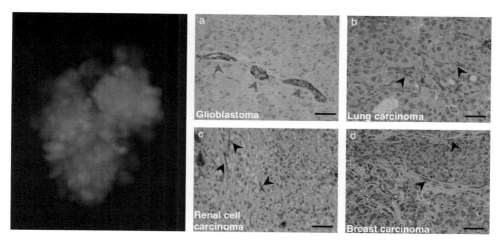

Figure 5.18 Metastatic seeding of tumour cells associated with stromal fibroblasts. Left: clump of cells shed from a kidney tumour isolated from circulating blood. The tumour cells (red) are associated with stromal cells (green). **Right: Clinical evidence for carryover of primary tumour stromal cells in human metastases.** Representative microscopy images: (a) glioblastoma (brain tumour), (b–d) human brain metastases originating from lung carcinoma (b), renal cell carcinoma (c) and breast carcinoma (d). Red arrow heads indicate tumour vessels after α-smooth muscle actin (αSMA)/CD31 double staining. In normal human brain and primary brain tumours only vessel associated pericytes and vascular smooth muscle cells are αSMA-positive. Vascular endothelial cells are CD31-positive. In the metastases, in addition to blood vessel staining, tumour-associated fibroblasts (αSMA-positive) are also present (black arrowheads). Fibroblasts are not present in normal brain tissue and must therefore have come from the original tumour. Scale bars: 50 μm. (Photos contributed by Rakesh Jain and Dan Duda (Duda *et al.*, 2010).) (See plate section for colour version of this figure.)

This type of metastasis may occur in at least some human cancers and, given that the cells doing the seeding have already jumped some of the major hurdles on the road to becoming a fully malignant tumour, may contribute to the aggressiveness of some cancers. This process could also occur even after a primary tumour has been surgically removed, giving a second mechanism for tumour recurrence in addition to incomplete removal of the primary tumour. There is, however, no conclusive evidence that this occurs in humans and it is a very difficult phenomenon to tackle experimentally.

How does cancer kill you?

Cancers are, of course, abnormal growths of cells and their damaging effects on the tissues they grow in can kill in a direct way. Thus, colon cancers, other tumours in the gastrointestinal tract and also ovarian carcinomas can obstruct the bowel, which would be fatal without surgical intervention. Similarly, lung tumours can be fatal if they block lung function (Fig. 5.19) and anaplastic thyroid tumours can, in effect,

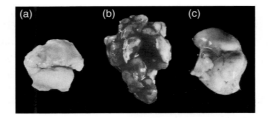

Figure 5.19 Metastatic tumour in mouse lung and the protective effect of combretastatin. (a) Normal mouse lung. (b) Extensive metastases as a result of which the weight of the lung has doubled, resulting in complete respiratory failure. (c) Lung of normal appearance in which metastases have been blocked by the administration of the drug combretastatin to the mouse. (See plate section for colour version of this figure.)

cause strangulation. A variant on this theme is contributed by leukaemias that result in a huge excess of white cells over red cells that so increases the viscosity of blood that circulation is drastically impaired. In general, however, human beings are remarkably resilient to organ damage. We can manage with half a kidney and if we lose two-thirds of our liver it will regenerate itself. It is therefore rare for cancer fatalities to be due to organ failure. Death is usually caused by secondary effects: principally infection. Cancer patients generally become increasingly susceptible to infection due to the decreased efficiency of their immune systems. Tumours that damage the walls of tissues can also increase vulnerability to infection. The agents are commonly bacteria (e.g. *E. coli, Pseudomonas*), which can overwhelm the host even with antibiotic treatment, but fungi are also significant contributors. Tumours can also cause damage to blood vessels, leading to haemorrhage, particularly in the liver.

About 40% of cancer deaths occur from malnutrition – a general condition of starvation and debilitation called **cachexia** (wasting syndrome) that develops in many other chronic diseases. Cancer cachexia is not understood and there are no satisfactory therapeutic treatments. Both chemotherapy and radiation therapy can also induce cachexia, and weight loss in turn reduces the efficacy of chemotherapy. Metastases may also suppress the immune system, e.g. if they are in the bone marrow or, if in the brain, raise intracranial pressure.

Conclusions

Tumour development is a complex process driven by multiple mutations and also by contributions from neighbouring host cells. Despite this complexity, the basic changes associated with the conversion from normal to tumour cell can be simply summarised (see key points below). In essence these reduce their dependence on environmental cues, avoid mechanisms that eliminate abnormal cells and recruit host cells to enable them to survive and spread to secondary sites. These remarkable achievements derive from the type of driver mutations we discussed in Chapter 4. In particular, constitutively activated RTKs promote proliferation independent of extracellular signals, and loss of function of key tumour suppressors (p53 and RB1) circumvents the central defence systems that protect against cancer, a topic we'll consider in greater detail in the next chapter. As these complex interactions are gradually unveiled they offer an increasing range of targets for therapeutic intervention, and we'll take that subject up again in Chapter 7.

Key points

- Cancers are diseases characterised by abnormal cell growth, that is, they are neoplasms.
- There are two main cancer categories: benign and malignant.
- The critical feature of most malignant cells is their capacity to spread (metastasise) from the site of origin (primary tumour) to other sites in the body (secondary tumours).
- The majority of human cancers are carcinomas. The remainder are sarcomas together with leukaemias and lymphomas.
- Eight properties characterise most cancers:
  (1) Growth signal autonomy – proliferation of normal cells is regulated by growth factors and nutrient availability. Cancer cells generally have a reduced requirement for growth factors.
  (2) Resistance to inhibitory growth signals – responsiveness to inhibitory signals that maintain normal tissue homeostasis is lost in cancer cells.
  (3) Resistance to apoptosis – the capacity for cell elimination (e.g. after DNA damage) is attenuated in cancer cells.
  (4) Unlimited replicative capacity – cancer cells maintain the length of their telomeres at the ends of chromosomes and thus escape the finite number of doublings of normal cells.
  (5) Induction of angiogenesis – cancer cells induce the formation of new blood vessels, essential for tumour development.
  (6) Abnormal metabolism – cancer cells obtain most of their energy from glucose using the glycolytic pathway, rather than from the oxidative breakdown of pyruvate in mitochondria.
  (7) Inflammation and activation of the immune system – tumour development is almost always associated with inflammation, which in turn leads to an immune response in the microenvironment of the tumour. This is initially an anti-tumour response. However, if it is overcome, the two-way signalling between cells of the tumour and neighbouring host cells becomes an integral component of tumour development.
  (8) Metastatic potential – cancer cells can spread via the lymphatic or blood circulatory systems from their primary site to other locations.
- Of these most important is the capacity to metastasise. This absolutely distinguishes malignant tumours from benign and it is the cause of mortality from most cancers.

Future directions

- Metastasis remains the critical problem in cancer in that there is no satisfactory treatment and it causes 90% of cancer deaths. The continued revelation of the intricacies by which tumour cells identify and establish themselves at secondary sites is beginning to provide additional therapeutic targets.

- The complexity of the tumour microenvironment is also being unveiled and with it non-tumour cell interactions (cells of the inflammatory and immune systems) that also offer promise for therapeutic intervention.

- The epithelial to mesenchymal transition (EMT) is a well established effect in cultured cells and animal models but whether it is precisely mirrored in human cancers remains controversial. In part this may be due to the difficulty of detecting cell markers before reversion sets in as cells establish themselves in a new site. Tracking labelled cells in humans is not feasible so this question will be difficult to resolve. However, the molecular minutiae are less relevant than the essential point: to emigrate cells change their adhesive and migratory properties in a process that is at least partially reversed as they settle into their new home. The EMT can also be stimulated in the bloodstream by direct interaction between circulating malignant cells and **platelets**. In the mouse models studied this requires platelet TGFβ and the importance of this additional facet of the transition is that potentially it offers a distinct anti-metastatic target.

- The cancer stem cell hypothesis continues to generate controversy in part because, if correct, the concept has important implications for drug therapy. The debate will be resolved with improved assignation of specific cell-surface markers and more refined fractionation of tumour cell populations. Regardless of the outcome, however, given that malignant cells are characterised by indefinite replication and unusual migratory properties, it is hardly surprising that some express markers associated with normal stem cells. The simplest explanation is that, rather than reflecting either a form of rogue stem cell or a consequence of reversion to an embryonic phenotype, the stem cell-like properties of some types of cancer cells reflect the oncogenic activity of MYC, a dominant 'driver' of tumourigenesis.

Further reading: reviews

CANCER CELL CHARACTERISTICS

Hanahan, D. and Weinberg, R. A. (2011). The hallmarks of cancer: the next generation. *Cell* **144**, 646–74.

Lazebnik, Y. (2010). What are the hallmarks of cancer? *Nature Reviews Cancer* **10**, 232–5.

Resistance to inhibitory growth signals

Heldin, C.-H., Landström, M. and Moustakas, A. (2009). Mechanism of TGF-β signalling to growth arrest, apoptosis, and epithelial–mesenchymal transition. *Current Opinion in Cell Biology* **21**, 166–76.

Further reading: reviews

Malumbres, M. and Barbacid, M. (2009). Cell cycle, CDKs and cancer: a changing paradigm. *Nature Reviews Cancer* 9, 153–66.

Resistance to cell death

Wyllie, A. H. (2010). 'Where, O death, is thy sting?' A brief review of apoptosis biology. *Molecular Neurobiology* 42, 4–9.

Unlimited replicative capacity

Artandi, S. E. and DePinho, R. A. (2010). Telomeres and telomerase in cancer. *Carcinogenesis* 31, 9–18.

Induction of angiogenesis

Adams, R. H. and Alitalo, K. (2007). Molecular regulation of angiogenesis and lymphangiogenesis. *Nature Reviews Molecular Cell Biology* 8, 464–78.

Carmeliet, P. and Jain, R. K. (2011). Molecular mechanisms and clinical applications of angiogenesis. *Nature* 473, 298–307.

Kerbel, R. S. (2008). Tumour angiogenesis. *New England Journal of Medicine* 358, 2039–49.

Oklu, R., Hesketh, R., Walker, T. G. and Wicky, S. (2010). Angiogenesis and current antiangiogenic strategies for the treatment of cancer. *Journal of Vascular and Interventional Radiology* 21, 1791–805.

ABNORMAL METABOLISM

How do cells sense oxygen levels?

Kroemer, G. and Pouyssegur, J. (2010). Tumor cell metabolism: cancer's Achilles' heel. *Cancer Cell* 13, 472–82.

Semenza, G. L. (2010a). Defining the role of hypoxia-inducible factor 1 in cancer biology and therapeutics. *Oncogene* 29, 625–34.

Semenza, G. L. (2010b). HIF-1: upstream and downstream of cancer metabolism. *Current Opinion in Genetics & Development* 20, 51–6.

HIF gives tumour vascular chaos a helping hand

Keith, B., Johnson, R. S. and Simon, M. C. (2012). HIF1α and HIF2α: sibling rivalry in hypoxic tumour growth and progression. *Nature Reviews Cancer* 12, 9–22.

Wilson, W. R. and Hay, M. P. (2011). Targeting hypoxia in cancer therapy. *Nature Reviews Cancer* 11, 393–410.

The return of Otto Warburg

Cairns, R. A., Harris, I. S. and Mak, T. W. (2011). Regulation of cancer cell metabolism. *Nature Reviews Cancer* 11, 85–95.

Grüning, N.-M. and Ralser, M. (2011). Cancer: sacrifice for survival. *Nature* 480, 190–1.

INFLAMMATION AND THE IMMUNE SYSTEM

de Visser, K. E., Eichten, A. and Coussens, L. M. (2006). Paradoxical roles of the immune system during cancer development. *Nature Reviews Cancer* 6, 24–37.

Dvorak, H. F. (1986). Tumours: wounds that do not heal. Similarities between tumour stroma generation and wound healing. *New England Journal of Medicine* 315, 1650–9.

What is a tumour?

Grivennikov, S. I., Greten, F. R. and Karin, M. (2010). Immunity, inflammation, and cancer. *Cell* **140**, 883–99.

Joyce, J. A. and Pollard, J. W. (2009). Microenvironmental regulation of metastasis. *Nature Reviews Cancer* **9**, 239–52.

Mueller, M. M. and Fusenig, N. E. (2004). Friends or foes: bipolar effects of the tumour stroma in cancer. *Nature Reviews Cancer* **4**, 839–49.

Steer, H. J., Lake, R. A., Nowak, A. K. and Robinson, B. W. S. (2010). Harnessing the immune response to treat cancer. *Oncogene* **29**, 6301–15.

METASTASIS AND METASTATIC POTENTIAL

Primary and malignant tumours

Valastyan, S. and Weinberg, R. A. (2011). Tumor metastasis: molecular insights and evolving paradigms. *Cell* **147**, 275–92.

Stem cells

Li, Y. and Laterra, J. (2012). Cancer stem cells: distinct entities or dynamically regulated phenotypes? *Cancer Research* **72**, 576–80.

Cancer signalling networks

The major 'drivers' of cancer development arise in pathways from receptor tyrosine kinases that ultimately control cell division. Thus, for example, aberrant activity of the RAS-MAPK pathway occurs frequently across a broad spectrum of cancer types. However, cancer-promoting mutations occur in other signalling pathways and, as we have seen, all aspects of cellular behaviour are ultimately affected. The molecular details of most of these pathways and how they can be subverted in cancers are now well established. The integration between diverse pathways creates a picture of a complex 'information network' rather than of discrete, linear systems. Despite the multiplicity of signalling pathways that can be involved, the tumour cells that emerge as the result of appropriate groups of mutations are phenotypically similar in that they share the characteristics discussed in the previous chapter. This suggests that the diverse pathways converge on a 'central axis' and in this chapter we'll look first at the central defenders that do indeed constitute the heart of our protection mechanisms against cancer. We'll then overlay the major pathways from receptor to nucleus to show how they impact on the central axis and then dissect from that complex map each of the pathways in turn to introduce the key players. The intention is not to show all known detail but to convey the principal features of the pathways with particular emphasis on identified (proto-)oncogenes and tumour suppressors.

Introduction

When we began this review of cancer cell and molecular biology back in Chapter 3 we focused on one pathway along which information is transmitted from membrane to nucleus. That was the RAS–MAPK pathway and we considered in some detail how the proteins in that signal relay talk to each other. In the following chapter we looked at the effects of the major oncogenic mutations that can affect that pathway, which, because it essentially regulates the cell cycle, have the effect of uncoupling cell proliferation from its normal controls. One other critical control on cell division is provided by the braking effect of tumour suppressors in the cell cycle, the loss of function of these being examples of the other main category of 'cancer mutation'.

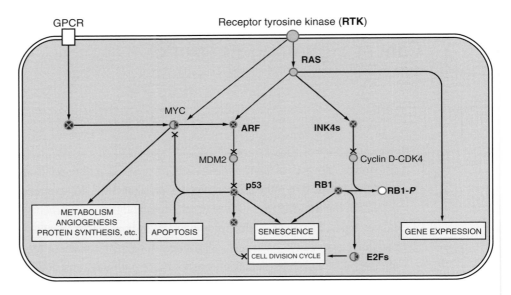

Figure 6.1 **The central axis of cancer signalling.** Many signalling pathways contribute to the tumour cell phenotype but essentially they converge on a core of four tumour suppressors (ARF, p53, INK4s and RB1) and two oncogenes (RAS and MYC). These normally control cell division and ensure that damaged cells either become senescent or undergo apoptosis. Loss of function of any of the tumour suppressors or oncogenic activation of RAS or MYC is therefore a major step towards tumour initiation. Discs: oncogenes; crosses: tumour suppressors.

We noted that there are multiple forms of receptors signalling through a considerable number of pathways and indeed one of the unresolved questions in cell biology is how cells make sense of a seemingly complex jumble of input signals. Mysterious though that is, cells clearly do make discrete decisions based on the integration of the messages they receive. An incipient tumour cell faces the problem of overcoming the defence systems built into these signalling pathways that have evolved either to permit repair of damaged DNA or to route irreparably stressed cells into apoptosis or senescence.

Fortunately it has emerged that the critical cancer resistance mechanisms reside in what might be called a 'central axis' into which all the main signalling pathways channel. The essential message is that if potentially oncogenic mutations subvert these pathways so that they become hyperactive, in general, all will be well if the central axis players retain their normal function. It is only when they become oncogenically activated or, in the case of the tumour suppressors, lose function that the defences are breached and the prospects for the cell become markedly brighter – brighter, that is, in terms of its chances of founding a tumour clone.

From our point of view this is very helpful because it means we can begin this chapter with a simple map that comprises only the central axis cast (Fig. 6.1). Two types of receptor are shown, and for the moment we will ignore the fact that multiple inputs are likely to be involved. Once we've considered how this bulwark operates we will take the plunge and put all the major pathways together, knowing that our map will then look somewhat intimidating. We can, however, pass quickly on from that by selecting each major pathway in turn and summarising the key players so that we'll have met most of the genes and proteins that have substantial roles in cancer. To put them back into context we'll end with an integrated picture of the phenotypic changes that occur when the defence systems have been overcome and a tumour cell has emerged.

The key players in the central axis of Fig. 6.1 are, of course, RAS, MYC, ARF, INK4s, p53 and RB1, together with the E2Fs that are regulated by RB1 (Chapter 4). These pathways, principally featuring p53 and RB1, lie at the heart of the defence against cancer and in general they work pretty well: despite the fact that cancers kill huge numbers, they are rare diseases at the molecular level and usually take a long time to develop into overt disease. Activation of either of the oncogenes (*MYC* and *RAS*) or loss of function of any of the tumour suppressors (ARF, INK4s, p53 and RB1) would constitute a major breach in the cancer defences and indeed almost all cancers feature at least one of these abnormalities.

The central axis: ARF, MDM2, p53, INK4, RB1

The pathways leading to both p53 and RB1 are nearly mirror images (Fig. 6.2). In the case of p53, the protein MDM2 functions as a ubiquitin ligase that inhibits p53-mediated transcription. The ubiquitylation of p53 marks it for degradation – a highly efficient way of removing a critical tumour suppressor. By neutralising p53, MDM2 acts as an oncoprotein and as such it is quite often over-expressed in human tumours. There is a second layer of control for p53 in the form of ARF: ARF binds to MDM2, even when it is bound to p53, and inhibits its activity. This makes ARF a

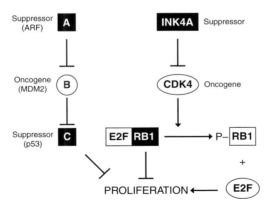

Figure 6.2 **Multiple controls of two key tumour suppressors.** The two arms show the major pathways regulating p53 and RB1. (Left) A and C inhibit tumour development; both can interact with B, which switches off the suppressor C: so B is a cancer promoter. A, however, suppresses the action of B on C. Thus loss of A would be a dominant contribution to cancer. (Right) Functionally the sequence controlling RB1 is the same but the intermediate oncoprotein is CDK4, which phosphorylates RB1 to release the proliferative activity of E2F.

tumour suppressor – the protector of p53 – and this key role makes it immensely important as a target for cancer-promoting mutations. Loss of ARF function is one of the most frequent aberrations in human cancers. As well as interacting with p53, MDM2 associates with RB1: the result is relief of RB1 suppression of E2F *trans*-activating function. Thus MDM2 not only releases a proliferative block by silencing p53 but drives proliferation by stimulating the transcription factor E2F.

The regulator of RB1 that corresponds to ARF for p53 is INK4A. ARF and INK4A are encoded by the same region of DNA so it's unsurprising that INK4A function is also lost in many cancers. INK4A inhibits cyclin D-CDK4: this is the kinase that phosphorylates RB1, thereby releasing the activity of E2F as a transcription factor and, in effect, switching on the cell cycle – so INK4A is a powerful cell cycle brake.

These pathways lie at the heart of the cellular defence so it is unsurprising that they feature sophisticated levels of control. Each arm includes a double negative in the shape of two tumour suppressors separated by a proto-oncogene. Oncogenic changes (e.g. expression of oncogenic MYC or RAS or loss of RB1 and hence activation of E2F) result in the up-regulation of ARF and hence cause cell cycle arrest or apoptosis. This is achieved not only by the indirect effect of ARF on p53 but also because ARF interacts directly with E2F to block its transcription factor activity. Thus ARF can be viewed as a dual-acting tumour suppressor protein in both the p53 and RB1 pathways – which is why its loss is such a disaster.

In Fig. 6.1, E2F1 is shown as a mixed oncoprotein/tumour suppressor. Up until now we've thought of E2F as a protein that switches on proliferation genes. However, it can also turn on apoptotic genes. The balance between proliferation and apoptosis is regulated by different signalling pathways, particularly those affecting CDK4/cyclin D and PI3K. In cancer cells this balance is perturbed, frequently by activating mutations in the PI3K pathway.

MYC and RAS

We noted earlier that the MYC protein is essential for normal cells to divide and that cancer cells very often lose control of MYC and make it in abnormal amounts, as much as one hundred times the normal level. That can happen when expression of MYC is driven through aberrant activation of a signalling pathway, although sometimes the *MYC* gene itself is mutated. Because over-expression of MYC is so common in cancers, it's perhaps a very good example of what has been called 'oncogene addiction' – the requirement for a specific function for the continued growth of a cancer. Unfortunately, quite often when a drug is used to block such an oncogene, cancer cells find a way round the block (they either mutate the target – again – or increase the level of *other* signalling molecules and use them instead). MYC may be an exception because of its unique role and it may be possible to make an effective inhibitor that works in humans without drastic side effects. We'll return to these important points in Chapters 7 and 8.

Remarkably, MYC can also be a tumour suppressor. How does it do that? Critical in this role is the fact that *ARF* is one of the many genes MYC can switch on. Thus when MYC is significantly over-expressed it activates the ARF–p53 axis. As we also noted above, over-expressed RAS acts similarly to activate both ARF and INK4A. So this sounds as though oncogenic MYC shouldn't be a threat. All the same, there are two reasons why it is. The first is that the ARF-driven protection system appears to be activated only by large increases in MYC expression: if the oncogenic stimulus only mildly increases MYC expression the defensive radar of the ARF system doesn't respond. So our main defence system, perhaps unsurprisingly, senses the *level* of damage and a small stimulus can persist undetected. The second reason why MYC is such a dominant force is that, if the ARF system is compromised (loss of function of ARF or p53), MYC now has a green light to drive unregulated cell proliferation.

Cancer and cell senescence

We saw earlier that cells can progress to a state of senescence that has been considered to be an irreversible, essentially vegetative, state. It is only in the last few years that senescence has been recognised as a cancer protection mechanism in animals. Unsurprisingly, p53 and RB1 play central roles in promoting senescence in vivo. Thus aberrant signals in pre-malignant cells activate either the ARF-p53 or the INK4A-RB1 pathway to force cells into senescence and prevent cancer progression. Notwithstanding, when the disease develops senescence is unblocked and proliferation activated: expression of INK4A and other senescence markers is lost.

In response to oncogenic signals (DNA damage, over-expression of RAS, short telomeres) signalling pathways can be activated that lead to the up-regulation of ARF and INK4A, the combined upshot of which is that p53 and un-phosphorylated RB1 combine to drive cells into senescence (Fig. 6.3).

Thus is revealed the central importance of ARF, INK4A, p53 and RB1 in our cancer defences. As long as they are present as functional genes we have two very effective strategies: apoptosis and senescence. Knock out even one and tumour development is almost inevitable. Encouragingly, recent transgenic mouse experiments reveal another route to senescence, and therefore cancer protection, that is independent of ARF and p53. This involves the cyclin-dependent kinases (CDKs) that regulate the cell cycle. We've seen (Chapter 3) that one way in which cell division is controlled is by the sequential synthesis and breakdown of key proteins – the cyclins and also the inhibitors of CDKs. Proteins to be broken down are tagged with ubiquitin by an enzyme complex (SKP2-SCF). Abnormally high expression of SKP2-SCF is found in some human cancers, constituting incriminating evidence that it may be cancer promoting. Knocking out SKP2-SCF activity might therefore be a useful therapeutic strategy. In transgenic mice at least this works: SKP2-SCF-defective mice are resistant to tumour development driven by oncogenic RAS or by loss of the tumour suppressor PTEN. Instead, SKP2-SCF inactivation triggers senescence: the mice have raised levels of the CDK inhibitors WAF1 and KIP1. The really startling result, however, is that this

Cancer signalling networks

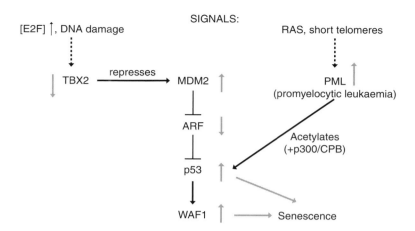

SIGNALS:

Figure 6.3 **Regulation of p53 activity to promote senescence.** Conversely to the promotion of senescence, somatic cells can be re-programmed into pluripotent stem cells by suppression of p53 or by enforced expression of MYC, KLF4, SOX2 and OCT4 in mouse fibroblasts, when the *Ink4a/Arf* locus is silenced. PML is the *promyelocytic leukaemia* protein (Chapter 4). TBX2 is a member of the T-box transcription factor family.

happens even if the ARF-p53 pathway is knocked out. In other words, even in the presence of very common cancer-causing mutations, interfering with the machinery of the cell cycle can protect against cancer. A slightly different approach generated CDK2-deficient mice and showed that they too are sensitised to senescence. So senescence, like apoptosis, is a protective mechanism that responds to the level at which key proteins are expressed. Even more exciting is the fact that drugs blocking the action of SKP2 or CDK2 can cause senescence in established mouse tumours, leading to regression.

Tumourigenic DNA viruses

Four types of DNA virus are associated with human cancers: human papillomaviruses (HPVs), hepatitis B virus (HBV), hepatitis C virus (HCV) and Epstein–Barr virus (EBV). Of the more than 100 types of HPV, 15 are oncogenic, of which the most important are types 16 and 18 that cause 70% of cervical cancers worldwide. They encode two proteins (E6 and E7) that target p53 and RB1 for degradation, thereby ablating two central tumour suppressors (Fig. 6.1). Hepatitis B virus, responsible for the majority of liver cancers that claim one million lives a year, also makes proteins that perturb normal cell signalling and promote genetic instability. The mechanism of HCV is unknown but it may interfere with DNA repair.

Burkitt's lymphoma, the most common childhood cancer in equatorial Africa, arises in B cells of the lymphatic system and commonly involves the jaw or other facial bones. It is caused by the Epstein–Barr virus (EBV, also called human herpesvirus 4

(HHV-4)). The EBV genome encodes a number of proteins that can 'immortalise' B cells in vitro, which is thought to reflect the way in which it promotes the human lymphoma. It can also cause nasopharyngeal cancer.

Signalling pathways that impact on the central axis

Having discussed the core pathways that determine cancer susceptibility, we should now take a deep breath and face the fact that we mentioned earlier: RAS-MAPK is not the only signal pathway that can contribute to cancer. Figure 6.4 overlays the other major pathways from cell surface receptors that play important roles in cancer. They're important because, as for RAS-MAPK, all their components have tumourigenic potential and many have been identified as oncoproteins or tumour suppressors. The first point to note about these pathways is that they work in exactly the same way as our MAPK model: they're protein relays – it's just that the performers are different. The salient feature of the map is that these pathways converge on the central axis. Thus, as long as the critical components of cancer defence are working, aberrant signalling by one of these pathways should activate a protection mechanism. It's only when a core control is lost that an abnormal signal is likely to lead to a tumour cell.

Somewhat paradoxically, although signalling overall is convergent, the pathways are frequently divergent – that is, they branch and 'cross-talk' with other pathways, as we mentioned in Chapter 3. A common means of divergence is via proteins that act as 'scaffolds' – they have multiple binding sites for other proteins and thus provide initiation points for branching routes. We've already met scaffolds as the cytosolic domains of RTKs in Chapter 3 and Fig. 6.4 shows that they can also activate phospholipase Cγ (PLCγ), SRC and JAKs. However, a striking point about RAS is that it initiates four more major pathways in addition to RAF-MAPK. RAS is extraordinary in that at least ten proteins can interact with a small domain in a not very large protein, making it a major point of pathway divergence. The pathways initiated from RAS regulate essentially every aspect of cellular behaviour: transcription, apoptosis, cytoskeletal structure and the cell cycle.

In addition to RAF-MAPK, the major pathways emanating from activated RAS are PI3K (which can also be activated directly by RTKs and controls survival), mTOR (which regulates proliferation, survival, metabolism and the cytoskeleton), JNK (stress-response pathway) and CDC42 (which with other small GTPases regulates the cytoskeleton). This feature of proteins acting as multi-pathway activators is widespread in cancer signalling and we shall see in the examples of major pathways that, although the scaffolds differ, the essential mantra of cancer signalling emerges as 'everyone talks to everyone else'.

The activity of many of these signalling pathways may be abnormal in cancers although the MAPK and the PI3K pathways appear to be the most frequently activated in a wide range of primary human tumours. To explain the principles clearly we have thus far looked only at the outlines of pathways and focused on a small number of genes and their encoded proteins as examples of the type of specific mutation that can

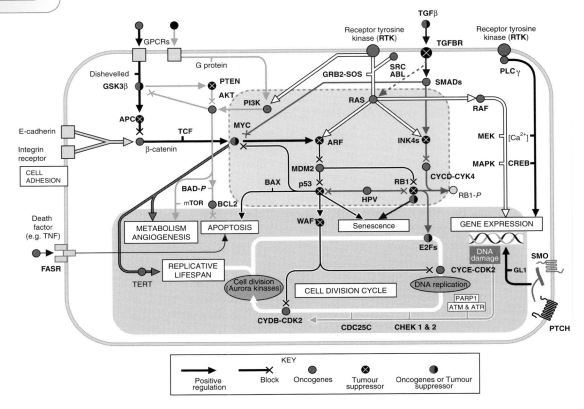

Figure 6.4 Major intracellular signalling pathways that can be involved in cancer. In addition to growth factors activating RTKs, pathways are shown that start from other types of receptors, namely from those involved in cell to cell contact (E-cadherin and integrins), from G-protein-coupled receptors (GPCRs) activated by WNT or by cytokines, and from the receptor complexes for transforming growth factor beta (TGFβ). TGFβ in normal cells signals via SMADs to inhibit MYC and activate INK4A, thereby inhibiting growth: the dotted line to RAS represents the switch to an oncogenic capacity that can occur in TGFβ signalling to activate the MAPK pathway. The PI3K pathway can be activated by RTKs and by GPCRs: AKT, activated by PI3K, promotes apoptosis (by phosphorylating BAD) and regulates glycogen synthase kinase 3β (GSK3β) by phosphorylation. GSK3β phosphorylates APC (*a*denomatous *p*olyposis *c*oli) and β-catenin, releasing β-catenin to form an active complex with transcription factors of the lymphoid enhancer factor (LEF1)/T cell factor (TCF) family. Genes activated by β-catenin/TCF include *MYC* and cyclin D1. GSK3β can also phosphorylate PTEN and MYC. As part of its widespread effects on cell behaviour due to the broad spectrum of genes that it controls, MYC can also activate the telomerase gene *TERT*, thereby increasing the replicative lifespan of the cell. Another transcriptional target of MYC is *ARF*, encoding a negative regulator of MDM2 – itself a negative regulator of p53. Thus the effect of increased ARF expression is to protect p53 in one of the major tumour suppression pathways. p53 and RB1 can drive cells into senescence. Both are targets for the oncoproteins of papillomaviruses that directly interact to degrade these key tumour suppressors. Members of the tumour necrosis factor (TNF) family can act through so-called death receptors to regulate the balance between apoptotic (BAX, BAD, BAK and BID) and survival (BCL2 and BCLx$_L$) factors in the cell. Other RAS-mediated pathways include the JUN N-terminal kinases (JNKs or stress-activated protein kinases, SAPKs), RHO/RAC GTPases (including CDC42, RAL, RHO and RAC) that are involved in the

affect pathway activity. The following summaries provide greater detail for the major pathways, together with a table listing the major 'cancer genes' or 'drivers' in which either germ line or somatic mutations have been identified. It should be borne in mind that these are a small selection from over 500 oncogenes and 100 tumour suppressor genes that have been identified. However, as we've noted, many of these appear to be either 'passengers' or possibly to exert a 'driving' capacity only in very specific cancers.

Finally, there is another group of genes in which mutations have not been identified but have been shown to be abnormally expressed in some tumour samples. Such data are usually obtained by immunostaining tissue sections but it is often difficult to obtain control tissues for comparison and quantitation of the amounts of protein present is technically challenging. With those reservations, many proteins have been reported to be abnormally expressed in a wide range of primary and secondary tumours and cell lines. Two notable examples, mentioned in Chapter 3, are the adaptor protein GRB2 and the exchange factor SOS to which it couples to regulate the activity of RAS. These proteins lie at the heart of one of the major 'cancer pathways' but, somewhat counter-intuitively, neither GRB2 nor SOS has been shown to undergo mutation in cancers although they have been reported to be independently over-expressed in some types of tumour cell.

Phosphatidylinositol 3-kinase (PI3K): survival and apoptosis signalling

Both RTKs and GPCRs can activate PI3K enzymes that are made up of a catalytic subunit (p100α, β or γ) with different regulatory subunits (p85s). PI3Ks phosphorylate the 3 position on the inositol ring of phosphatidylinositol (PI) to produce phosphatidylinositol 3,4,5-trisphosphate (PtdIns(3,4,5)P_3) to which a number of plekstrin homology domain proteins bind, leading to the activation of AKT kinases (Fig. 6.5). AKTs (AKT1, 2 and 3) are multi-functional enzymes that regulate a range of transcription factors and other proteins, the general effect of which is to promote cell survival. Thus, for example, one target is the BCL2 family protein BAD, which when un-phosphorylated binds to the anti-apoptotic BCL2. Phosphorylation inactivates BAD, releasing BCL2 to suppress the intrinsic apoptotic pathway.

B-cell lymphoma 2 (BCL2), first identified in chromosomal translocations occurring in follicular lymphomas, is the founding member of a large family of regulators of **caspase** activity and apoptosis characterised by three or four regions of homology

---

(**Figure 6.4** caption continued from overleaf):

organization of the actin cytoskeleton, cell cycle regulation and membrane trafficking, and mTOR. RTKs may also activate, directly or indirectly, SRC, phosphatidylinositol 3-kinase (PI3K) and an anti-apoptotic pathway (via the AKT serine kinase and BAD), and phospholipase Cγ (PLCγ) leading to hydrolysis of phosphatidylinositol 4,5-bisphosphate (PIP$_2$) and elevation of the free, intracellular concentration of calcium ([Ca^{2+}]$_i$). Changes in [Ca^{2+}]$_i$ are sensed by calmodulin and calmodulin-dependent kinases can phosphorylate the cAMP response element-binding protein (CBP) to regulate transcription and enhance the anti-apoptotic activity of AKT. The mTORC1 complex includes RAPTOR (regulatory associated protein of mTOR), mLST8/GβL and PRAS40, that of mTORC2 RICTOR (rapamycin-insensitive companion of mTOR), GβL and mSIN1.

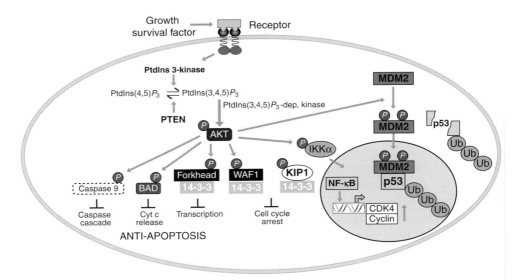

Figure 6.5 Pathways activated by phosphatidylinositol 3-kinase. PtdIns(3,4,5)P_3 (PIP$_3$) recruits phosphoinositide-dependent protein kinase 1 (PDK1) to the plasma membrane, which in turn activates the serine/threonine kinase AKT (also called PKB). PIP$_3$ is de-phosphorylated by the phosphatase PTEN, a tumour suppressor. AKT has multiple targets: it acts as a survival (anti-apoptosis) signal (by phosphorylating BAD and caspase 9), as a transcriptional regulator (by phosphorylating forkhead transcription factors and IKKα), to promote cell cycle progression (by phosphorylating WAF1 and KIP1), and it regulates glycogen synthase kinase 3β (GSK3β) in the APC/β-catenin pathway by phosphorylation. Another AKT target is MDM2, a critical negative regulator of p53 activity because it promotes ubiquitination and degradation of p53. In addition, MDM2 actively promotes proliferation because it stimulates E2F1 *trans*-activation. PTEN can interact with p53 to attenuate MDM2-mediated p53 inhibition. *MDM2* is transcriptionally activated by p53 in a negative feedback loop. Prolonged expression of p53 activity can also turn on transcription of *PTEN*.

(*BCL2 Homology* domains). The family comprises anti-apoptotic (e.g. BCL2, BCLx$_L$) and pro-apoptotic proteins (e.g. BAX, BAK) of which a sub-group has only the short (9–16 amino acid) BH3 domain (BAD, BID, BIM, NOXA, PUMA, Beclin-1).

Aberrant activity of the PI3K pathway can obviously be caused by anomalous 'upstream' activation of receptors or RAS, but mutations in genes encoding pathway components themselves are among the most frequent in solid tumours, the cumulative result being that this pathway is constitutively active in about half of all human cancers. Pathway mutations include a number of point mutations in p85, particularly in pancreatic, colon and brain tumours; amplification of p110α in ovarian and breast cancer; and loss-of-function of the tumour suppressor PTEN either by deletion, promoter silencing or point mutations particularly in brain and pancreatic tumours. These pathway-activating mutations promote cell survival, invasiveness and tumourigenesis.

The mammalian target of rapamycin (mTOR: also called FK506 binding protein 12-rapamycin associated protein 1, encoded by *FRAP1*), a multi-subunit serine/threonine kinase, is part of the PI3K-AKT pathway that acts as a central regulator of cell growth through its capacity to modulate metabolism, macromolecular synthesis and **autophagy**. The mTORC1 complex is inhibited by low nutrient levels, growth factor deprivation, reductive stress or rapamycin. It is activated when the TSC1/2 complex is inhibited by AKT phosphorylation when it mainly phosphorylates the S6K1 kinase and the eIF4E binding protein (4E-BP1). mTORC2 is regulated by growth factors and nutrient levels, phosphorylating AKT, SGK1 and protein kinase Cα (PKCα) and also regulating the cytoskeleton. Thus mTORC2 provides an additional level of AKT activation, promoting mTORC1 activity. mTOR is negatively regulated by several tumour suppressors (TSC1, TSC2, PTEN and LKB1/STK11). LKB1 is less frequently mutated than PTEN but it is often inactivated in non-small cell lung cancer. Deletion of LKB1 in lung cancers results in activation of both AKT and SRC/FAK signalling (Fig. 6.8): this not only reverses the negative effect of LKB1 on mTOR but promotes cell migration and hence metastasis. In addition, activating point mutations in mTOR have been detected in several cancers (lung, skin, ovary, bowel, brain and kidney). mTOR signalling is also enhanced by GOLPH3, an oncoprotein that is amplified in a variety of human cancers. These multiple roles suggest that mTORC2 may emerge as an important target for cancer therapy.

During autophagy the action of a PI3K and beclin-1 (also known as autophagy-related gene (*Atg6*)) initiates the formation of **autophagosomes**. Beclin-1 is another member of the BH3 family that is released from the interaction with BCL2 proteins in the stress response. It is in fact a **haploinsufficient** tumour suppressor with a role in both autophagy and apoptosis.

Autophagy is emerging as a complex factor in cancer in that, for some types at least, it can suppress initiation but may also provide an essential support for tumour progression. Thus mice with disrupted autophagic genes (*Atg5* and *Atg7*) develop benign growths that do not become malignant, consistent with a normal role for autophagy of preventing tissue damage and cancer progression. **Heterozygous** deletion of beclin-1 also causes tumours in a variety of tissues although these do become malignant, probably because beclin has other activities in addition to its autophagic role. The evidence that autophagy actually promotes tumourigenesis is particularly strong for pancreatic cancers for which, in both genetic mouse models and xenografts of human tumour cells, inhibition of autophagy prevents growth and causes regression. This is presumed to be due to an increase in reactive oxygen species and DNA damage, together with decreased oxidative phosphorylation, as a response to inhibition. Autophagy plays a prominent role in tumours with activating mutations in RAS (prevalent in lung, pancreatic and colon cancers), as indicated by the high basal autophagy in cell lines derived from such cancers. In these aggressive tumours the autophagic response sustains the level of mitochondrial function necessary for growth, a requirement that has been termed '**autophagy addiction**' by Eileen White and her colleagues.

Death receptors and the extrinsic pathway of apoptosis

The tumour necrosis factor receptor (TNFR) family respond to cell death signals by promoting apoptosis in a variety of cells. The activating ligands are trimeric complexes

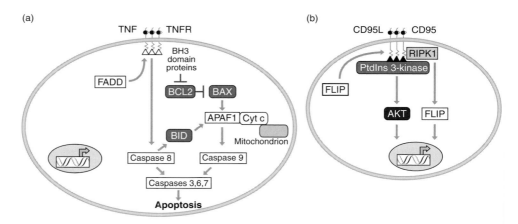

Figure 6.6 The apoptosis pathway. (a) For apoptosis, the adaptor protein FADD is recruited to the exposed death domain of the activated receptor. FADD then recruits the initiator procaspase 8, which cleaves itself to form the active caspase 8 – an enzyme that promotes apoptosis. (b) When cells express FLIP, rather than apoptosis, CD95 activates other pathways, including NFKB1 (NF-κB) and PI3K, that promote tumour development. NF-κB is activated by a variety of signals including TNFα and IL1β and can contribute to cancer by inhibiting apoptosis and promoting proliferation.

that interact with three monomeric TNFR chains. Activation of these cell-surface receptors by their ligands induces apoptotic death in many cell types (Fig. 6.6).

Remarkably, however, in tumour cells that express CD95 (also known as FAS and APO1), CD95 can act in an autocrine manner (i.e. it binds to receptors on the same type of cell from which it is secreted) to promote tumour development. This can happen because the intracellular signal pathways activated by CD95L/CD95 depend on the cell type.

The expression of TRAIL (tumour necrosis factor-related apoptosis inducing ligand) receptors is high on some tumour cells (e.g. from colorectal carcinoma) and **monoclonal antibodies** that selectively activate TRAIL show promise as a means of inducing tumour cell apoptosis.

The JAK-STAT pathway

In Chapter 3 we noted the importance of the JAK-STAT signalling in hematopoietic cell development. Consistent with this role, activating mutations in JAK2 and JAK3 occur in some leukaemias and in myelofibrosis (abnormal proliferation of bone marrow cells) and the resultant constitutive effect is presumed to drive abnormal cell proliferation. A selective inhibitor of JAKs has been developed (INCB018424) that gives sustained remission of myelofibrosis. An inactivating mutation in JAK1 has been identified in a rare smooth muscle tumour: this appears to inactivate interferon-γ signalling and thus promote evasion of the immune system by tumour cells. In addition, over-expression of members of the STAT family has been detected in a variety of tumours, indicating the impact that aberrant JAK-STAT signalling may have.

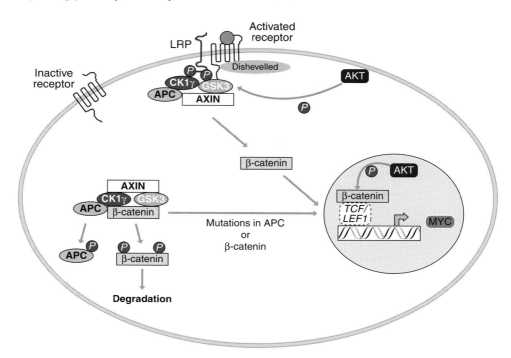

Figure 6.7 WNT signalling. In the absence of WNT the complex containing glycogen synthase kinase 3 (GSK3) and casein kinase 1 (CK1) phosphorylates β-catenin and adenomatous polyposis coli (APC) proteins, promoting their degradation. Activation by WNT of the receptor complex (Frizzled associated with the membrane protein LRP and Dishevelled) permits phosphorylation of LRP by GSK3 and CK1 leading to the degradation of axin and the release of un-phosphorylated β-catenin. β-catenin binds to TCF/LEF1, displacing a co-repressor, to activate transcription of *MYC*, cyclin D1 and the metalloproteinase matrilysin. Mutations in APC or β-catenin occur in colon carcinoma that block degradation and cause anomalous transcription. The PI3K/AKT pathway can affect WNT signalling in two opposing ways. When AKT is associated with the plasma membrane it can phosphorylate GSK3 to stimulate β-catenin-dependent transcription. However, AKT can also translocate to the nucleus where, by phosphorylating β-catenin and an associated protein, it can repress the transcriptional activity of β-catenin.

WNT and GPCR signalling

The WNT family are secreted morphogenic ligands that are involved in development. Amplification or over-expression of several WNTs has been detected in a variety of cancers and reduced expression of specific WNT receptors (members of the Frizzled family of GPCRs) has also been reported (e.g. Frizzled 5 in some prostate carcinomas). In the canonical WNT pathway (Fig. 6.7) a complex of casein kinase 1, glycogen synthase kinase 3, axin and the APC (*a*denomatous *p*olyposis *c*oli) protein regulate the fate of β-catenin – whether it is ubiquitylated or released to activate transcription of key cell proliferation genes (notably *MYC*).

The majority of mutations identified in *APC* inactivate the gene product by truncation and, because APC is involved mitosis, this promotes chromosome instability. Mutant forms of APC are unable to associate with β-catenin and hence cannot down-regulate its *trans*-activation function. Colorectal tumours with normal APC generally have mutations in β-catenin, the latter being the most frequently mutated gene in colorectal cancer. Hence loss of APC promotes anomalously high expression of genes driving cell proliferation, an alternative way of inducing genetic instability. APC/β-catenin complexes also mediate signal transduction from E-cadherin cell surface adhesion proteins. The prolyl isomerase PIN1 regulates β-catenin turnover and subcellular localisation by interfering with its interaction with APC. Inhibition of PIN1 induces apoptosis but it is strongly over-expressed in a sub-set of human tumours and this increases the transcription of β-catenin target genes. Depending on whether AKT is attached to the plasma membrane or located in the nucleus it can exert opposing effects on β-catenin, repressing its transcriptional activity when nuclear.

In addition to the WNT system, activating mutations occur in a number of other GPCR pathways (e.g. in the thyroid stimulating hormone receptor in thyroid tumours) and in G_α subunits (e.g. in $G_{\alpha s}$ in pituitary adenomas and in $G_{\alpha i}$ in ovarian tumours). G_q subunit coupling to phospholipase Cβ (Chapter 3) activates the protein kinase C (PKC) family, of which there are about ten isoforms. Mutations have been identified in each of the PKC isozymes and they appear to be particularly associated with pancreatic tumours and glioblastomas. These enzymes therefore number among the $\sim$150 of the human genome kinase complement of 539 that can be mutated in cancer. As we have seen, kinase mutations are almost always activating, i.e. they confer gain-of-function. However, the PKC family and some other kinases (e.g. MAPK2K4 and DAPK3) are exceptional in that the mutations are inhibitory, that is, the normal form of these enzymes acts as a tumour suppressor. This means of course that PKC inhibitors will not be useful therapeutic agents.

Cell adhesion: cadherin signalling

The superfamily of cadherins comprises nearly 200 trans-membrane proteins that are related by the presence in their extracellular domains of multiple copies of the cadherin motif. A sub-group (T-cadherins) attach to the plasma membrane via a **glycosylphosphatidylinositol** (GPI) anchor. Through the interactions of identical or closely related sub-types they mediate cell–cell adhesion (in contrast to integrins that mainly associate with the extracellular matrix). This means that they play important roles in development, e.g. during epithelial–mesenchymal transitions in embryonic growth, and we have already noted changes in the expression of endothelial cadherin during the corresponding pathological change as epithelial cells become malignant. In addition, cadherins can signal intercellular status to intracellular pathways. This is best understood for E-cadherin, expressed on many types of epithelial cell, which binds β-catenin via a phosphorylated intracellular domain. Through its interaction with α-catenin, β-catenin can regulate the actin cytoskeleton but it is also an indirect regulator of *MYC* transcription. These activities may explain why loss of E-cadherin

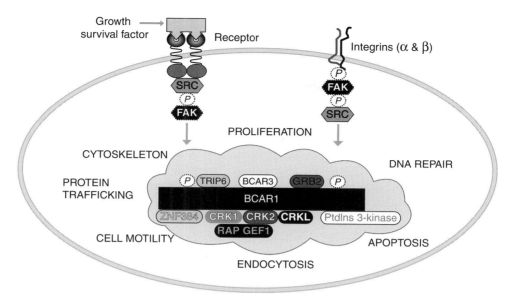

Figure 6.8 Signal integration by an integrin scaffold protein. The central role of the SRC and FAK kinases is to activate by tyrosine phosphorylation the scaffold protein BCAR1 (also known as p130 CAS, CRK-associated substrate or CAS scaffolding protein family member 1). Some of the proteins that bind to tyrosine phosphorylated BCAR1 are shown including CRKs (that activate small GTPases and regulate motility), GRB2 (activating RAS-MAPK and proliferation), PI3K (activating AKT to promote cell survival), TRIP6 that mediates thyroid hormone signalling and the transcription factor ZNF384 that activates expression of matrix metalloproteinases (MMPs). The broad effects of integrin receptor engagement reflect not only the interaction with the extracellular matrix and hence cell motility and proliferation but can also include modulation of apoptosis, endocytosis and protein transport. In addition, integrin signalling can protect against some forms of DNA damage.

function is associated with tumour cell metastasis and why it is one of the 'invasion suppressor' genes referred to in Chapter 5.

Integrin signalling

The integrins are a family of over 20 trans-membrane proteins that bind specifically to extracellular matrix proteins (e.g. laminin or fibronectin) as α/β heterodimers (Fig. 6.8). They respond to signals from the extracellular matrix to modulate cell shape, motility and division. Integrins do not possess enzymatic activity but when activated they recruit cytosolic molecules and can transmit signals bidirectionally. The intracellular tail of integrin-β links to filamentous actin via talin. Talin (or paxillin for some integrin-α subunits) recruits focal adhesion kinase (FAK), a tyrosine kinase that acts as a major scaffold protein for relaying integrin signals. FAK has docking sites for SRC, GRB2, SHC, PLCγ and PI3K and can thus interact with all the major signalling pathways. This includes the angiogenic system of which integrins $\alpha_v\beta_3$ and $\alpha_v\beta_5$ are a part and they too signal primarily through the RAS-RAF-ERK1/2 and PI3K pathways.

Integrin signalling is principally mediated by the effectors BCAR1 (p130^{CAS}), NEDD9, CRK, SRCIN1 (p140^{CAP}) and the IPP complex (comprising ILK, LIMS1/pinch and parvin). These have different tissue expression patterns but they are all scaffold proteins with multiple binding sites for signal propagation. A striking feature of these adaptors is that they receive signals not only from integrins but also from RTKs, cytokine receptors and steroid hormones. They thus integrate a wide range of extracellular signals. The emerging picture of integrin signalling in cancers is of aberrant expression of receptors and adaptors, rather than of mutational changes. Thus, for example, integrins α_2, α_3, α_4, α_5, β_1 and β_2 are over-expressed in melanomas and high levels of BCAR1, CRK and ILK have been detected in a range of tumours.

FAK (*PTK2*) over-expression has been found in many types of cancers, particularly in carcinomas but also in haematological malignancies, and there is considerable evidence associating this with metastatic disease. The stimulation of integrin-FAK signalling activates SRC that in turn phosphorylates BCAR1, a step that is required for metastasis to the lung of human breast cancer cells expressing oncogenic RAS or PI3K. This is consistent with the role of FAK/SRC signalling in LKB1-deficient tumours. However, SRC may also be activated by disrupted integrin-FAK signalling (e.g. by ablation of FAK). Although SRC can promote tumourigenesis, as with other oncoproteins, unrestrained SRC signalling can lead to cell death. This implies that cell fate is determined by precise regulation of SRC in focal adhesions (protein complexes that connect the extracellular matrix with the cytoskeleton). In part this is achieved through the action of CBL acting as a receptor for SRC, targeting it for autophagy in a process that involves proteins of the ATG family, mentioned earlier. The sequestration and destruction of SRC thereby permits cell survival and tumour development as a response to compromised signalling through the integrin-FAK pathway. Inhibition of the autophagic pathway leads to accumulation of active SRC at focal adhesions, which is toxic to the cell. These effects point to the promise of therapeutic combinations that include SRC/FAK inhibitors.

The integrative nature of these systems is illustrated by their direct effects on RTK signalling. Thus, for example, $\alpha_5\beta_1$-integrin regulates re-cycling of the EGFR to the cell surface: the rate is increased through the transcriptional effects of a mutant form of p53, thereby enhancing signalling from the RTK and increasing invasive potential.

BCR-ABL1

In Chapter 4 we saw that the *BCR-ABL1* translocation occurs in almost all cases of CML and in about 25% of adult ALL, generating a fusion protein that is a constitutively active tyrosine kinase. In addition, BCR-ABL1 acts as a scaffold for a sequential assembly of protein complexes that is regulated by phosphorylation (Fig. 6.9). This generates a large network of interacting proteins emanating from three sets of adaptors that bind directly to BCR-ABL1: the GRB2, DOK and CRK complexes. Adaptor proteins in the core complexes may also act as scaffolds (e.g. GAB2 binds PI3K and INPPL1/SHIP2) so that BCR-ABL1 may be viewed as a 'super scaffold'. It is notable that some of the players are familiar from the numerous other scaffolds that we have already encountered and so the story is essentially one of variations on a theme: these

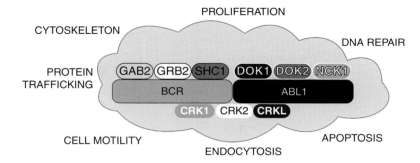

Figure 6.9 **BCR–ABL1 signalling.** Three adaptor complexes bind directly to BCR–ABL1 as part of a network of over 60 interacting proteins. These can modulate essentially all the major signalling systems including RAS-MAPK, PI3K, JAK-STAT, NF-κB and small GTPases, consistent with the pleiotropic effects of this fusion protein on haematopoietic cells. The three core adaptor complexes are GRB2 (GRB2, GAB2, SHC1), DOK (DOK1, DOK2, NCK1) and CRK (CRK1, CRK2, CRKL). Other proteins (e.g. CBL, UBASH3B/STS1, INPPL1/SHIP2 and p85) that modulate growth factor signalling can also associate directly with BCR–ABL1. Many of these adaptors are common to other 'super scaffolds' and the corresponding responses are a general feature of the central signalling systems involved in cancer.

signalling platforms activate pathways that can affect all aspects of cellular function. Signalling from the GRB2 complex generally promotes proliferation whereas that from DOK is inhibitory. SHC1 interacts with both the DOK and GRB2 complexes, accounting for the evidence that it can exert opposing effects on haematopoietic cell proliferation depending on cellular context.

The Hedgehog pathway and GLI signalling

Sonic hedgehog (SHH) is a ~45 kDa precursor auto-catalytically cleaved to release the N-terminal signalling domain (SHH-N, ~20 kDa) and a ~25 kDa C-terminal domain with no known signalling role (Fig. 6.10). During processing cholesterol is attached to the carboxyl end of the N-terminal domain. This modification mediates trafficking, secretion and receptor interaction of the ligand. SHH can signal in an autocrine manner or, after secretion, by a paracrine mechanism (i.e. after secretion it binds to receptors on nearby cells) that requires the participation of the 12-trans-membrane protein Dispatched.

SHH binds to the Patched receptors (PTCH1, PTCH2) that, in the absence of ligand, inhibit Smoothened (SMO), a member of the Frizzled family of GPCRs that do not interact directly with their endogenous ligand, SHH. In the unstimulated state SMO interacts with and is repressed by PTCH. When SHH interacts with PTCH, SMO is released and phosphorylated to activate a signalling cascade that controls the activity of GLI transcription factors.

Hedgehog signalling plays a major role in development but it is abnormally activated in a variety of cancers. In a familiar pattern, the activity of GLI signalling is positively regulated by both the RAS-MAPK and AKT pathways. Genes transcriptionally

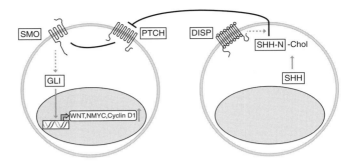

Figure 6.10 Sonic hedgehog (SHH) signalling. The SHH precursor is auto-catalytically cleaved and cholesterol is added to the N-terminus. Secretion and paracrine signalling requires the trans-membrane protein Dispatched. SHH binds PTCH on its target cell that relieves the inhibition by PTCH of SMO, which can then activate GLI proteins, a family of three zinc-finger transcription factors (GLI1, GLI2 and GLI3), to promote proliferation. Abnormal signalling can occur in tumour cells in an autocrine manner: paracrine signalling can also occur in which SHH, secreted by tumour cells, activates adjacent stromal cells.

activated by Hedgehog signalling include insulin-like growth factor-binding protein, cyclin D2 and osteopontin. The first two of these contribute to cell proliferation: osteopontin promotes malignancy in melanoma cells. Ninety per cent of basal cell carcinomas, the most common skin cancer in humans, have PTCH mutations: in the remaining 10% SMO mutations block repression by PTCH. PTCH and SMO mutations also occur in 20% of the malignant brain tumour medulloblastoma. Mutations have not been reported in pancreatic, prostate, colorectal, oestrogen receptor α-negative breast and ovarian cancers but Hedgehog signalling is commonly up-regulated in these cancers.

The small molecule inhibitor GDC-0449 that blocks the PTCH/SMO receptors is in **clinical trials** for a range of cancers. Drug resistant metastases have arisen after GDC-0449 treatment as a result of the acquisition of SMO coding mutations that were not present in the primary tumour.

Transforming growth factor beta (TGFβ)

The term 'transforming growth factor' was coined when two completely unrelated TGFs (TGFα and TGFβ) were discovered in the medium of transformed cells growing in culture. Transformation had been by infection with a sarcoma virus and the TGFs released could transform normal fibroblasts – that is, make these 'normal' cells lose contact inhibition and the requirement for substrate attachment, so that they formed colonies in soft agar. In the fullness of time TGFβ has emerged as one of the most pleiotropic of all molecules – i.e. it causes a wide range of responses in essentially all the major cell lineages. Thus, despite its name, it is a potent inhibitor of epithelial cell growth and its capacity to cause cell cycle arrest, differentiation and apoptosis enables it to play a key role in development and tissue homeostasis. In these contexts TGFβ behaves essentially as a tumour suppressor. In addition, TGFβ regulates the response to

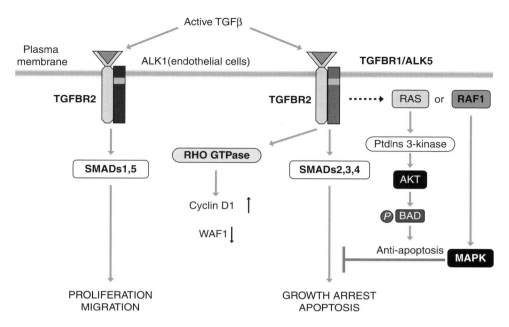

Figure 6.11 **TGFβ signalling in endothelial cells.** TGFβ (there are three isoforms, TGFβ₁, TGFβ₂ and TGFβ₃) binds to the constitutively active serine/threonine kinase TGFBR2, which then activates TGFBR1 and the canonical signalling pathway mediated by SMADs 2, 3 and 4 that causes cell cycle arrest and apoptosis in normal cells. In transformed cells other pathways may be activated by TGFβ that inhibit apoptosis and promote cell migration and invasion. The ALK5 form of TGFBR1 is ubiquitously expressed. In endothelial cells, a second type 1 receptor is expressed, ALK1, that signals via SMADs 1 and 5 to promote proliferation and migration.

injury, one component of which is activation of the epithelial to mesenchymal transition (EMT) and the associated activation of cell invasiveness and motility. This step is also generally considered to be part of the induction of metastasis and thus reflects the capacity of TGFβ to promote carcinogenesis. So TGFβ can switch from tumour suppressor to promoter when its growth arrest and apoptotic functions are lost but other signalling responses are retained. This is consistent with the fact that most tumour cells are refractory to growth arrest by TGFβ. How does it do that?

Like all protein hormones TGFβ binds to its cognate trans-membrane receptors (Fig. 6.11). Rather than RTKs they are serine/threonine kinases that phosphorylate a family of signal proteins, SMADs, that are specific to TGFβ receptors (the canonical signalling pathway). The SMAD family are transcription factors and, depending on the proteins they interact with, they can regulate the expression of at least 100 genes. In normal cells two major targets of this pathway are *MYC*, which is repressed, and *INK4B*, which is turned on by TGFβ signalling. As MYC is essential for cell proliferation and INK4B is a CDK inhibitor, two complementary mechanisms for bringing about cell cycle arrest and indeed apoptosis are activated. Thus TGFβ is a tumour suppressor. However, in contrast with this benign, protective behaviour, abnormal activity of the TGFβ signalling system is widespread in cancer. TGFβ is secreted by a variety of tumour cells and somatic mutations occur in the receptors and in SMADs. Mutations in TGFBR1 are rare but there is a significant incidence of TGFBR2 mutations in liver and ovarian tumours and in colorectal and gastric carcinomas that are replication error positive (fail to accurately replicate repetitive nucleotide sequences). Colorectal and gastric tumours also have a substantial frequency of SMAD4 mutations. Mutations in this pathway have not been reported in primary breast tumours but have been found in relapses after tamoxifen therapy.

Although some receptor mutations inhibit catalytic activity, their overall effect is to redirect pathway signalling rather than to block it. Thus secreted TGFβ is essential for maintenance of the EMT. Transformed cells retain an active SMAD signalling pathway, consistent with their continued requirement for TGFβ, and expression of SMAD2 is required for tumour extravasation and metastasis. The switch in signalling activates transcription of, for example, matrix metalloproteinases (MMPs) and, in endothelial cells, of angiogenic factors, while selectively inhibiting responses associated with tumour suppressor activity. One mechanism appears to be that TGFβ can also switch on the RAS-MAPK pathway (see Figs. 6.1 and 6.11) and the changes in cancer may shift the balance in favour of MAPK so that growth inhibitory/apoptotic signalling is overwhelmed by an invasive/metastatic phenotype. Accordingly, TGFβ is also shown in Fig. 6.11 as a mixed tumour suppressor and oncogene.

Vascular endothelial growth factors (VEGFs) and Notch signalling

Figure 6.4 does not include VEGFs even though they are the most important angiogenic drivers. The reason for its omission is that VEGFs signal via RTKs and so, from a signalling point of view, they are just a variation on a theme. They're quite a complicated variation (there are five different VEGF ligands (cytokines) and three types of VEGF RTK): that's probably because they control vasculogenesis, angiogenesis and lymphangiogenesis (Fig. 6.12). Nevertheless, although that means they offer additional drug targets for treating cancer, they are indeed in our story no more than a variant of RTK signalling.

An important target of VEGF is the Notch signalling system (Fig. 6.13). In the developing embryo endothelial cells lining blood vessels express a protein called delta-like 4 (DLL4). These cells additionally carry Notch receptors to which DLL4 binds to promote cell proliferation. Notch is also expressed on at least some tumour cells, e.g. in some carcinomas and on white blood cells in T-cell acute lymphoblastic leukaemia (ALL), and mutations in one form of Notch occur in >50% of human T-ALL. These activate Notch signalling, as does DLL4, so that either mutations in the tumour cell gene or activation via DLL4/Notch in normal endothelial cells in the tumour environment can drive tumour development.

VEGFA is a powerful activator of Notch signalling because it induces expression of DLL4 and thereby promotes angiogenesis. When expressed by tumour cells DLL4 can activate Notch signalling in host stromal cells, improving vascular function. Conversely, DLL4 expressed in endothelial cells acts via Notch 3 expressed on adjacent colorectal cancer cells

Signalling pathways that impact on the central axis

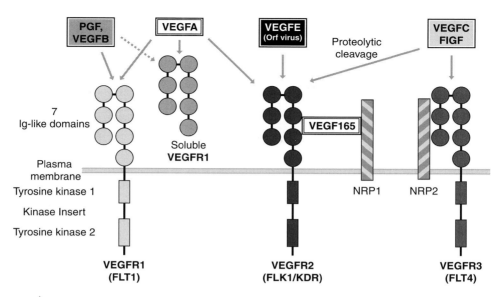

Figure 6.12 Vascular endothelial growth factor ligands, receptors and co-receptors. The family of VEGFs (VEGFA, VEGFB, VEGFC, FIGF (*FOS* induced *growth factor*, formerly VEGFD)), and placenta growth factor (PGF/PLGF) bind as dimers with differing specificities to three types of receptor; binding promotes *trans*-phosphorylation of receptor dimers. Neuropilins (NRP1 and 2) are membrane-bound co-receptors for VEGFs and for semaphorins. There are five principal, alternatively spliced isoforms of VEGFA denoted by the number of amino acids (121/145/165/189/206). Gradients of VEGF isoforms regulate blood vessel branching. Proteolytic cleavage of VEGFC and VEGFD increases their affinity for VEGFR3 and for some VEGFR2s. An additional form, VEGFE, is encoded by *parapoxvirus ovis* (PPVO or Orf virus) and VEGFE, like VEGFA, binds with high affinity to VEGFR2. The three isoforms of PGF activate VEGFR1 causing *trans*-phosphorylation of VEGFR2 and amplifying VEGF-driven angiogenesis. PGF2 also binds to NRP1.

or T-ALL cells to promote the switch from dormancy to growth. Blockade of DLL4-mediated Notch signalling has been shown to inhibit tumour growth in several mouse models. The Notch pathway thus joins the list of potential therapeutic targets.

We have already noted that integrin signalling pathways are of importance in angiogenesis but another receptor-mediated signalling pathway is also significant. This is activated by **sphingosine 1-phosphate** binding to the G-protein-coupled receptor S1PR1. S1PR1 is highly expressed in endothelial cells and it activates cAMP, RHO and RAC GTPases and phosphatidylinositol signalling pathways. S1PR1 signalling is important in vascular development as evidenced by the death in utero of *S1p1*-null embryos due to incomplete vascularisation and the inhibition of angiogenesis by small interfering RNA-mediated silencing. Another member of the receptor family, S1PR2, expressed mainly in endothelium and vascular smooth muscle cells, regulates negative feedback to the S1PR1 pathway. Accordingly tumour growth is accelerated in S1PR2-deficient mice opening the possibility that activators of S1PR2 signalling might function as anti-angiogenic agents. The

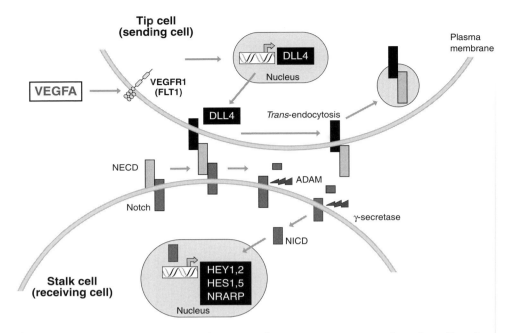

Figure 6.13 The Notch signalling pathway. Notch receptors are expressed on the cell surface as a heterodimeric receptor. The extracellular domain (NECD, *Notch extracellular domain*) and the membrane-bound domain (Notch intracellular domain) associate via non-covalent interactions. The DLL4 ligand promotes endocytosis and non-enzymatic dissociation of the Notch heterodimer. NECD is *trans*-endocytosed into the signal-sending cell, exposing Notch to ADAM and γ-secretase proteolysis for release of the Notch intracellular domain (NICD), which translocates to the nucleus to trigger transcriptional activation of Notch target genes that include HEY1, 2, HES1, 5 and NRARP.

inter-dependence of diverse signalling pathways that contributes to the control of angiogenesis is emphasised by the fact that S1P receptors are also *trans*-activated by RTKs (specifically the PDGFR).

Cellular responses during tumour development

When all the elaborate defence systems that have evolved fail, what is the effect of all the perturbations of cell signalling pathways we have discussed on the cancer cell as a whole? All facets of cellular behaviour can be changed (Fig. 6.14). The cell may synthesise and secrete growth factors that further promote expansion of the tumour cell clone, notably insulin-like growth factors (IGFs). These pathways may also regulate the cell's own cytoskeleton, which plays a role in cell division, and affect cell adhesion. The release of enzymes that degrade the extracellular matrix, which is essential for migration and invasion, is a first step to metastasis. The release of factors that drive the synthesis of new blood vessels (angiogenesis) is an essential component of tumour expansion and, finally, the overall metabolic activity of the cell is enhanced to support all these responses.

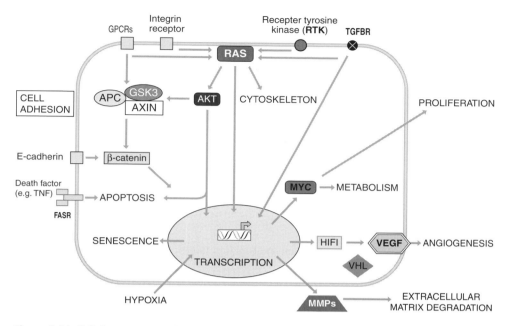

Figure 6.14 Cellular responses during tumour development.

In short, the effects are to coordinate the modulation of all cellular responses to equip a cancer cell to thrive independently of normal controls and to have the capacity to colonise new locations, ultimately the most lethal of all the properties of a tumour cell.

Signalling and systems biology

This review of the major pathways associated with cancer confirms what we suspected from the outset: signalling is very complicated. The complexity arises not because the individual signal pathways are difficult to understand but because, seemingly, almost every component can indeed communicate with any other. So the concept emerges that, rather than thinking of linear pathways, we should view the entire system as an 'information network'. Astonishing though it is, cells clearly manage to process the huge amounts of information conveyed by such networks and come up with a decisive response – to divide or to differentiate, for example. We know that individual cells don't dither from elegant experiments using fluorescent reporters to monitor the activation of MAPK pathways in yeast. Simultaneous exposure to two signals (e.g. that activate stress and mating responses) produces *a mixed response in the population as a whole* but individual cells are quite decisive: they mount *either* a stress *or* a mating response. For mammalian cells that receive a plethora of signals, although a huge amount is known about the detail of pathways, we have little idea about how they achieve the critical, integrative decision-making bit.

This is an attractive subject for mathematical modelling and an instructive example uses Boolean logic to describe the status of each of 300 proteins ('nodes') that have

Cancer signalling networks

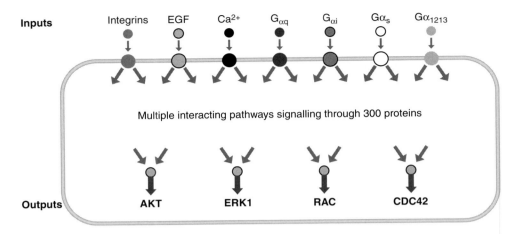

Figure 6.15 Boolean signal transduction model. (Helikar *et al.*, 2008.)

been identified as participating in pathways emanating from three major receptor families: receptor tyrosine kinases, G-protein-coupled receptors and integrins (Fig. 6.15). The model then simulates exposure of the network to tens of thousands of random combinations of inputs representing typical ligands and analyses the combined dynamics of multiple outputs. One way of doing this is to select four outputs: AKT (survival), ERK1 (transcription), RAC and CDC42 (cytoskeletal regulation), and define just three output ranges (weak, intermediate and strong response). This reduces the output to a possible 3^4 (= 81) states.

This approach reveals two particularly striking results. Firstly, far from requiring 81 outputs, a mere 15 can account for 94% of the 10,000 inputs. Secondly, almost half the inputs lead to just two output combinations. These outputs represent quiescent states in which the outputs ERK1, RAC and CDC42 are inactive, in contrast to AKT that must remain active to suppress apoptosis.

This model indicates that intracellular signalling pathways cluster widely varying input combinations into classes of cellular responses. It is possible to increase the noise input to the system: even when this is done the model shows that the cell's network can perform non-fuzzy classifications, i.e. act decisively.

Conclusions

Integrating the main signalling pathways of cancer into one diagram conveys a vivid impression of the complexity of intracellular communication. However, these pathways converge on a central axis of the most critical oncogenes and tumour suppressors. This represents the main defence against tumour development, which, when fully functional, can not only arrest cell cycle progression but divert cells into either senescence or apoptosis. These roles explain why the tumour suppressors involved (p53, RB1, ARF and INK4) are the most frequent to suffer loss-of-function mutations in human cancers.

Future Directions

The diverse pathways that lead from receptors to the central axis have oncogenic potential – that is mutation in any component could lead to aberrant signalling. Such abnormal signalling may be countered by the central defence, but the combination of hyperactive signalling and a defensive defect normally represents a recipe for cancer. We have therefore looked in some detail at the proteins in these pathways to provide a reasonably comprehensive picture of the players that can contribute to tumour development. This knowledge offers potential therapeutic targets and we now turn to the story of how drug treatments have evolved over the last 50 years.

Key points

- Numerous ligands signal to mammalian cells by activating multiple intracellular pathways that control cell behaviour.
- In principle, abnormal function of any of the components can contribute to cancer.
- In practice, the majority of cancers have mutations concentrated in core signalling pathways, particularly RAS-MAPK and phosphatidylinositol 3-kinase (PI3K).
- These focus on a central axis that includes the major tumour suppressor genes *ARF*, *INK4*, *P53* and *RB1*. Signalling by this network determines whether the cell response is proliferation, apoptosis or senescence.
- Although ligands specifically activate their cognate receptors, for the receptor tyrosine kinase (RTK) family in particular, this results in the activation of multiple pathways. This occurs in part because receptor cytosolic domains act as 'scaffold' proteins that have multiple binding sites for intracellular proteins.
- Several intracellular proteins also act as scaffolds or points of signalling divergence.
- The activity of many pathways with the capacity to 'cross-talk' implies that signalling through diverse types of receptor exerts effects on a similar, broad spectrum of cellular responses. The precise cellular response reflects integration of these overlapping signals.

Future directions

- The delineation of aberrant signalling in cancer cells has revealed critical genes and pathways that are mutated with high frequency. The activated pathways therefore offer attractive targets for therapy. However, because these pathways impact on the regulation of the cell cycle, targeting their components raises the problem of the effect on normal cells.

Further reading: reviews

INTRODUCTION

Vogelstein, B. and Kinzler, K. W. (2004). Cancer genes and the pathways they control. *Nature Medicine* 10, 789–99.

THE CENTRAL AXIS: ARF, MDM2, P53, INK4, RB1

Sherr, C. J. (2001). The INK4a/ARF network in tumour suppression. *Nature Reviews Molecular Cell Biology* 2, 731–7.

MYC AND RAS

Herold, S., Herkert, B. and Eilers, M. (2009). Facilitating replication under stress: an oncogenic function of MYC? *Nature Reviews Cancer* 9, 441–4.

Rothenberg, M. E., Clarke, M. F. and Diehn, M. (2010). The Myc connection: ES cells and cancer. *Cell* 143, 184–6.

CANCER AND CELL SENESCENCE

Collado, M. and Serrano, M. (2010). Senescence in tumours: evidence from mice and humans. *Nature Reviews Cancer* 10, 51–7.

Hydbring, P. and Larsson, L.-G. (2011). Tipping the balance: Cdk2 enables Myc to suppress senescence. *Cancer Research* 70, 6687–91.

Ishikawa, F. (2000). Aging clock: the watchmaker's masterpiece. *Cellular and Molecular Life Sciences* 57, 698–704.

7

The future of cancer prevention, diagnosis and treatment

The science of treating cancer with specific chemicals as an adjunct to surgery and radiotherapy developed in the second half of the twentieth century, from the use of single agents to the administration of drug cocktails, with spectacular results for some types of cancer. By the beginning of the next millennium these had been joined by the first 'targeted' agents, namely kinase inhibitors and the first monoclonal antibodies. These advances have significantly expanded treatment options but one of the biggest obstacles remains that of drug resistance. The astonishing success of vaccines that prevent infection by human papillomaviruses (HPVs) is already making a major impact on the incidence of cervical carcinoma and other types of cancer caused by oncogenic forms of these viruses.

In parallel with these therapeutic and preventative advances, developments in imaging methods, both for tumour detection and monitoring response to therapy, are showing promise. The pressing need to define tumour biomarkers that can be detected at the earliest possible stage has brought the emerging sciences of proteomics and metabolomics to the fore while gene expression profiling has already made substantial contributions to the diagnosis and classification of tumours.

The development of anti-cancer drugs

Major developments in surgical methods essentially established the basis for cancer treatment in the early part of the twentieth century. The second half of that century saw the gradual introduction of chemicals for the treatment of disease and the foundation of the science of chemotherapy. In the most general sense chemotherapy means the chemical treatment of disease, something that has been practised since ancient times when the Egyptians used arsenic. However, we are here considering more recent developments, an example being taking penicillin for a bacterial infection that is indeed a course of chemotherapy. In common usage and in the present context the term refers to the use of chemicals to treat cancer, these agents being taken either singly or in combination, with the aim of killing tumour cells. These cytotoxic drugs are normally administered systemically, meaning that they circulate in the bloodstream and can, in principle, affect every cell in the body. With very few exceptions drugs are not specific, any targeting that may occur arising from, for example, abnormally rapid division rates of tumour cells. However, for the most part cancer cells proliferate slowly relative to many normal cells and, moreover, their proliferation rates vary widely between different types of tumour. This problem has focused much effort on producing toxic agents directed more specifically to tumour cells, either through binding to proteins that are either exclusively, or at least predominantly, present on the surface of tumour cells or by targeting a signalling pathway that strongly contributes to tumour development. The terms hormone therapy, targeted therapy and gene therapy are essentially variants of chemotherapy in which these methods are used to enhance specificity for tumour cells.

One of the earliest demonstrations that chemicals could act as pharmacological inhibitors of cancers came in the early 1940s when Charles Huggins showed that the growth of prostate tumours could be modulated by hormonal treatment. The ensuing 20 years saw the emergence of the anti-folate **aminopterin**, first used by Sidney Farber in 1947, and then of the less toxic **methotrexate** in 1948 as effective agents against childhood leukaemia. These were followed by **6-thioguanine** and **6-mercaptopurine** for acute leukaemia and of 5-**fluorouracil** as the first treatment for solid tumours. It was not until the 1960s, however, that the now common practice of using combinations (cocktails) of drugs was first tested in humans. The results were spectacular in showing that drug combinations could be much more effective than single agents. The use of methotrexate, **vincristine**, 6-mercaptopurine and **prednisone** increased the five-year survival rate for childhood acute lymphoblastic leukaemia (ALL) from essentially zero to 60%. By 1996, the development of combination therapy had raised the five-year survival rate to 81%. Currently, children with acute myelogenous leukaemia (AML) have survival rates between 50% and 70%, and over 80% of children with acute promyelocytic leukaemia, a subtype of AML, are cured. By 1980, combination strategies had made similar impacts on Hodgkin's disease and testicular cancer, which now, provided they are detected at an early stage, have essentially a 100% survival rate.

Nowadays combinations of drugs are often used in a chemotherapy regimen, the idea being to hit different targets in the cancer cell, thereby increasing the 'fractional kill' – the proportion of tumour cells eliminated. Chemotherapy regimens are often denoted by acronyms, some of which are fairly logical, e.g. FOLFOX: **folinic acid** (FOL), **fluorouracil** (F) and **oxaliplatin** (OX), used to treat bowel cancer; and FOLFIRI: folinic acid, fluorouracil and **irinotecan** (IRI) (sometimes also with **cetuximab**) used for colorectal cancer. Others, however, are pretty gnomic (e.g. Stanford V) and it is fortunate that these days one can refer to www.chemocare.com/bio/list_by_acronym.asp for enlightenment.

In the following sections we will briefly summarise the considerable range of chemotherapeutic strategies that have emerged since the 1980s (Fig. 7.1) before considering their overall impact on human cancers and how cancer detection and therapy may evolve as we move through the twenty-first century. Specific chemical agents mentioned in these sections are only included as examples. A comprehensive table of currently available agents and their cellular targets is given in Appendix B.

Chemotherapeutic strategies for cancer

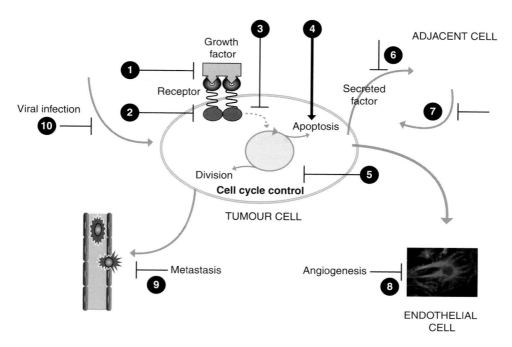

Figure 7.1 Potential chemotherapeutic targets for the prevention or inhibition of cancers.
(1) Inhibition of growth factors. (2) Receptor inhibition. (3) Signal pathway inhibition (blockade of transcription activation). Note also that drugs inhibiting, for example, mTOR or PI3K indirectly target tumour metabolism. (4) Stimulation of apoptosis. (5) Cell cycle inhibition. (6) Inhibition of tumour cell secreted factors. (7) Inhibition of stromal signals. (8) Inhibition of angiogenesis. (9) Inhibition of metastasis. (10) Inhibition of infection by oncogenic viruses.

Chemotherapeutic strategies for cancer

Kinase inhibitors

We have seen the significance in cancer development of abnormal kinase activities, particularly of members of the receptor tyrosine kinase family. Because phosphorylation is such a central mechanism for regulating cell behaviour (there are approximately 518 kinases encoded in the human genome), the kinase family has been the major focus for the design of specific inhibitors. Currently there are 11 Food and Drug Administration (FDA) approved kinase inhibitors with a further 80 in **clinical trials**.

Kinases catalyse the removal of the terminal (gamma) phosphate of ATP and its transfer to the amino acids serine, threonine or tyrosine. All kinases have a conserved cleft that contains the binding site for ATP. The majority of kinase inhibitors directly block ATP binding (e.g. PD166326) although those with the greatest selectivity among the kinase family are **allosteric** inhibitors (e.g. CI-1040). The availability of X-ray crystallographic structures for many kinases has provided a basis for the rational design of inhibitors.

As described previously, the major categories of kinase relevant to cancer are (1) abnormally active as a result of mutation (either within the gene, by amplification or by translocation (e.g. EGFR, ERBB2, BCR-ABL1); (2) key signalling proteins that are rarely mutated but are abnormally activated by upstream regulators (e.g. MEK1/MAP2K1, MEK2/MAP2K2, mTOR); and (3) proteins that support tumour growth, for example, by promoting angiogenesis (e.g. VEGFRs). Food and Drug Administration approved kinase inhibitors include erlotinib (versus EGFR), lapatinib (ERBB2), sorafenib (VEGFRs, PDGFRB and RAF1), sunitinib (VEGFRs, PDGFRB and RET), temsirolimus (mTOR), imatinib (Gleevec® or Glivec®) and nilotinib (BCR-ABL1).

Kinase inhibitors are non specific in that they cannot distinguish between tumour cells and normal cells, with the exception of imatinib, which targets a mutation in ABL1 (i.e. BCR-ABL1), although it also inhibits KIT and PDGFR. Imatinib has been highly effective, raising the five-year survival rate for chronic myeloid leukaemia (CML) to 89% for newly diagnosed patients in the chronic phase of the disease. It is, however, surprising that the effects of kinase inhibitors on normal cells are sufficiently well tolerated to make them viable therapeutic agents. Some kinase inhibitors have high specificity (e.g. gefitinib for EGFR and imatinib) but, remarkably, some relatively non-specific inhibitors are sufficiently effective to have received FDA approval (e.g. dasatinib for use against CML despite its inhibitory action against members of the SRC family, KIT and PDGFR as well as BCR-ABL1). Dasatinib is a second-generation inhibitor developed to control the expansion of subclones that carry mutations conferring imatinib resistance in patients with CML or gastrointestinal stromal tumours after imatinib treatment. Nevertheless, a relatively common mutation in BCR-ABL1 (at threonine 315) is resistant to all three inhibitors because it locks the molecule in the ABL1-active conformation. New (third-generation) inhibitors targeting the flexible regions involved in the switch between inactive and active conformations are under development. Despite the success of imatinib in treating CML, it is much less effective against BCR-ABL1-positive ALL. The main reason for this is the presence of deletions of *INK4A-ARF* in CML. It should also be borne in mind that, in addition to mutations, alternative resistance mechanisms (gene amplification or the up-regulation of alternative kinase pathways) play a substantial part in neutralising the efficacy of kinase inhibitors.

Selective oestrogen receptor modulators (SERMs)

The importance of oestrogen and hence anti-oestrogen therapy, particularly in breast cancer, was discussed in Chapter 3. SERMs are selective because some are agonists in all tissues (oestrogen), some are mixed or partial agonists/antagonists in that their effect is tissue dependent (tamoxifen) and some are pure antagonists, acting as such on all tissues (fulvestrant). SERMs are also used in the treatment of osteoporosis.

Monoclonal antibodies

Monoclonal antibodies are usually considered as a category distinct from conventional chemotherapy drugs. The first monoclonal antibodies for cancer treatment received FDA approval in the late 1990s. These were rituximab (Rituxan®) and tras-tuzumab (Herceptin®), humanised monoclonal antibodies binding to CD20 and the ERBB2 receptor, respectively. CD20 is a B cell surface protein and rituximab made an immediate impact on the treatment of chemotherapy-resistant forms of non-Hodgkin's lymphoma and remains in use for B cell lymphomas. Herceptin® is used to treat ERBB2-positive metastatic breast cancer in combination with conventional chemo-therapy. Even so, a high proportion of ERBB2+ tumours are unresponsive and drug resistance develops rapidly in responders.

Several other monoclonal antibodies now have FDA approval for cancer use, notably cetuximab (anti-EGFR) for colorectal and head and neck cancers. Monoclonal antibodies have also been used for the targeted delivery of anti-cancer vehicles in (1) radioimmunotherapy (antibodies against cell antigens linked to a radionuclide, so that the target cell is killed by radiation); (2) antibody-directed enzyme pro-drug therapy (ADEPT: monoclonal antibodies linked to an enzyme with pro-drug-activating capacity); and (3) immunoliposomes (liposomes being synthetic lipid vesicles that can be formed so that they contain drugs or radionuclides: by incorporating antibodies in the membrane they can be directed to tumour cells). In general, attempts at targeting have been unsuccessful.

Immunotherapy

The principle of immunotherapy in cancer is to stimulate the activity of the immune system against tumour cells. Two cytokines have FDA approval for use as anti-cancer drugs: interleukin 2 (IL2) and interferon alpha 2b (IFNA2). Interleukins stimulate the proliferation of sub-sets of T and B cells and interferons activate particularly natural killer cells and macrophages. IL2 is approved for the treatment of metastatic kidney cancer and metastatic melanoma, although the response rate of melanomas is only about 20%. Interferon alpha is also used for the treatment of melanoma as well as for some leukaemias and AIDS-related Kaposi's sarcoma. Nonetheless, neither of these is well tolerated, causing side-effects that can include seizures and liver damage.

Anti-angiogenic agents

In 2004, the humanised version of a monoclonal antibody to vascular endothelial growth factor (VEGFA), bevacizumab, better known as Avastin®, became the first FDA approved anti-angiogenic drug. The antibody recognises all isoforms of VEGFA and has a long circulating half-life of up to 21 days after intravenous infusion. Initial clinical trials showed that when used alone Avastin® was generally ineffective but that it had significant effects when combined with other drugs. It was therefore initially approved for metastatic colorectal cancer in combination with 5-fluorouracil. Subsequently it has been approved for treatment of unresectable, recurrent or meta-static non-squamous-cell lung cancer, breast cancer and **glioblastoma multiforme** and is currently undergoing extensive clinical trials for use against many types of tumours including melanoma, ovarian carcinoma, renal cell carcinoma, gastric carcinoma and prostate cancer. Although the precise mechanism of action is unknown, the anti-VEGF antibody may play a role in the 'normalisation' of tumour vasculature thereby making it more susceptible to drugs administered subsequently. Despite its promise, Avastin® received a set-back in 2010 when the FDA rescinded its approval for use against breast cancer because cumulative evidence has shown that it does not prolong life and has serious side effects. Avastin® retains approval for the treatment of other cancers.

In addition, Avastin® is also being evaluated in phase 3 trials for treating the vascular damage associated with age-related macular degeneration, recent phase 2 trials having shown an improvement in vision in patients. Ranibizumab (Lucentis®), another monoclonal antibody recognising $VEGFA_{165}$ (an isoform of VEGF having 165 amino acids) and pegaptanib (Macugen), a single strand nucleic acid **aptamer**, bind to the heparin-binding domain of $VEGFA_{165}$. These are in use for treating the 'wet' type of age-related macular degeneration (ARMD).

The future of anti-angiogenic agents

Currently 12 anti-angiogenic agents have received FDA approval although their effects have been relatively limited and there is at least anecdotal evidence that recurrence of some cancers after anti-angiogenic treatment may come in the form of more aggressive tumours than arise in patients not so treated. This puzzling observation is beginning to be resolved by studies in immunosuppressed mice of intravenously injected human metastatic breast cancer or melanoma cells as a model for metastatic seeding. Treatment either before or after tumour cell injection with VEGFR inhibitors (sorafenib, sunitinib or SU10944) accelerated the formation of metastases by approximately ten-fold with a corresponding decrease in the median survival time. This occurred notwithstanding the fact that sunitinib, for example, strongly inhibits the growth of established primary tumours in mice. Similarly, anti-angiogenic drugs (anti-VEGFR2 antibody DC101) or tumour cell deletion of *Vegfa* may inhibit primary tumour growth but increase rates of invasion and metastasis. Another anti-angiogenic drug, cilengitide, which blocks α_v integrins and is in phase 3 clinical trials, also has the anomalous effects at low doses of stimulating angiogenesis and tumour growth. Non-specific effects of these inhibitors may contribute to these responses. On the other hand, these findings strongly suggest that VEGF inhibitors can promote signalling events that, for example, stimulate the release of bone marrow cells to act as markers for tumour cell adhesion in pre-conditioned niches. This in turn indicates that both drug combinations and their administration regimes will be critical in evolving effective anti-angiogenic strategies. The options for administration include continuous versus discontinuous, **adjuvant** or neo-adjuvant (i.e. before surgery) therapy, and metronomic chemotherapy (frequent administration of very low doses over prolonged periods).

It may also be noted that most of the currently approved anti-angiogenics are directed at either VEGF or RTKs. That is, they target the growth factors (by antibodies and soluble forms of their receptors, so called 'ligand traps'), the ligand binding domains of their receptors (by antibodies), and the activated receptors (by kinase inhibitors). As the complexity of the molecular biology is gradually unveiled other targets present themselves. Thus, for example, HIF1 and PHD proteins, the regulation of VEGF by TIS11B and VHL, and the diverse functions of the four PI3K isoforms have come to the fore as potential targets for the treatment of cancers and ischaemic diseases. The major role of RAS in signalling from almost all RTKs makes it a less attractive target for angiogenic therapy and attempts to develop RAS inhibitors have not been notably successful. However, it has recently been shown that NF-κB signalling, promoted by RAS acting via RALGDS, RALB and TBK1, is required for tumour formation in a mouse model of RAS-induced lung cancer. Inhibition of TBK1 selectively kills RAS-mutant cells in a **synthetic lethal** interaction.

Vascular targeting agents (VTAs)

This broad category of drugs damage the vasculature of the circulatory system, which can lead to tumour nutrient starvation. They are administered directly into the circulation so that their target is essentially the monolayer of endothelial cells that lines the vessels. Within those cells VTAs bind to cytoskeletal microtubules, destabilising them and thus inhibiting cell division. They do not, therefore, have specificity for tumour cells. Nevertheless VTAs have been widely used in the treatment of cancer, particularly the taxanes (which include paclitaxel and docetaxel) and combretastatin.

An alternative approach has been to target matrix metalloproteinases (MMPs), a large family of zinc-dependent endopeptidases that degrade extracellular matrix proteins and cell surface receptors. Because of their capacity for tissue remodelling they are important in angiogenesis and metastasis and high levels of MMPs correlate with invasion and metastasis in a variety of human cancers. More than 50 MMP inhibitors have undergone clinical trials for the treatment of a variety of cancers but the results have thus far been disappointing.

Liposomal therapy

One of the major problems of chemotherapeutics is that drug delivery to the target cells is inefficient. To compensate, more of the drug is administered, which in turn increases the incidence of severe side effects. To circumvent these problems much effort has been focused on drug delivery by liposomes that can be targeted to specific sites where their drug content is released in a concentrated burst. A variety of ingenious targeting strategies have been used, the most common being the incorporation of receptor-targeting antibodies in liposomal membranes.

A notable example of this method has used liposomes to deliver doxorubicin (an anthracycline antibiotic that intercalates between the bases of DNA). The drug is pegylated, that is, covalently attached to a polyethylene glycol polymer to reduce its immunogenicity, generating what have been called 'stealth' liposomes. This preparation has been used in the treatment of ovarian cancer and AIDS-related Kaposi's sarcoma. A pegylated form of cisplatin (Platinol®) is currently being tested in vivo. The same design of vehicle has also been used in a combination of an anti-EGFR monoclonal antibody and either vinorelbine, gemcitabine or paclitaxel.

Drug delivery from inert capsules

A method under development uses minute plastic spheres injected into the blood-stream to deliver drugs to tumours. These drug-encapsulating beads are made of polyvinyl alcohol hydrogel modified by the addition of sulphonate groups and are loaded with anthracycline drugs (e.g. doxorubicin) before injection. The administration procedure is called transarterial chemoembolisation (TACE): the loaded beads accumulate in the abnormal vasculature of the tumour, forming an embolism, from which the encapsulated drug slowly elutes.

Gene therapy

This term refers to any therapeutic method in which exogenous genes are expressed. This means that the gene concerned has to be administered in a vector. Viral vectors are often used, taking advantage of their capacity to infect cells, following which the inserted therapeutic gene is transcribed. Viruses, of course, do not specifically target tumour cells and targeting remains a problem, although it is possible to engineer viruses to express proteins that bind to specific cell surface receptors. The first gene replacement therapy was carried out in 1990, when the T cells of a four-year-old girl were exposed outside her body to retroviruses containing an RNA copy of a normal adenosine deaminase gene. Injection of the engineered cells allowed her immune system to begin functioning.

An elegant example of the viral strategy is the modified adenovirus ONYX 015 that can destroy tumour cells that have lost the *P53* tumour suppressor gene while having no significant effect on normal cells. The promise of approaches based on genetic engineering is enormous but technical difficulties have meant that progress thus far has been slow.

Therapeutic (treatment) vaccines

The strategy driving work on these vaccines is to boost the immune response of the patient and they could be used not only to inhibit the growth of tumours but to complement conventional therapies. Although progress in this field has been slow, sipuleucel-T (Provenge®) was approved by the FDA in 2010 for prostate cancer. Sipuleucel-T elicits an immune response and has prolonged the life of patients with the advanced stage of hormone-refractory prostate cancer.

Prophylactic (preventive) vaccines

These are designed to prevent the initiation of cancer and work in a similar manner to traditional vaccines by activating an immune response in the recipient, the antibodies thus produced acting to protect against infection. There are FDA approved vaccines for the DNA human papillomaviruses (HPVs) and for hepatitis B virus.

Approximately 70% of cervical cancers and about 5% of *all* cancers worldwide are caused by HPVs and one of the great triumphs of science in the battle against cancer has been the development of vaccines (Cervarix® and Gardasil®) that appear to give almost complete protection against infection by the tumour-promoting HPVs. The vaccines are artificially synthesised, non-infectious, virus-like particles that induce a strong immune response. The antibodies thus produced block binding of HPV to the basal cells of stratified epithelium and hence prevent infection. They have no effect as therapeutic vaccines. Recombivax HB® and Engerix-B® are FDA approved vaccines for HBV.

Peptide vaccination

In principle, vaccination with protein fragments that act as immunogens to prime the immune system is the simplest method for enhancing natural defences against tumour development. The protein fragments used are derived from tumour-associated antigens, a considerable number of which have now been identified. Because these are small peptides, they can bind directly to MHC class I or II molecules and induce specific effector and memory T cells. A number of peptide vaccines have been tested in small numbers of patients against cancers that include acute myeloid leukaemia, myelodysplastic syndrome, multiple myeloma, breast cancer and colorectal cancer. Usually such vaccines have been administered together with an 'adjuvant' that enhances the T cell response of the recipient. Despite some positive results, peptide vaccination in general has been disappointing and the focus of research has shifted to adoptive cell transfer (ACT) therapy. In ACT, lymphocytes that have invaded a tumour, and are therefore antigen specific, are removed and grown in vitro to generate expanded clones of tumour-reactive T cells. The patient undergoes lymphocyte depletion by chemotherapy before the activated cells are infused back into them. Genetic engineering of the antigen-specific T-cell receptor has also been incorporated in this method. Adoptive cell transfer has been particularly effective in patients with metastatic melanoma.

Drug resistance

Drug resistance is a major limitation to the effectiveness of chemotherapy and most cancer deaths occur because of the inadequacy of the available drugs. The effect is very variable with some tumours having generally low levels of resistance (e.g. Hodgkin's lymphoma, childhood acute leukaemia). Other types of tumour usually respond to initial treatment but eventually acquire resistance (e.g. non-small cell lung cancer) and some types are inherently resistant (e.g. melanoma). Several mechanisms can promote resistance including either increased expression or loss of the target. For example, methotrexate treatment can cause amplification of the dihydrofolate reductase gene so that eventually the maximum dose that can be administered is ineffective against the high levels of enzyme that are expressed. Conversely, tamoxifen can be rendered ineffective by loss of the oestrogen receptor from breast or ovarian cells. However, the most significant drug resistance mechanisms are the acquisition of mutations by tumour cells that render the drug ineffective together with an increased capacity for exporting the drug whence it came. These include the effects of imatinib on BCR-ABL1 and KIT referred to earlier but may extend to switching to an alternative receptor-driven pathway. An example of the latter is the amplification of *MET* in lung cancers that develop resistance to the EGFR inhibitor gefitinib.

Drug efflux is mediated by members of the **ABC transporter superfamily** of ATP-driven, trans-membrane pumps, notably ABCB1 (also known as MDR1) or the permeability glycoprotein (P-gp). P-gp has broad specificity and it is expressed by a wide variety of cell types. Chemotherapeutic agents may increase the expression of P-gp and, in particular, resistance to doxorubicin, the vinca alkaloids (vinblastine, vincristine, vindesine and vinorelbine) and epipodophyllotoxins (e.g. etoposide) derives from enhanced P-gp activity.

The problem of drug resistance is one the greatest cancer challenges. Even with new drugs directed against the increasing spectrum of targets revealed by whole genome sequencing (Chapter 8), maintaining patients in sustained remission with complex cocktails may prove problematic. These difficulties have been illustrated in the treatment of tuberculosis for which a standard drug combination cures 95% of cases. Resistance to this first-line combination defines multidrug-resistant TB: second-line drugs are available but these are more expensive, require more prolonged administration and have severe side effects. These problems also arise with cancer but the multi-factorial nature of these diseases may offer alternative routes for killing tumour cells, as exemplified by recent developments in ovarian carcinoma therapy. These tumours have a poor prognosis because they have generally metastasised by the time they are diagnosed. Treatment is usually by surgery followed by cisplatin chemotherapy. The majority of patients respond to this drug but will eventually succumb as resistance develops. A novel approach has used a sphingosine analogue, FTY720 (fingolimod, Gilenya®), that after phosphorylation, binds to S1PR1 members of the GPCR family, important in angiogenesis (Chapter 6). The cells turn on an autophagic response as if to a stress and, although autophagy is a survival mechanism, eventually the cells die. Combination of FTY720 with inhibitors of autophagy accelerates the rate of necrosis. This strategy targets a normal cellular response and may therefore be promising when combined with agents that, for example, promote tumour cell apoptosis.

Non-specific effects

A further problem is that among the non-specific targets of drugs will be cells of the immune system, so that one of the innate defence systems will be compromised by agents designed to kill tumour cells. Nevertheless, under some circumstances cytotoxic drugs can stimulate an immune response. This occurs by a variety of mechanisms including activation of dendritic cells, inhibition of immunosuppressive cells and modulation of surface protein expression and hence targeting. Consistent with this is accumulating evidence that immunotherapy (i.e. the use of a vaccine or therapeutic antibodies to enhance the activity of the immune system against tumours) can be combined with conventional chemotherapy (e.g. administration of docetaxel) to generate a more effective anti-tumour response than is generated by either method on its own.

The efficacy of chemotherapy

Despite these problems the prospects for patients with a range of cancers have increased significantly over the latter part of the twentieth century, as we noted earlier. The overall five-year survival rate for white Americans diagnosed between 1996 and 2004 with breast cancer was 91%; for prostate cancer, non-Hodgkin's lymphoma and leukaemia the figures were 99%, 66% and 52%, respectively. These statistics are part of a long-term trend of increasingly effective cancer treatment and there is no doubt that the advances in chemotherapy summarised above are a contributory factor. Nonetheless, the precise contribution of drug treatments remains controversial and impossible to disentangle quantitatively from other significant factors, notably earlier detection and improved surgical and radiological methods. On the cautionary side we should also note that for metastatic (breast) cancers there has been little change in the survival rates and that for breast cancer, although chemotherapeutic advances have reduced the rate of recurrence, cancer does recur in a substantial proportion of cases.

Cancer detection: tumour imaging and molecular imaging

The diagnosis of cancer has historically relied on clinical examination and imaging methods together with histological examination and immunohistochemical staining of tumour samples. Imaging mainly refers to the use of X-rays to create two-dimensional images but also includes scintigraphy, wherein two-dimensional images are acquired after the administration of radioactively-labelled agents preferentially taken up by specific organs, and thermography, used particularly for the detection of surface temperature increases associated with the abnormal metabolic activity and neoangiogenesis occurring in breast tumours. In addition, ultrasound imaging has been used, in particular for the detection of breast and prostate cancers.

Molecular imaging

Positron emission tomography (PET)

Aside from conventional X-rays, perhaps the most familiar imaging method is positron emission tomography (PET), which produces computer-generated images that represent sites of specific biological activity within the body. It is widely used to investigative a range of neurological disorders (e.g. Alzheimer's disease) as well as in diagnosis and monitoring response to the treatment of cancers. PET involves injection into the bloodstream of a small amount of a positron-emitting isotope incorporated in

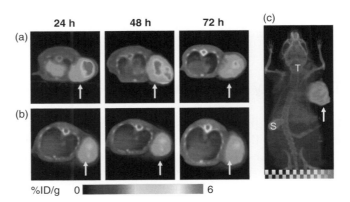

Figure 7.2 SPECT/CT of chemokine receptor 4 (CXCR4) expression in xenografts of a human glioblastoma cell line detected using ^{125}I-labelled antibodies. Severe combined immunodeficient mice bearing subcutaneous tumours were injected with: (a) ^{125}I-anti-CXCR4 antibody or (b) ^{125}I-IgG$_{2A}$: sections through the tumour are shown. (c) Whole mouse imaged 48 h after injection. The images show radioactivity accumulating in the tumours over 48 h and beginning to clear by 72 h. IgG$_{2A}$ (b) is a control that also shows some accumulation. %ID/g = percentage of injected dose per gram of tissue; arrows indicate tumour; S = spleen; T = thyroid. (Nimmagadda et al., 2009.) (See plate section for colour version of this figure.)

an organic substance (usually ^{18}F-fluorodeoxyglucose (FDG: a glucose analogue)). Positrons (electric charge +1: the anti-matter equivalent of an electron) travel only a short distance before being annihilated by colliding with an electron: this releases two or more gamma ray photons that can be detected. ^{18}F-fluorodeoxyglucose accumulates preferentially in tumour cells because of their abnormal metabolism and this is revealed in a 'PET scan'.

Single photon-emission computed tomography (SPECT) is similar to PET although the radioisotope is usually attached to the ligand for a specific cell surface receptor and the gamma radiation is measured directly. These differences decrease the resolution, relative to PET scans, but they also reduce the cost. PET or SPECT can be used together with computed tomography (CT) to provide one of the most powerful methods for locating precisely the position of a tumour (Fig. 7.2). Computed tomography uses X-rays to acquire two-dimensional images but from a large number of such images, taken as the radiation beam moves through the body, a three-dimensional picture can be pieced together. Because this permits whole organs to be visualised it has become an immensely powerful diagnostic tool since its introduction in the early 1970s. SPECT/ CT cameras comprise a gamma detector (the commonly used radioisotopes are thallium (^{201}Tl), technetium-99m (^{99m}Tc), iodine (^{123}I) and gallium (^{68}Ga)) and a CT scanner. PET/SPECT-CT therefore combines the metabolic activity signal from PET with the anatomical image generated by CT.

Because, after injection, FDG becomes trapped in any cell that sequesters it, tissues with high glucose uptake, such as the brain, the liver and many cancers, give a strong radiolabelling signal. Not all cancer cells take up significant amounts of FDG but PET is commonly used to detect Hodgkin's disease, non-Hodgkin's lymphoma, lung,

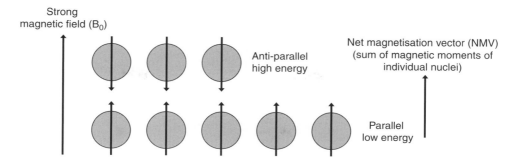

Figure 7.3 Representation of the concept of individual nuclei aligning in a strong magnetic field. An excess align parallel to the field to create a net magnetisation vector.

pancreatic and oesophageal cancers. The standardised uptake value (SUV) is a pseudo-quantitative measure of FDG uptake but a number of factors make the SUV, and hence the interpretation of PET data, a somewhat uncertain business. These include the level of glucose in the patient's bloodstream (because this competes with FDG), the time of day (which affects metabolism) and even the scanner used.

Response Evaluation Criteria in Solid Tumours (RECIST) is an internationally accepted set of guidelines for the measurement of the response of cancer patients to treatment – in other words, whether they improve ('respond'), stay the same ('stabilise'), or worsen ('progression') during treatment.

Although PET scanning is considered to be non-invasive, the subject is exposed to ionising radiation. For ^{18}F-FDG the typical dose is 8 mSv and for whole body CT scans the range is 7 to 30 mSv. The risk of such exposures leading to cancer is small but it is not negligible and the estimate is that about 1.5% of cancers in the USA might be attributable to single CT scans.

In the quest for more sensitive methods, other radiolabels are being investigated, for example, ^{18}F-sodium fluoride for bone metastases from renal cell carcinoma. ^{11}C-acetate and technetium-99m methylene diphosphonate ((^{99m}Tc)MDP) are also being developed.

Magnetic resonance imaging (MRI)

The earliest report that tumours could be distinguished from normal tissue in vivo by nuclear magnetic resonance (NMR) was in 1971, and by the end of that decade the first whole body scans had been carried out. From these beginnings the field of magnetic resonance imaging (MRI) has evolved to become a widely used radiological method to construct two- or three-dimensional images of the body. The technique uses powerful magnetic fields that align (polarise) atomic nuclear spins with the field (Fig. 7.3), followed by a brief pulse of an electromagnetic field (usually of radio frequency (RF)) that flips the spin on the aligned nuclei. The energy of the RF pulse is absorbed and then radiated at a specific resonance frequency. The resonance frequency of a nuclear spin depends on the chemical environment of the nucleus and changes therein, referred to as chemical shifts, are reported in ppm.

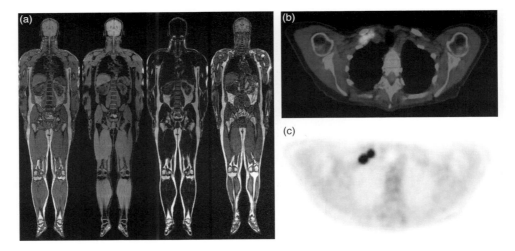

Figure 7.4 MRI and PET. (a) Whole-body MRI imaging sections showing high soft-tissue contrast. (Images kindly contributed by Dr Peter Börnert, Philips Technologie GmbH.) (b) and (c) Two images showing a section through the pelvis. (b) Fluorine-18-fluorodeoxyglucose PET-CT imaging shows mainly bone structures and the more active muscles. (c) Fluorine-18-fluorodeoxyglucose PET image. Both these scans are of a patient with metastatic melanoma: the two, small tumours are clearly visible in both scans. (Images kindly contributed by Dr John Buscombe, Division of Nuclear Medicine, Addenbrooke's Hospital, Cambridge.)

Magnetic resonance imaging usually images the hydrogen atoms (protons) in water and imaging is possible because the protons in different tissues return to their equilibrium state at different rates. A typical MRI scan comprises about 20 sequences that can be designed to give T_1-weighted or T_2-weighted images depending on the selected values of the echo time (T_E) and the repetition time (T_R). T_1-weighted images show water and fluid within tissues as dark regions with fatty tissues being light; T_2-weighted images highlight water.

In addition to proton signals, MRI can utilise other nuclei, for example, endogenous ^{31}P or ^{23}Na, and protein NMR utilises heteronuclear signals for the determination of structures. Exogenous compounds labelled with ^{19}F, ^{13}C or ^{17}O can be used in vivo by injecting reagents into the bloodstream to improve the contrast in tissues being imaged. Compared to PET, MRI gives greatly enhanced soft-tissue contrast and resolution that is particularly valuable in delineating tumours (Fig. 7.4). A development of NMR called high-resolution magic-angle-spinning ^{1}H NMR spectroscopy permits quantitative analysis of metabolites in tissues and has been applied to the characterisation of a range of tumours in vivo. In contrast to PET and CT these NMR methods do not use ionising radiation and they are considered to be non-invasive and very safe.

Images can be obtained of slices in essentially any orientation through a tissue. This has led to the development of three-dimensional scanning methods, giving MRI a valuable role in delineating solid tumours prior to surgery.

In addition to anatomical imaging, Fourier transformation of NMR signals permits the identification of specific metabolites (Fig. 7.5). Thus, for example, the **choline**

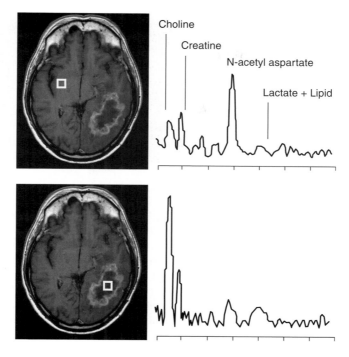

Figure 7.5 Proton MRI imaging of the human brain. Left: magnetic resonance image showing an astrocytoma (glioblastoma multiforme), delineated by a white border. (Image contributed by Tom Booth, Cambridge Research Institute.) **Right:** in vivo metabolite profiling of the brain by proton NMR. The spectra are from: **Top:** a region of normal brain. **Lower:** a brain tumour (also an astrocytoma). The spectra show lower N-acetyl aspartate and elevated choline, lactate/lipid resonances in the tumour. (Spectra reproduced with permission from Gillies and Morse, 2005.)

signal is increased in several types of tumour whereas there are lower levels of N-acetyl aspartate in brain tumours compared with corresponding normal tissue. Changes in phospholipid metabolism have also been detected by ^{31}P MR spectroscopy and raised levels of phosphocholine in breast tumours have been associated with tumour malignancy. Broadly speaking, MR has shown that the levels of lipids may reflect the rate of cell proliferation, metastatic capacity and the acquisition of drug resistance. However, thus far proton NMR has been relatively limited in both its diagnostic capacity and in monitoring early tumour responses to drugs.

Despite these limitations, the rapidly developing field of tumour imaging is of potentially great importance in both detection and monitoring response to treatment. The combination of MRI (or CT) scans with PET images (a strategy called image fusion or co-registration) is used to produce the most refined tumour maps. This has led to the design of programmes that precisely match the beams of therapeutic radiation to the contours of a tumour so as to minimise exposure of healthy tissue. Together with developments in radiation methods, these advances mean that radiotherapy is steadily improving in terms of dose refinement and radiation targeting.

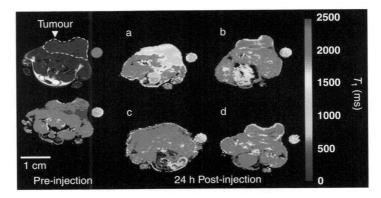

Figure 7.6 Magnetic resonance imaging of the accumulation of a phosphatidylserine-binding contrast agent in the tumours of drug-treated animals 24 hours after injection. The smart contrast agent used was gadolinium conjugated to a protein that binds to phosphatidylserine. The position of the tumour is indicated on the grey-scale image (left). Colour scale indicates T_1 values for image voxels. Drug-treated and untreated tumours are shown in (a) and (c), respectively. (b) and (d) show controls using a non-phosphatidylserine binding reagent, drug-treated and untreated, respectively. The treated tumour (a) shows marked accumulation of the PS-active contrast agent, illustrating the potential of this approach for monitoring responses to therapy (Krishnan *et al.*, 2008). (See plate section for colour version of this figure.)

Smart contrast agents

A current problem is the inadequacy of imaging methods in detecting early tumour responses to drugs. This has prompted the development of smart magnetic resonance sensors based on gadolinium (Gd^{3+}) that respond to changes in their immediate environment (Fig. 7.6). One promising method for identifying apoptotic cells uses a conjugated Gd^{3+} contrast reagent that binds phosphatidylserine expressed on the surface of apoptotic cells. This and similar approaches using technetium-99m labelled reagents have shown promise in detecting cell death in mouse tumours.

^{13}C hyperpolarisation

Many non-hydrogen nuclei with an intrinsic magnetic moment occur at very low abundance in animals and are therefore only detectable at low sensitivity by NMR. Considerable effort has been directed towards increasing the MRI signal that can be obtained from, for example, ^{13}C and ^{3}He. Although thus far confined to mice, the method of hyperpolarised ^{13}C-MRI enhances by over 50,000-fold the signals that can be obtained from ^{13}C NMR. In this method dynamic nuclear polarisation of ^{13}C-labelled metabolic substrates is carried out at low temperature (~ 1 K), the polarised mixture is then rapidly dissolved in buffer at room temperature to produce a hyperpolarised solution that is injected into mice in an MR scanner for spectral recording. The ^{13}C-labelled substrates (alanine, lactate, pyruvate, fumarate and malate) permit quantitation of the differential metabolism that characterises cancer cells. This essentially non-invasive method has shown potential for grading mouse models of

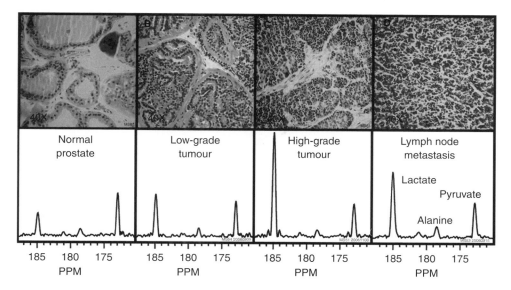

Figure 7.7 Stained sections and hyperpolarised ^{13}C spectra for histologically defined groups of a transgenic mouse model for prostate adenocarcinoma. The hyperpolarised ^{13}C spectra illustrate the strong correlation that exists between the amount of hyperpolarised ^{13}C lactate and the progression of the disease from normal prostate through low-grade primary tumours to high-grade primary tumours (Albers *et al.*, 2008). (See plate section for colour version of this figure.)

prostate cancer and also for three-dimensional ^{13}C spectroscopic imaging of tumours (Fig. 7.7).

The same approach applied to lymphomas implanted in mice has shown that conversion of fumarate to malate, reflecting tumour cell necrosis, is increased by administration of etoposide, the topoisomerase II inhibitor that promotes apoptosis of cancer cells (Fig. 7.8). There is a pressing requirement for methods to detect early responses to chemotherapy and these and related findings indicate the potential of hyperpolarised ^{13}C-MRI in this respect.

A currently promising development combines hyperpolarisation with superparamagnetic iron oxide nanoparticles (SPIONs). SPIONs are a class of contrast reagents that can be coated with ligands to target them to cells (Fig. 7.9). For example, coating nanoparticles with peptides that are over-expressed on some tumour cells can promote binding to integrin receptors. The magnetic field gradient generated by SPIONs attenuates the MR signal so that localisation of the nanoparticles at metastases produces dark spots. The sensitivity is such that single cells can be detected and the method has, for example, detected early-stage lung metastases from human breast and prostate carcinoma tumours implanted in mice.

Optical imaging

The principal medical optical imaging methods are fluorescent imaging and bioluminescent imaging. In the former the entity to be visualised (small molecule, protein or

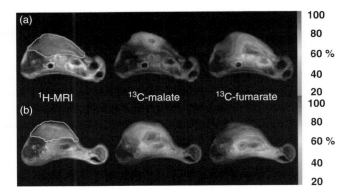

Figure 7.8 Representative transverse images from (a) untreated and (b) etoposide-treated mice with implanted lymphoma tumours. The four false colour images on the right represent hyperpolarised [1,4-^{13}C]malate and [1,4-^{13}C]fumarate following intravenous injection of hyperpolarised fumarate. The increased malate signal in the drug-treated tumour is thought to be due to tumour cell necrosis. Image reproduced by permission of Dr Ferdia Gallagher, Prof. Kevin Brindle and colleagues (Gallagher *et al.*, 2009). (See plate section for colour version of this figure.)

cell) is labelled with a fluorophore and injected into an animal. Upon excitation by light of the appropriate wavelength the emission signal from the fluorophore may be collected as a measure of the reporter's tissue localisation.

Bioluminescence signals are generated from cells transfected with a reporter construct that usually expresses luciferase under the control of an appropriate promoter. Thus, for example, stably transfected tumour cells may be grown in a mouse as a model system (Fig. 7.10). The administration of the substrate (D-luciferin, which is oxidised with the emission of yellow-green light (~575 nm)) generates a bioluminescence signal from the transfected cells. Because the signal can be generated repeatedly, this method can be used to measure progressive tumour growth and drug responsiveness in individual animals. Compared to fluorescence there is virtually no background bioluminescence signal but the in vivo use of both methods is limited by signal attenuation and scattering. Bioluminescence is increasingly used in mouse models of cancer development and therapeutic responsiveness but the technicalities of these two methods mean that they are unlikely to see significant use in humans.

Proteomics

The term proteome refers to the entire complement of proteins present in a cell population at any instant. Proteomics usually relies on two-dimensional difference gel electrophoresis (DiGE: Fig. 7.11) coupled with **mass spectrometry** (e.g. MALDI-TOF and a variety of methods for the fragmentation and separation of compounds). The recent development of protein microarray systems (forward or reverse phase arrays) permits determination of the phosphorylation state of individual proteins (the 'phosphoproteome'). The ultimate aim of these methods is to provide a complete picture of the protein expression profile including all post-translational modifications

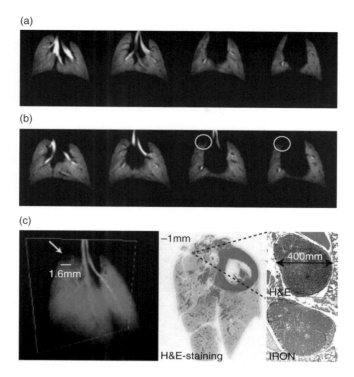

Figure 7.9 Lymph node metastasis of human prostate carcinoma cells in a mouse detected by MRI. This study used luteinising hormone-releasing hormone (LHRH) as the ligand to target tumour cells in lung metastases followed by high-resolution MRI of hyperpolarised helium (^{3}H). (a) No LHRH-SPION injection. (b) LHRH-SPION injection showing a lymph node metastasis (circled). (c) Three-dimensional representation of the lungs showing the site of the metastasis and a magnified view of the excised tissue after staining for iron and with haematoxylin and eosin, showing the high density of SPIONs in the metastasis. Reproduced with permission (Branca *et al.*, 2010). (See plate section for colour version of this figure.)

(phosphorylation, ubiquitination, glycosylation, methylation, acetylation, oxidation, nitrosylation, etc.).

In DiGE two protein samples, usually from a tumour and corresponding normal tissue, are labelled with different fluorophores, pooled and separated on the same gel (Fig. 7.12). The fluorescence signal from individual spots indicates whether the expression of that protein has increased, decreased or remained constant between the two samples. The aim of these strategies is to define tumour biomarkers that permit classification and prognosis. Both tumour tissue itself and serum proteins can be informative in these respects. In serum samples protein signatures have been reported that give a strong correlation with metastatic relapse in primary breast cancer patients.

The problems of cost and automation of DiGE have prompted development of PROTOMAP, which requires only one-dimensional gel electrophoresis but, following this separation into groups of proteins, all proteins are sequenced by mass spectrometry (a procedure sometimes referred to as shotgun proteomics). This method

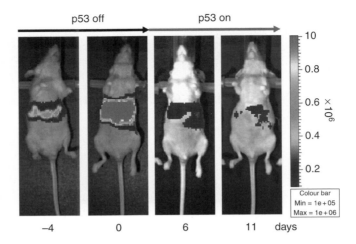

Figure 7.10 Reactivation of p53 in p53-deficient tumours causes regression. In this transgenic mouse experiment bioluminescence imaging monitors the size of the lung tumour that arises in *P53*-negative mice (days –4 to 0) but regresses (days 0 to 11) when p53 expression is activated. (Images contributed by Dr Scott Lowe, Sloan-Kettering Institute (Xue *et al.*, 2007).) (See plate section for colour version of this figure.)

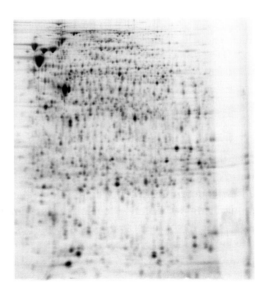

Figure 7.11 Two-dimensional difference gel electrophoresis (DiGE) separation of proteins. In this method proteins are first separated by isoelectric point in a pH gradient (isoelectric focusing) and then in a second dimension by sodium dodecyl sulphate polyacrylamide gel electrophoresis (SDS-PAGE) on the basis of their denatured relative molecular mass (M_r). This type of method is suitable for any soluble proteome. (Image kindly contributed by Kathryn Lilley, Cambridge Centre for Proteomics.)

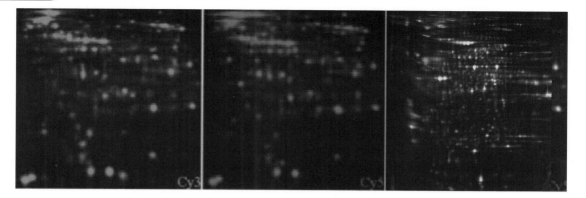

Figure 7.12 An overlay image from a two-dimensional DiGE gel of proteins labelled with two fluorescent cyanine dyes. In this experiment 50 μg of wild-type *Xenopus* embryo lysate was labelled with Cy3, which minimally labels lysine residues. The same amount of protein from a corresponding sample from a mutant strain was labelled with Cy5. Both samples were pooled prior to two-dimensional electrophoresis. Fluorescent images were captured using excitation and emission parameters unique to Cy3 and Cy5. The two resulting images were overlayed and false coloured (Cy3 green, Cy5 red). Proteins present in both samples at similar abundances will directly overlay and appear yellow on this false colour image. The ratio of Cy3:Cy5 intensity for any spot reflects the relative abundance of the protein(s) present in that spot between the two samples. (Image kindly contributed by Kathryn Lilley, Cambridge Centre for Proteomics.) (See plate section for colour version of this figure.)

can incorporate stable isotope labelling to permit relative quantification of thousands of proteins in one experiment by using isotope-coded affinity tags (ICAT) or isobaric tag relative absolute protein quantitation (iTRAC).

Metabolomics

The term metabolomics refers to the analysis of the metabolites present in a biological sample (cell, tissue or fluid) at a given time. The 'metabolic profile' obtained essentially reflects the upshot of the actions of the proteome, which is a function of the transcriptome, itself dependent on the genomic profile of the cell, tissue or organism. The term 'metabonomics' is sometimes used to distinguish a comparison of metabolic profiles that result from a specific perturbation (e.g. change in diet). The major steps in generating a metabolic profile are separation of components after they have been extracted from the sample, followed by detection. Separation can utilise gas chromatography, high performance liquid chromatography, capillary electrophoresis or combinations thereof. Detection is usually by NMR spectroscopy and mass spectrometry. As long ago as 1984 metabolic analysis by ^{1}H NMR was shown to have the capacity to identify patients with diabetes mellitus and this approach is still commonly used. However, because a complete metabolome includes not only metabolic substrates but lipids, peptides, vitamins and cofactors, a combination of methods is often necessary to define a metabolome. Alternatively, target analyses can be performed on specific pathways or classes of compounds. This complexity means that metabolomics is still a developing field both technically and in the statistical methods used for data analysis. Because metabolomics can, in principle, detect metabolic perturbation, considerable effort is being directed to its use for identifying biomarkers for tumours and for their response to treatment. A current programme evolving from these methods is the eTUMOUR study (www. etumour.net) that is developing methods of pattern recognition in an accumulating database of ^{1}H NMR spectra to provide automated diagnosis of brain tumours.

Gene expression profiling

The rate of advance of chemotherapeutics in the twentieth century was in part limited by being confined to those cancer-associated genes that had been identified and sequenced by the slow and laborious methods then available. Thus, for example, from the 1970s it was possible to measure RNA expression by Northern blotting but only of one, previously sequenced, gene at a time. The field was revolutionised by the arrival of DNA microarrays, a method derived from Southern blotting, that utilises ~10,000 oligonucleotide probes attached to glass. Hybridisation to cDNA samples is usually detected by fluorescence.

Expression profiling began to be applied to the analysis of tumour samples at the end of the 1990s (Fig. 7.13). One of the most influential studies identified a signature of 70 expressed genes that strongly predicted the development of metastases from primary breast tumours. This provided a much more robust indicator than had previously been available from studies based on one or a few genes. The results had an immediate impact on treatment strategies because they showed that about three quarters of the patients with lymph node-negative breast cancer who would have been candidates for adjuvant therapy would not have gone on to develop distant metastases. That is, they would have been given a treatment from which they could not have benefited and that might have caused side effects. The 70-gene 'poor prognosis' signature has subsequently been developed as a molecular diagnostic test (MammaPrint) for breast cancer and has received FDA approval. Similar approaches have developed a 76-gene assay (also known as the Rotterdam signature) and defined a panel of 21 genes (using the Oncotype DX assay) to predict the probability of distant recurrence in patients with lymph node-negative, oestrogen-receptor-positive breast cancer. A 2010 survey of four American centres revealed that the data provided by the 21-gene assay had caused a change in the treatment recommended for nearly one third of patients, a noteworthy impact in the six years since the test was introduced. Similar prognostic tests for colon cancer, acute myeloid leukaemia, diffuse large-B-cell lymphoma and Burkitt's lymphoma have followed.

Protein imaging

The application of the amazing technical developments summarised in this chapter (gene expression arrays, proteomics, metabolomics and tumour imaging) is propelling

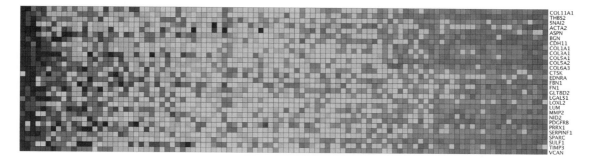

Figure 7.13 Gene expression profiling. This method permits the simultaneous determination of the levels of expression of thousands of genes in a cell sample. Heat maps show the variation in the relative amounts of mRNA between samples for individual genes. Each row represents the expression level of an individual gene: each column is a different sample. In the above example red indicates high and blue reduced expression. One of the first studies to use oligo-nucleotide microarrays for expression profiling screened 6,827 mRNAs from 72 patients and showed that the patterns of 50 genes permitted differentiation between classes of acute leukae-mia – ALL and AML (Golub *et al.*, 1999). Numerous subsequent studies have utilised the power of this technique to classify sub-types of cancer and to identify gene expression 'signatures' that, for example, are prognostic indicators for metastatic development. The heat map shown above is from a recent study in which RNA was profiled from xenografts of individual human cancer cell lines. This identified a signature that is significantly triggered only when a cancer has reached a particular stage of invasiveness. The heat map shown is for colon cancers but the invasion-associated signature occurs in many other human cancers (e.g. breast, lung, ovarian). It includes many genes associated with the epithelial to mesenchymal transition (Chapter 5) including COL11A1 (collagen type XI), SNAI2 (slug), ACTA2 (α-smooth muscle actin), CDH11 (cadherin 11), FN1 (fibronectin 1), PDGFRB (platelet-derived growth factor receptor β). (Image kindly contributed by Dimitris Anastassiou, Columbia University; Anastassiou *et al.*, 2011.) (See plate section for colour version of this figure.)

cancer diagnosis and treatment into realms that were almost unimaginable before the twenty-first century. Yet another pioneering step has for the first time provided the capacity to track changes in the concentration and location of proteins in a living cell. This 'dynamic-proteomics' method involves infecting cells with retroviruses carrying fluorescent protein genes. Called 'central dogma tagging' (CD-tagging), this relies on the random integration of retroviral cDNA into the host cell genome: appropriate integration within a gene results in the synthesis of a fluorescently tagged protein. By this means over 1,000 individual cells were isolated, each expressing a different, labelled protein, from which clones were generated. By making time-lapse movies of each clone, the response of the 1,000 proteins to the topoisomerase 1 inhibitor camptothecin was monitored (Fig. 7.14). Perhaps surprisingly, the majority of the proteins underwent significant, drug-induced changes in concentration with most (76%) decreasing while a small fraction (7%) showed increased expression. About 2% of the proteins underwent localisation changes, including translocation of the drug target (TOP1) to the cytoplasm. While this method for the quantitation of proteome dynamics has so far covered only about one twentieth of the expression capacity of the genome and can only be carried

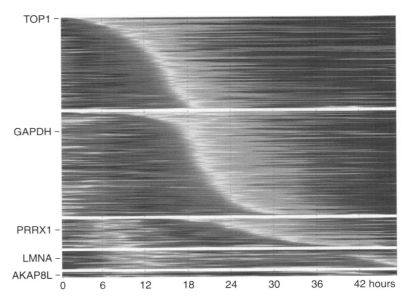

Figure 7.14 **Temporal patterns of protein fluorescence intensity in response to a drug.** Each row corresponds to one protein averaged over all cells in the movie at each time point. Proteins were clustered according to their dynamics and show waves of accumulation and fall in intensity (Cohen *et al.*, 2008). Among the most rapid responders of the 1,000 tagged proteins was the target of the drug (camptothecin) – topoisomerase-1 (TOP1) that is degraded within a few minutes. Other classes of dynamic behaviour are represented by GAPDH (level increases over 12 h before declining), PRRX1, LMNA and AKAP8L (increases after 30 h). (See plate section for colour version of this figure.)

out on one cell type at a time, it provides a new vista by which we may be able to approach the molecular behaviour of cells and their response to therapeutic drugs.

Nanotubes, graphene and nanocells

The continuing development of this vast range of therapeutic approaches and monitoring methods offers much promise as the twenty-first century unfolds. Furthermore, it is already clear that they will be joined by exciting new technologies from the world of physics, most notably in the form of carbon nanotubes and graphene (Fig. 7.15). Carbon nanotubes, members of the **fullerene** family, are cylindrical carbon molecules – essentially rolled up sheets of carbon hexagons – that are taken up by cells and can be coated with biological agents so that they can be both targeted and act as carriers. Folic acid coating has been used to direct nanotubes to the relatively high density of folic acid receptors present on some cancer cells. These small molecules are taken up by cells and, once inside, infrared laser light, directed at the target tissue, is specifically absorbed by the nanotubes, raising their temperature and thus killing the cells.

Graphene is a novel material comprising a planar sheet of carbon just one atom thick. These sheets can be made biocompatible by conjugation with branched

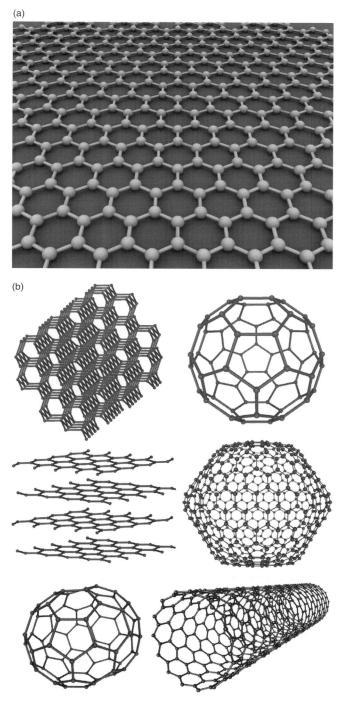

Figure 7.15 Structures of a graphene sheet (a) and allotropes of carbon (b). Bottom right is a single-walled carbon nanotube to which chemotherapeutic drugs (e.g. paclitaxel) have been conjugated to provide efficient in vivo delivery.

polyethylene glycol (PEG) polymers. Like carbon nanotubes, graphene can be coated with targeting molecules and drugs. As it can also be rendered luminescent, graphene offers the capacity to deliver and load drugs into cancer cells and to image the process simultaneously. This approach has already been used for selective targeting of B cell lymphoma cells by using adsorbed Rituxan® (the chimeric monoclonal antibody against the B cell surface protein CD20).

Other drugs have already been delivered using graphene although the method still has to overcome one of the most besetting problems of cancer therapy, namely how to achieve specific targeting of more than a very limited number of tumours. Nevertheless, these new technologies hold great promise. They also carry a degree of irony because graphene first came to the fore, albeit not by name, as a constituent of soot that caused cancer of the scrotum in eighteenth-century chimney sweeps. Indeed there is evidence that, if inhaled in large quantities, carbon nanotubes could cause lung cancer in a similar manner to the action of asbestos fibres in causing mesothelioma. However, these hazards, now well understood, should not be a barrier to the continued development of these ingenious methods.

Other synthetic drug delivery systems include nanocells comprising a core nanoparticle coated with polyethylene glycol that confers solubility in water and is non-immunogenic. Nanoparticles have been fabricated from poly(lactic-co-glycolic) acid (PLGA), a copolymer that is both biocompatible and biodegradable. Nanocells have been used to deliver drug combinations, notably an anti-angiogenic agent together with a chemotherapeutic drug. The vascular collapse caused by the anti-angiogenic agent traps nanocells within the tumour. In mice this has been shown to inhibit tumour growth and, by concentrating the chemotherapeutic drug, give an increased therapeutic index. As with other delivery vehicles, nanocells can be targeted by coating with specific ligands and can also carry not only drugs but siRNAs (to act through **RNA interference**) or MRI contrast agents.

Conclusions

Surgery and radiation continue to be the mainstays of cancer treatment. However, since the introduction of methotrexate in 1950, they have been joined by an ever-expanding armoury of chemotherapies. The development of drug combinations has produced some dramatic improvements in survival rates, notably for some forms of leukaemia and for testicular cancer. Nevertheless, most of the large number of drugs currently in use for cancer are non-specific – that is, they do not differentiate tumour from normal in the cells they target. This has led to much emphasis in developing targeted therapies based on knowledge of the underlying molecular biology. These include monoclonal antibodies, engineered proteins (e.g. targeting ligand linked to cytotoxic agent), and small molecule kinase inhibitors with selectivity for specific mutants. However, only 22 have so far achieved FDA approval.

The development of preventive vaccines that have proved to be almost completely effective in blocking infection by papillomaviruses has been one of the great success stories in medicine. This offers the means to prevent 70% of the deaths from cervical carcinoma, the seventh highest cause of cancer worldwide.

In parallel with these advances, increasingly sophisticated methods for detecting tumours and for monitoring their response to treatment are under development, principally based on imaging by PET and MRI. The progressive expansion of the 'omics' repertoire to include protein expression (the proteome), metabolites (the metabolome) and the complete profile of expressed genes is transforming the scale of cancer detection. In the final chapter we will review the genomic revolution that is revealing many more therapy targets and introducing the concept of tailoring the design of therapeutic strategies to the molecular signature of individual cancers.

Key points

- The development of modern cancer chemotherapy began in the second half of the twentieth century with the use of methotrexate for the treatment of childhood leukaemia and purine biosynthesis inhibitors for solid tumours.
- The greatest chemotherapeutic successes have been some of the tyrosine kinase inhibitors (e.g. imatinib (Gleevec®) for chronic myeloid leukaemia), selective oestrogen receptor modulators (e.g. tamoxifen for breast cancer) and monoclonal antibodies (e.g. Herceptin® for breast cancer).
- The most successful vaccines thus far developed (Cervarix® and Gardasil®) confer almost complete protection against human papillomavirus (HPV), the major cause of cervical cancer.
- The major barrier to effective cancer chemotherapy is drug resistance that can arise through the activation of efflux mechanisms or by modulation of the target.
- In parallel with chemotherapeutic developments, major technical advances are improving tumour detection and the capacity to monitor treatment response, most notably using positron emission tomography (PET) and magnetic resonance imaging (MRI).

Future directions

- Innovative developments for the non-invasive imaging of tumours offer the promise of being able to monitor drug response from the earliest stages of treatment. The requirement for advances in this field is indicated by the fact that the current detection limit for clinical imaging of solid tumours is a diameter of approximately 1 cm (representing between 10^8 and 10^9 cells). This represents about 30 doublings of the initiating cell. For primary breast tumours the median volume doubling time has been estimated as 157 days (for women aged 50 to 70 years; significantly shorter in younger women), meaning that tumours have typically been growing for about 13 years before they are detected.

- Thus far the limiting problem of drug resistance has proved insoluble and appropriate strategies are urgently required. Recent evidence, summarised in Chapter 5, indicates that the proportions of macrophages and cytotoxic T cells in tumours can serve as an indicator of prognosis and that macrophage populations infiltrating tumours can affect the response to chemotherapy given before surgery. This highlights the importance of refining methods for quantifying the leucocyte population in tumours.

Further reading: reviews

THE DEVELOPMENT OF ANTI-CANCER DRUGS

Pavet, V., Portal, M. M., Moulin, J. C., Herbrecht, R. and Gronemeyer, H. (2011). Towards novel paradigms for cancer therapy. *Oncogene* **30**, 1–20.

CHEMOTHERAPEUTIC STRATEGIES FOR CANCER

Aggarwal, S. (2010). Targeted cancer therapies. *Nature Reviews Drug Discovery* **9**, 427–8.

Kaitin, K. I. and DiMasi, J. A. (2011). Pharmaceutical innovation in the 21st century: new drug approvals in the first decade, 2000–2009. *Clinical Pharmacology & Therapeutics* **89**, 183–8.

Yap, T. A., Carden, C. P. and Kaye, S. B. (2011). Beyond chemotherapy: targeted therapies in ovarian cancer. *Nature Reviews Cancer* **9**, 167–81.

KINASE INHIBITORS

Ott, P. A. and Adams, S. (2011). Small-molecule protein kinase inhibitors and their effects on the immune system: implications for cancer treatment. *Immunotherapy* **3**, 213–27.

INHIBITORS OF METABOLISM

Vander Heiden, M. G. (2011). Targeting cancer metabolism: a therapeutic window opens. *Nature Reviews Drug Discovery* **10**, 671–84.

SELECTIVE OESTROGEN RECEPTOR MODULATORS (SERMS)

Ahmad, N. and Kumar, R. (2011). Steroid hormone receptors in cancer development. A target for cancer therapeutics. *Cancer Letters* **300**, 1–9.

MONOCLONAL ANTIBODIES

Nelson, A. L., Dhimolea, E. and Reichert, J. M. (2010). Development trends for human monoclonal antibody therapeutics. *Nature Reviews Drug Discovery* **9**, 767–74.

IMMUNOTHERAPY

Stewart, T. J. and Smyth, M. J. (2011). Improving cancer immunotherapy by targeting tumour-induced immune suppression. *Cancer and Metastasis Reviews* **30**, 125–40.

ANTI-ANGIOGENIC AGENTS

Oklu, R., Walker, T. G., Wicky, S. and Hesketh, R. (2010). Angiogenesis and current antiangiogenic strategies for the treatment of cancer. *Journal of Vascular and Interventional Radiology* **21**, 1791–805.

VASCULAR TARGETING AGENTS (VTAS)

Kanthou, C. and Tozer, G. M. (2007). Tumour targeting by microtubule-depolymerising vascular disrupting agents. *Expert Opinion on Therapeutic Targets* 11, 1443–57.

LIPOSOMAL THERAPY

Chen, Z. (2010). Small-molecule delivery by nanoparticles for anticancer therapy. *Trends in Molecular Medicine* 16, 594–602.

DRUG DELIVERY FROM INERT CAPSULES

Damiano, G. C. and Girolamo, R. (2011). Trans-arterial chemoembolisation as a therapy for liver tumours: new clinical developments and suggestions for combination with angiogenesis inhibitors. *Critical Reviews in Oncology Hematology* 80, 40–53.

GENE THERAPY

Cattaneo, R., Miest, T., Shashkova, E. V. and Barry, M. A. (2008). Reprogrammed viruses as cancer therapeutics: targeted, armed and shielded. *Nature Reviews Microbiology* 6, 529–40.

Jiang, H., Gomez-Manzano, C., Lang, F. F., Alemany, R. and Fueyo, J. (2009). Oncolytic adenovirus: preclinical and clinical studies in patients with human malignant gliomas. *Current Gene Therapy* 9, 422–7.

THERAPEUTIC (TREATMENT) VACCINES

Klebanoff, C. A., Acquavella, N., Yu, Z. Y. and Restifo, Ni. (2011). Therapeutic cancer vaccines: are we there yet? *Immunological Reviews* 239, 27–44.

PROPHYLACTIC (PREVENTIVE) VACCINES

Julius, J. M., Ramondeta, L., Tipton, K. A. *et al.* (2011). Clinical perspectives on the role of the human papillomavirus vaccine in the prevention of cancer. *Pharmacotherapy* 31, 280–97.

Liu, M. A. (2011). DNA vaccines: an historical perspective and view to the future. *Immunological Reviews* 239, 62–84.

Stanley, M. (2010). Prospects for new human papillomavirus vaccines. *Current Opinion in Infectious Diseases* 23, 70–5.

PEPTIDE VACCINATION

Klug, F., Miller, M., Schmidt, H. H. and Stevanovic, S. (2009). Characterization of MHC ligands for peptide based tumour vaccination. *Current Pharmaceutical Design* 15, 3221–36.

DRUG RESISTANCE

Engelman, J. A. and Settleman, J. (2008). Acquired resistance to tyrosine kinase inhibitors during cancer therapy. *Current Opinion in Genetics and Development* 18, 73–9.

Singh, A. and Settleman, J. (2010). EMT, cancer stem cells and drug resistance: an emerging axis of evil in the war on cancer. *Oncogene* 29, 4741–51.

MOLECULAR IMAGING

POSITRON EMISSION TOMOGRAPHY (PET)

Alberini, J.-L., Edeline, V., Giraudet, A. L. *et al.* (2011). Single photon emission tomography/ computed tomography (SPET/CT) and positron emission tomography/computed tomography (PET/CT) to image cancer. *Journal of Surgical Oncology* **103**, Special Issue: 602–6.

MAGNETIC RESONANCE IMAGING (MRI)

Gore, J. C., Manning, H. C., Quarles, C. C., Waddell, K. W. and Yankeelov, T. E. (2011). Magnetic resonance in the era of molecular imaging of cancer. *Magnetic Resonance Imaging* **29**, 587–600.

PROTEOMICS

Hu, H., Deng, C., Yang, T. *et al.* (2011). Proteomics revisits the cancer metabolome. *Expert Review of Proteomics* **8**, 505–33.

METABOLOMICS

Blekherman G., Laubenbacher, R., Cortes Diego F. *et al.* (2011). Bioinformatics tools for cancer metabolomics. *Metabolomics* **7**, 329–43.

NANOTUBES, GRAPHENE AND NANOCELLS

MacDiarmid, J. A., Madrid-Weiss, J., Amaro-Mugridge, N. B., Phillips, L. and Brahmbhatt, H. (2007). Bacterially-derived nanocells for tumour-targeted delivery of chemotherapeutics and cell cycle inhibitors. *Cell Cycle* **6**, 2099–105.

Wang, C., Cheng, L. A. and Liu, Z. A. (2011). Drug delivery with upconversion nanoparticles for multi-functional targeted cancer cell imaging and therapy. *Biomaterials* **32**, 1110–20.

Yang, X. Y., Wang, Y. S., Huang, X. *et al.* (2011). Multi-functionalized graphene oxide based anticancer drug-carrier with dual-targeting function and pH-sensitivity. *Journal of Materials Chemistry* **21**, 3448–54.

WEBSITES

Chemotherapy index: www.chemocare.com/bio/list_by_acronym.asp

eTUMOUR study: www.etumour.net/

The future of cancer in the post-genomic era

The unveiling of the sequence of DNA in the human genome in 2003 was one of the most dramatic milestones in the history of science. Nevertheless, even in the immediate aftermath of that event it would have been difficult to predict the extraordinary advances of the following eight years that now permit individual genomes to be sequenced with great rapidity at low cost and have prompted an endeavour to compile a database of 10,000 complete cancer genomes. Within that period whole genome sequencing has revealed new cancer genes and promoted the development of novel drugs. It has illuminated hitherto unsuspected flexibility in human DNA, provided alternative strategies for the classification of tumours and already begun to change the treatment regimes that are offered to patients by clinicians. Taken together with the advances described in the previous chapter, an armoury of great breadth and sophistication can now be deployed for the detection, classification and treatment of cancer.

Human genome sequencing

The draft sequence of the human genome was largely completed by April 2003 in a phenomenal achievement that required quite stunning developments both of sequencing machines, robotics to handle clones and computing power to process the data and make it easily usable by the scientific community. Since 2003, equally dramatic technical advances have produced an almost unbelievable increase in the rate at which sequences can be obtained. These new technologies are called 'next-generation' or 'second-generation' sequencing and permit rapid, so-called 'massively parallel' sequencing of complete genomes in a single experiment. Efficient though second-generation sequencing is, it may be about to be overhauled by 'third-generation' sequencing in which an accurate sequence can be obtained without sequencing thousands of copies (Box 8.1).

Box 8.1 Whole genome sequencing

Sanger dideoxy sequencing

Up until about 2005 all sequencing was done, in essence, by the 'dideoxy' method devised by Fred Sanger. In 1977, this had yielded the first sequence of a DNA genome (the bacteriophage phi X 174 that has just over 5,000 bases) followed by human mitochondrial DNA (about 17,000 bases). The replacement of slab gels by a capillary system giving automated readout permitted sequencing of the 3,000 million base pairs in the human genome in 2003.

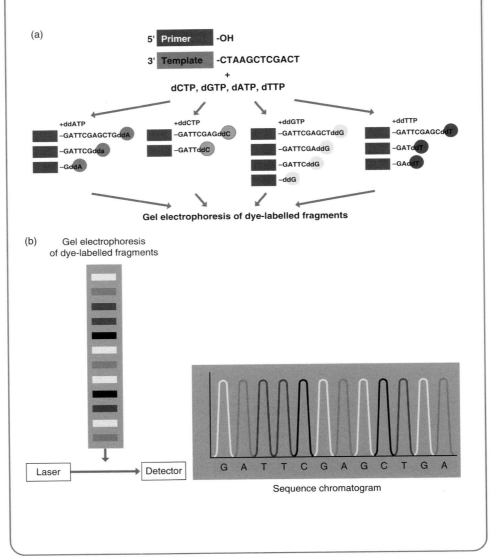

Box 8.1 cont'd

Box 8.1a (*cont.*) **Sanger dideoxy sequencing.** (a) This method utilises the mechanism by which DNA is replicated in cells. DNA polymerases require a short oligonucleotide (a primer) that binds to the template strand: a deoxynucleoside triphosphate (dNTP) is added to the 3′-hydroxyl group of the primer to form a phosphodiester bond. The reaction is repeated so that a new strand, complementary in sequence to the template, is synthesised in the 5′ to 3′ direction. The sequencing reaction uses dideoxynucleoside triphosphate (ddNTP) analogues that halt strand elongation because they lack the 3′-hydroxyl group required for formation of the next phosphodiester bond. By selecting suitable relative concentrations of normal dNTPs and chain terminating ddNTPs (the polymerase incorporates these randomly) fragments terminating at every base over the length of the template are made. (b) Labelling ddNTPs with different fluorophores followed by electrophoretic separation of the fragments on the basis of size permits the sequence to be read.

Next-generation sequencing

In the ten years since the finished sequence of the human genome was published technical developments have transformed sequencing rates from those possible using the Sanger dideoxy method. These technologies are referred to as 'second-generation' sequencing (or 'next-generation') and 'third-generation' sequencing and they permit rapid sequencing of complete genomes in a single experiment.

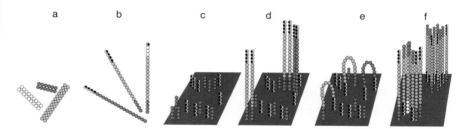

a   b   c   d   e   f

Box 8.1b **Principle of a second-generation DNA sequencing platform.** The scheme represents the elements of the Illumina system: (a) genomic DNA is fragmented; (b) common adaptors are ligated to each end; (c) forward and reverse primers corresponding in sequence to the adaptors are tethered via a flexible linker oligonucleotide to a solid surface; (d) adaptors bind to primers and unique DNA inserts are PCR amplified: un-tethered DNA is washed away leaving tethered amplification products; (e) free ends anneal to tethered primers to permit bridge amplification; (f) the reverse strands are cleaved and washed away, leaving clusters of ~1,000 copies of a single member of the template library. The flow cells used contain 100s of millions of clusters that are sequenced simultaneously ('massively parallel sequencing'). The first sequencing cycle begins by adding four labelled reversible terminators, primers and DNA polymerase. The products are washed away and the cycle repeated until sequencing is complete.

Box 8.1 cont'd

Second-generation sequencing systems use amplification of DNA (by PCR) to permit cyclic (i.e. iterative) sequencing in parallel of large numbers of identical template clusters (clonal amplicons). These may be attached to beads or tethered to substrates. As in the Sanger method, sequence detection is by fluorescent labelling, e.g. using reversible dye-labelled terminators. The dramatic speed and cost improvements in these methods, together with the massive parallelisation of sequencing that is important for the quality of the 'read', have come by the use of fully automated, high throughput flow cells.

Pyrosequencing

This method differs from those based on Sanger sequencing in detecting the release of pyrophosphate that accompanies nucleotide incorporation, rather than chain termination. This requires, in addition to reagents for DNA synthesis, the inclusion of enzymes and substrates that result in the production of light in the form of chemiluminescence, detectable by a charge coupled device (CCD) chip. The template DNA is tethered in a flow cell, solutions of dNTPs (dATP, dCTP, dGTP and dTTP) being sequentially added and removed and, as with other second-generation methods, sequencing is in a massively parallel fashion.

Third-generation sequencing

This category refers to the miniaturisation that is possible if sequences can be determined directly from a single DNA molecule without the requirement for DNA amplification. A variety of the approaches that have been developed are summarised below.

Heliscope single molecule sequencing (Helicos Biosciences)

In this method poly(A) tails are attached to DNA fragments, which then hybridise to poly(T) oligonucleotides bound to a flow cell. Sequencing is by DNA extension using fluorescently labelled nucleotides and billions of single DNA or RNA molecules can be sequenced in one run.

Single molecule real time sequencing (Pacific Biosciences)

This technology uses electron-beam lithography to fabricate substrates comprising 150,000 zero mode waveguides (ZMWs). These tiny chambers hold 1 zepto-litre (10^{-21} litre) and each contains one immobilised DNA polymerase molecule. Synthesis of DNA complementary to a bound template uses nucleotides labelled with a fluorophore at the terminal phosphate moiety (a different colour for each dNTP). When a cognate nucleotide is incorporated a fluorescence pulse is emitted: on cleavage the labelled pyrophosphate group rapidly diffuses from the chamber. The ZMW is illuminated from below by laser light and DNA sequencing is therefore monitored in real time.

Box 8.1 cont'd

Nanopore sequencing
In contrast to all the previous methods this technology does not rely on the detection of labelled nucleotides but measures the changes in electrical conductivity as single strands of DNA are drawn through artificial nanopores (~1 nanometre diameter). The apertures can either take the form of specific transmembrane proteins (e.g. α-haemolysin) or be made by etching on a silicon wafer (by using a beam of ions – 'ion-beam sculpting'). The change in specific conductance with DNA movement depends on which base is restricting the current flow.

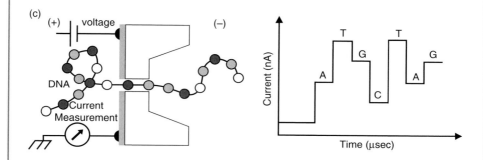

Box 8.1c Nanopore sequencing. Schematic of design and output signal (from Gierhart *et al.*, 2008).

Ion torrent sequencing
As each nucleotide is incorporated into a strand of DNA a hydrogen ion is released, changing the pH of the solution. Ion torrent sequencing uses a high-density array of micro-wells beneath which an ion-sensitive semiconductor layer detects pH change. A different DNA template is anchored to each well and the chip is sequentially flooded with one nucleotide after another, permitting massively parallel sequencing.

Footnote
In addition to those summarised above a number of other ingenious sequencing methods are under development including the use of an engineered DNA polymerase to permit nucleotide determination by fluorescent resonant energy transfer (VisiGen Biotechnologies) and multiplex polony sequencing in which discrete clonal amplifications of a single DNA molecule (polymerase colonies) are grown in a gel matrix.

The technical progress embodied in the above brief summaries is astonishing. The cost estimate for sequencing in the 1980s was $1 per base: the current (2012) figure is $30/Gb. The first sequencing machines in the early 1980s managed

> **Box 8.1** cont'd
>
> 10 kb (10,000 bases) a day; the present rate is heading towards 100 million kb per day. So the cost has gone down by a factor of ten thousand million and the speed has gone up by 10 million.
>
> In addition to their massive impact on cancer, these developments have led to the '1,000 genomes project' that is ultimately intended to provide a comprehensive map of worldwide genetic variation between multiple populations that will revolutionise the study of disease susceptibility (www.genome.gov/27541917).

These advances have resulted in huge increases in both the speed and the precision of sequencing and ushered in the era of 'personalised medicine', meaning that individual genomes can be sequenced in a day for a cost approaching US$1,000. This has the revolutionary implication that the mutation pattern of an individual tumour can be used to design a therapeutic strategy before treatment is started. We will see shortly when considering tumour biomarkers that genome sequencing also offers a method both for monitoring tumour response to treatment and potentially for tumour detection from DNA in blood long before there are any clinical signs (palpable lumps, bleeding, etc.). From this it is evident that the science of genomics is poised to make the greatest impact on medical science of any advance in our history and the following examples illustrate some of the major advances that have already occurred in the first decade of the twenty-first century.

Following the full sequencing of the human genome, Michael Stratton, working at the Sanger Centre, completed a study that would have been inconceivable just a few years earlier. Inconceivable because it required the sequence of the human genome but also because it relied on the fantastic technical developments that permitted massive lengths of DNA to be sequenced at high speed. Five hundred kinase enzymes were selected that were known or thought to be involved in cell signalling pathways that controlled growth. A large number of tumour samples were then screened by sequencing to see if there were any mutations in the kinases. As some kinases were already known to be frequently mutated in human cancers, the appearance of some of these was anticipated. What came as a surprise, however, was that one kinase gene, *BRAF*, had the same single base mutation in about two thirds of the tumours (V600E). The tumour they had chosen was melanoma – the most rapidly increasing cancer in frequency of occurrence in the UK population – about which virtually nothing was known at the molecular level. Stratton's group had discovered a new 'cancer gene' that played a major role in a very prevalent cancer. This rapidly led to the synthesis of a drug that is very efficient at blocking the action of the mutant form of *BRAF* (see below) and thus offers a chemotherapeutic approach to treating melanoma. In the end, of course, *BRAF*'s role in melanoma would have been discovered by the same slow and painstaking methods that had previously identified many cancer genes, but that might have taken years or even decades. Whole genome sequencing and the era of genomics had made an almost immediate impact on medical science.

Whole genome sequencing (WGS) and cancer

In 2007, the J. Craig Venter Institute released the first individual genome sequence, that of the institute's founder. This was followed almost immediately by that of James Watson, from the Human Genome Sequencing Center in Houston, the first full genome to be sequenced using next-generation rapid-sequencing technology. Shortly after that three complete sequences of individual human genomes were published simultaneously in 2008. These were from a male Yoruba from Ibadan, Nigeria, a male Han Chinese and a female who had died from acute myeloid leukaemia (AML). The new technology of massively parallel sequencing permitted repeated determinations to give an 'average depth' of over 30 times, greatly increasing the confidence with which variants could be assigned. Each of these sequences revealed several million single nucleotide variants (SNVs) when compared with a human reference assembly, for example, the single nucleotide polymorphism database (dbSNP) run by the National Center for Biotechnology Information (NCBI) that includes a range of molecular variation in addition to SNPs (Fig. 8.1). The AML study identified nearly four million tumour SNVs. Somewhat surprisingly, the majority of these are also present in the reference genome or in the Venter or Watson genomes and, after their subtraction, 31,632 new SNVs remained that were unique to the tumour genome (Fig. 8.1). Most of these were in intronic regions or untranslated regions but 14 were validated as germ line SNVs (i.e. SNPs) and eight were somatically acquired, non-synonymous mutations (in *CDH24, SLC15A1, KNDC1, PTPRT, GRINL1B, GPR123, EBI2* and *PCLKC*). Mutations in *FLT3* and *NPM1* that had been identified previously were also detected.

The extensive overlap between SNPs in the AML tumour and in the reference genomes (Fig. 8.1) highlights the astonishingly dynamic nature of human DNA and indicates that the definition of 'normality' with regard to sequence is somewhat arbitrary. This ever-changing background might suggest the impossibility of detecting

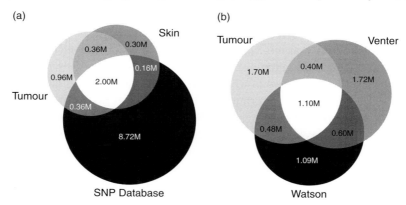

(a)   Skin
Tumour   0.36M   0.30M
0.96M   0.16M
2.00M
0.36M
8.72M
SNP Database

(b)   Venter
Tumour   0.40M
1.70M   1.72M
1.10M
0.48M   0.60M
1.09M
Watson

Figure 8.1 Genomic fluidity illustrated by the overlap of SNPs between the genomes from different tissues in one individual and between individuals. (a) Skin and tumour (acute myeloid leukaemia) genomes from the first complete sequences. (b) Overlap between SNPs in the tumour genome and the genomes of J. D. Watson and J. C. Venter (Fig. 1: Ley *et al.*, 2008).

individual polymorphisms that make small contributions to tumour development. But as we saw in Chapter 5 with the example of two intronic SNPs that increase expression of the *FGFR2* gene and promote ER$^+$ breast cancers, such associations can be identified provided the numbers of cases and controls studied are sufficiently large.

Subsequent WGS has characterised copy-number alterations in primary lung adenocarcinomas, identifying 57 significantly recurrent events in 371 tumours. These included 24 amplifications and 7 homozygous deletions, 25 of these gross changes not having been previously associated with lung cancer. Also 26 of 39 chromosome arms had large-scale copy number gain or loss. The mutational signature in a small cell lung cancer genome arising from carcinogens in tobacco smoke has also been resolved revealing 22,910 somatic mutations in the genome of one individual, of which 134 were in coding exons. This permitted the estimation that the cells of the tumour had acquired, on average, one mutation for every 15 cigarettes smoked.

Somatic rearrangements generating chimeric genes are a familiar mechanism of oncogenic activation in leukaemias, as we saw in Chapter 4. However, Stephens *et al.* (2009), using paired-end sequencing of 65,000,000 randomly generated (500 bp) DNA fragments (i.e. sequencing both ends), showed that this process also plays an important role in breast cancer (Fig. 8.2). In 24 primary breast tumours there was an average of 38 such rearrangements per tumour with over 200 present in some tumours. Rather than generating a fusion gene, the most common rearrangement was tandem duplication. The frequency showed an unexpected variation between tumours – from none to over 100 with a size range of duplicated segments from 3 kb to >1 Mb. From a prognostic viewpoint the most exciting finding was that one of the four main categories of breast cancer, namely basal-like cancers, generally had large numbers of tandem duplications, fewer rearrangements being associated with luminal-A and luminal-B types.

In an alternative approach to breast cancer, the genome and the transcriptome of a breast cancer metastasis were sequenced and the genomic sequence compared with that of DNA from the primary tumour that had arisen nine years earlier. This identified 32 non-synonymous coding mutations that fell into three mutational patterns: (1) mutations in five genes present in both the primary and the metastasis; (2) six mutations that appeared to be present only in minor clones of the tumour; and (3) 19 metastasis mutations not present in the primary. This study also identified 75 RNA **editing** events, two of which were novel changes, affecting 12 loci that had occurred in the metastatic transcriptome. The detection of RNA editing illustrates the importance of integrating genomic and RNA sequencing and, taken together, the data reveal the extent of evolution that can occur during metastatic progression.

Turning to pancreatic carcinomas, 1,562 somatic mutations have been identified in a screen of 24 tumours. From these emerged a set of 12 intracellular signalling pathways, six of which had a mutation in one of its constituent genes in all the 24 tumours screened. The other six pathways were also mutated in over half the tumours. There was, nonetheless, a diversity in the specific genes involved so that, for example, mutations in four different genes could contribute to disruption of the TGFβ signalling pathway. This kind of genomic snapshot revealing critical pathways can only be obtained through sequencing the entire genome.

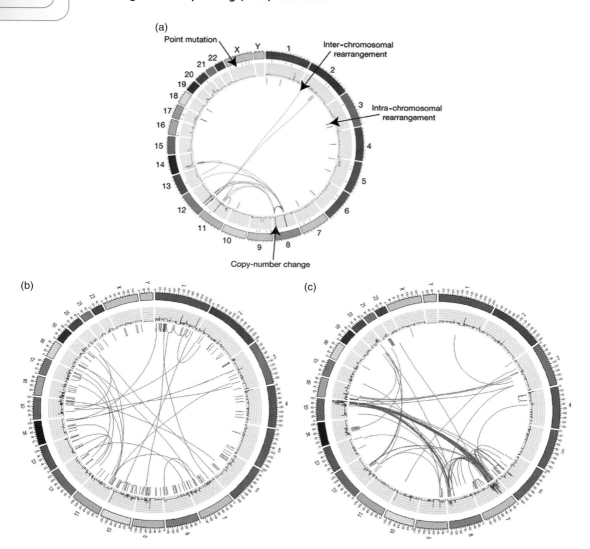

Figure 8.2 Somatic mutations in breast cancer represented as genome-wide Circos plots. (a) The mutational information included in the plots. An ideogram of a normal **karyotype** is shown in the outer ring. The blue lines show copy number variation, green denotes an intra-chromosomal rearrangement and each purple line an inter-chromosomal rearrangement. (b) and (c) show somatic rearrangements in two primary breast cancers. (b) A triple negative tumour (oestrogen receptor (ER⁻), progesterone receptor (PR⁻) and ERBB2 negative) in which hardly any gene amplification has occurred but there are about 150 tandem duplications. (c) An ER⁺, PR⁺, ERBB2⁻ tumour with nearly 200 amplified genes but very few tandem duplications (see Appendix E: Breast cancer). (Reproduced with permission from Stratton *et al.*, 2009 and Stephens *et al.*, 2009. (See plate section for colour version of this figure.)

Chromothripsis

These and a continuing stream of studies are revealing the immense complexity of mutational events that make every tumour unique at the detailed molecular level, even though major patterns of 'drivers' may characterise specific types of cancer. Bewildering though the complexity may be, it is at least consistent with the notion of protracted accumulation of mutations as clonal evolution takes its course and for the majority of cancers that picture is probably accurate. In a startling development, the application of paired-end sequencing to multiple tumours has revealed that perhaps 3% of cancers, of widely varying type, arise through a completely different mechanism of genetic instability. This takes the form of a single cataclysmic event in which localised regions of a limited number of chromosomes shatter into fragments. DNA repair systems are activated that repair the damage as best they can, mainly by non-homologous end-joining, in what appears to be a random process giving rise to every type of inversion and juxtaposition of the fragments that are rescued (Fig. 8.3).

Stephens and colleagues (2011) have called this process **chromothripsis** (Greek: thripsis – shattering into pieces) and suggest that it arises from a mitotic defect when the condensed structure of chromatin would predispose to the clustering of breaks within limited segments.

The highly focused pattern of mutations that characterises chromothripsis is evident from the genome-wide profile of rearrangements in a case of chronic lymphocytic leukaemia (Fig. 8.4). Other types of cancer from which evidence for chromothripsis has been obtained include melanoma, small cell lung cancer, non-small cell lung cancer, glioma, synovial sarcoma, and oesophageal, colorectal, renal and thyroid tumours.

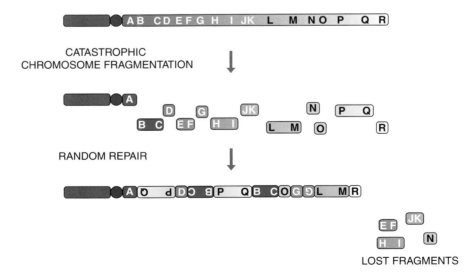

Figure 8.3 Chromosome shattering and fragment assembly in the process of chromothripsis (Stephens *et al.*, 2011).

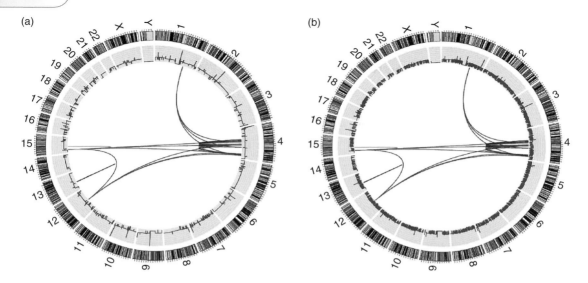

(a)

(b)

Figure 8.4 **Genomic rearrangements localised to chromosome 4q in a chronic lymphocytic leukaemia.** The two plots represent the genome-wide profile of rearrangements before (a) and at relapse 31 months after chemotherapy (b). The patterns are not significantly different, indicating that in this cancer no further evolution had occurred (Stephens *et al.*, 2011: Fig. 1c,d). (See plate section for colour version of this figure.)

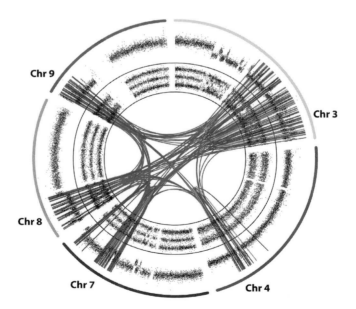

Figure 8.5 **Chromothripsis involving five chromosomes.** This plot represents 147 rearrangements between chromosomes 3q, 4q, 7q, 8p and 9p that have occurred in a chordoma, a rare type of bone tumour (Stephens *et al.*, 2011: Fig. 4a). (See plate section for colour version of this figure.)

One feature of chromothripsis is that, although alteration of gene copy number occurs at numerous locations in affected regions, the almost invariable result is only one or two copies. This finding, together with the retention of **heterozygosity** and the intensive clustering of the alterations within discrete regions of the affected chromosome, provide strong evidence that the majority of the rearrangements occur in a single cellular catastrophe (Fig. 8.5).

One particularly telling example of how chromothripsis can promote cancer development has come from a small cell lung tumour sample that had huge amplification of the *MYC* gene on chromosome 8, producing up to 200 copies per cell. A combination of sequencing and FISH revealed one normal copy of chromosome 8 together with two massively rearranged derivative chromosomes. The latter had arisen by the random stitching of fragments from the shattered regions of the other chromosome, followed by chromosomal duplication. In addition, 15 other fragments had been pieced together to form a **double minute** chromosome of about 1 Mb that had subsequently undergone massive amplification. The fragments included *MYC*, thereby revealing the mechanism of hyper-expression of this potent proliferation driver. In parallel with specific oncogenic events, chromothripsis has also been shown to cause either disruption of tumour suppressor genes or their complete loss if they are carried by fragments that fail to be incorporated into the derivative mosaic by the DNA repair machinery.

The almost incredible event of chromothripsis may give birth to hundreds of chromosomal rearrangements in a single event. Not the least amazing aspect of this nuclear breakdown is that cells, albeit a tiny minority, can survive as functional units among which some now have a selective advantage in terms of growth and hence their capacity to evolve into a tumour clone.

These examples illustrate the bewildering complexity of instability in cancer genomes, only now being fully unveiled through the power of second-generation sequencing. Although these are early steps in cancer genome sequencing, they have already revealed not only new 'cancer genes' that are potential therapeutic targets but pathways that are high frequency mutational targets in specific cancers together with patterns of mutational evolution during progression from primary tumour to metastasis.

Genomic partitioning

It is self-evident that a complete understanding of the molecular signature of a cancer requires the complete genomic sequence. Desirable though that may be, the identification of over 2,000 chromosomal rearrangements in 24 breast cancer samples (Fig. 8.2) underlines the difficulty of delineating driver mutations from the large background of passengers. Furthermore, although massively parallel methods now permit complete sequencing of individual genomes at high speed, there is, of course, a cost and time component that prompts consideration of more selective approaches. The choices include (1) sequencing selected candidate genes or a specific region of the genome already implicated in a pathway (as used to reveal *BRAF*); (2) sequencing only protein coding sequences, i.e. exons; and (3) sequencing common variants, i.e. SNPs. This

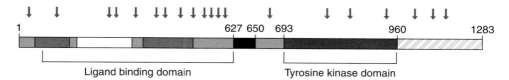

Figure 8.6 **Distribution of mutations in ERBB4.** These new mutations (arrows) were identified in melanomas by sequencing the tyrosine kinome (Prickett *et al.*, 2009).

strategy of 'genomic partitioning' therefore uses a variety of methods to focus on discrete regions of interest.

An example of the first of these selective approaches focused on the 86 members of the protein tyrosine kinase superfamily by first sequencing 593 exons of the tyrosine kinome in 29 melanoma samples. This yielded 30 somatic mutations in 19 genes that were then selected for sequencing in a total of 79 melanoma samples, yielding 99 non-synonymous somatic mutations, most of these not previously known to be associated with melanoma. It was found that 19% of the melanomas had mutations in the EGFR family member ERBB4 that increased the kinase activity and transforming capacity of the protein (Fig. 8.6). Mutations in ERBB4 had not previously been identified and they offer new targets for therapeutic intervention.

Protein-coding sequences make up only ~1% of the human genome (the exome) and the strategy of exome sequencing involves only ~5% of the sequencing necessary for the whole genome. The negative aspect is that it may miss up to 20% of exonic sequences and repeated sequencing of at least eight-fold ('depth of coverage') is required to provide statistically reliable data on variants. Despite these problems, exome sequencing applied to small numbers of individuals has identified the causes of several monogenic diseases.

As the melanoma example shows, even selective sequencing generates large amounts of information. Systems approaches are currently being developed to integrate 'omic' data to classify individual cancers more precisely and thus accelerate the implementation of effective therapies. One example of this approach focuses on the problem of identifying drivers in the enormous number of genomic aberrations present in individual tumours. The high frequency of copy number aberrations is a particular problem in this context because the phenotype is driven by the expression level of a gene, i.e. the amount of mRNA and protein made, which may not directly reflect its copy number. The application of an algorithm that integrates copy number and gene expression levels (CONEXIC) to a sample of melanomas has identified known drivers and has also revealed two novel genes necessary for melanoma growth.

In a similar approach combining DNA sequencing, gene expression (mRNA) profiling and functional proteomics, the most probable drivers of breast and colorectal tumours have been resolved and are represented by the interconnecting pathways shown in Fig. 8.7.

These studies revealed that breast and colorectal cancers typically have point mutations that alter about 80 genes and have major copy number changes (either deletion or

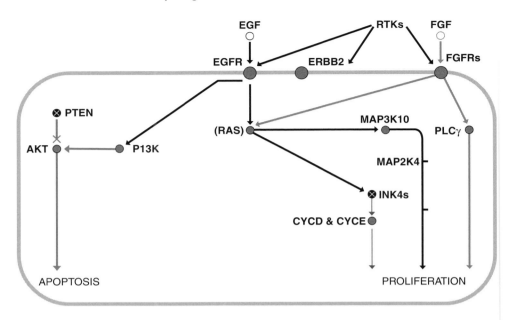

Figure 8.7 Systems biology applied to genomic and proteomic data. The method is designed to define drivers of individual cancers, permitting rational therapeutic design. An analysis of breast and colorectal cancers revealed alterations in the central signalling pathways emanating from FGFR1, FGFR2 (see Fig. 4.10) and the EGFR. Mutations in each of those receptors as well as in ERBB2 were identified in both types of cancer. In addition mutations in PI3K and PTEN were present in both types of cancer and of AKT in breast cancers. Mutations in RAS (shown in brackets) were not present in these tumours. Other alterations included mutations in MAP3Ks and MAP2Ks in colorectal tumours and in PLCγ (Fig. 6.4) in a breast tumour. (Gonzalez–Angulo *et al.*, 2010; Leary *et al.*, 2008).

at least 12 copies) in 17 genes. Predictably, many of the affected genes are those we have encountered as major oncogenes or tumour suppressors. A notable finding, consistent with some of the whole genome sequencing data summarised earlier, is that the majority can be grouped into functional pathways, for example, those controlling proliferation and survival, the cell cycle and cell adhesion (Fig. 8.8). Within one pathway specific genes may be deleted or amplified, indicating that anomalous signalling can arise from aberrations in either positive or negative regulators. Thus, for example, point mutations and copy number changes occurred in pathways driven by EGFR and ERBB2 and involving PI3K in both types of tumour. Point mutations and homozygous deletions affecting *P53*, *SMAD2*, *SMAD3* and *PTEN* characterised colon cancers, together with amplification of *MYC* and *EGFR*. Breast tumours showed homozygous deletions of INK4 genes (*CDKN2A* and *CDKN2B*) and amplifications of cyclins D1, D3 and E3. Additional genes not previously known to undergo copy number change in these tumours were also resolved.

Although the 'cancer landscape' that emerges from such studies is complex, the prominent genes and pathways represent major targets for therapeutic intervention.

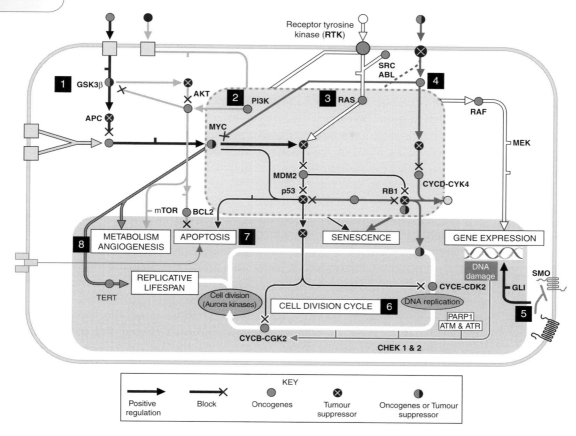

Figure 8.8 Major cancer signalling pathways showing proteins for which there are specific inhibitors. 1. WNT/Notch. 2. PI3K/AKT. 3. RAS. 4. TGFβ 5. Hedgehog. 6. Cell cycle. 7. Apoptosis. 8. Angiogenesis. In one example of the impact of whole genome sequencing 33 mutations were identified in pancreatic carcinomas distributed over six intracellular signalling pathways such that each of the pathways was affected in all of the 24 tumours examined (Jones *et al.*, 2008).

A combination of drugs capable of inhibiting all or most of these might produce effective remission, even allowing for the mutational fluidity of heterogeneous tumours described in Chapter 5.

Genome-wide association (GWA) studies and inherited cancer predisposition

As we noted in Chapter 4, about 10% to 15% of cancers are hereditary, depending on the type of cancer. Thus, for example, approximately 10% of breast cancers and ~35% of colorectal cancers arise from an inherited susceptibility. *BRCA1* and *BRCA2* are two familiar major susceptibility genes for breast cancer: inherited mutations in either conferring a high risk of developing the disease. Variants in some other identified genes also increase the risk but, taken together, these known mutations account for less than 25% of the overall familial risk of breast cancer. High-risk germ line mutations for colon cancer occur in *APC* and in mismatch repair genes (*MLH1* and *MSH2*) but, as with breast cancer, the known loci are responsible for only a small proportion (<6%) of cases. These and similar observations have led to the conclusion that susceptibility to hereditary cancers comes mainly from the combined effects of many loci with each individual variant conferring only a small increase in relative risk. A succession of GWA studies have addressed this problem by comparing the genomic sequences of several thousand individuals with the disease with those of a similar number of normal control individuals so that sufficient statistical power is generated to identify SNPs that predispose to specific cancers. These have typically used the Illumina HumanHap550 BeadChip Array (e.g. the Colorectal Tumour Gene Identification (CoRGI) Consortium and the Breast Cancer Association Consortium). This approach has identified alleles that confer low-penetrance susceptibility to most major cancers (breast, lung, ovarian, pancreatic, prostate, testicular germ cell, thyroid, urinary bladder and colorectal cancers, glioma, follicular lymphoma, neuroblastoma, childhood ALL, CLL and melanoma) and to a variety of other diseases (Alzheimer's disease, Crohn's disease, amyotrophic lateral sclerosis, atrial fibrillation, bipolar disorder, coronary artery disease, rheumatoid arthritis, diabetes). For breast cancer GWA methods have identified SNPs (or at least **haplotypes**) in four genes (*FGFR2*, *TNRC9*, *MAP3K1* and *LSP1*) that account for 3.6% of the familial risk in a first step to identifying the complete complement of low-penetrance variants that can contribute to breast cancer.

The identification of cancer-associated SNPs raises the problem of determining their function prior to developing therapeutic strategies. For *FGFR2* the two SNPs in intron 2 mentioned earlier have been shown to alter the binding affinity of transcription factors (OCT1/RUNX2 and C/EBPβ). The result is an increase in *FGFR2* expression, thereby increasing the risk of developing breast cancer. Despite this finding, determining the functional effect of SNPs in general is technically very challenging.

The facility with which individual genomes can now be analysed also permits the identification of mutations responsible for relatively rare diseases and the genomes of a number of individuals with a variety of genetic diseases have already been sequenced (familial hypercholesterolemia, familial neuropathy, etc.).

Therapeutic strategies for cancer driven by the sequencing revolution

Whole genome sequencing of individual tumours has typically revealed thousands of somatic mutations, for example more than 30,000 in a metastasis of malignant melanoma and, as noted earlier, in AML. These mutations affect, as expected, major cancer genes (i.e. 'drivers', e.g. *MYC*). The majority are clearly 'passengers' but, from analysis of such data, the following points have already emerged for a variety of cancers: (1) although the major mutations are in genes already known to be 'drivers', mutations are also being detected in genes not previously known to be cancer associated; (2) this type of screen has shown that what is often critical is aberrant signalling in a pathway, which can be caused by mutations in a number of different genes – without whole genome sequencing such 'core signalling pathways' would not be identified; and (3) tumours sub-classified on the basis of mutational signature may respond differently to treatment. Even with currently available drugs, the latter is important because that information could spare patients from chemotherapy treatments that won't work.

As this approach is extended to all the major cancers it should greatly increase the range of diagnostic and predictive biomarkers. We have already seen an example of genomics providing biomarkers through the application of gene expression profiling to breast cancer. However, WGS has advantages over expression profiling in that it is easier to obtain DNA than RNA and the complete sequence reveals all mutations. This includes amplifications, translocations, tandem duplications and copy number variations, the extent of which is only just beginning to be recognised.

Tailoring therapy on the basis of whole genome sequencing

High throughput sequencing has already made important contributions to cancer diagnosis and prognosis, and as part of that identified novel cancer genes. The capacity to identify rapidly not only the genes but the precise mutations that they carry is critical in the rational design of therapy regimes. This point is exemplified by the EGFR inhibitor erlotinib, which has been approved for the treatment of non-small cell lung carcinoma. However, the efficacy of the drug depends on the precise EGFR mutation. Patients with the exon 19 mutation have a significantly better response to this drug than those with other EGFR mutations. Thus an important outcome of WGS will be the acquisition of biomarkers predictive of response so that individual tumour genotypes can be matched with appropriate targeted agents to maximise patient benefit and minimise pointless treatment.

The identification of novel cancer genes will prompt efforts to design specific therapies for each that emerges. For the products of oncogenes – the majority of 'cancer genes' – that means inhibitors and indeed the vast majority of targeted chemotherapy focuses on knocking out oncoproteins. The re-activation of tumour suppressor genes is a much more intractable problem and, apart from gene therapy referred to earlier, can only be approached if the gene product is part of a combination that shows **synthetic lethality** such that inactivation of the other component causes cell death. An example of the potential of this approach is the anti-tumour activity of olaparib, an inhibitor of poly (ADP-ribose) polymerase (PARP) in preliminary trials on patients with BRCA1 or BRCA2 mutations.

In the previous chapter we reviewed the general field of chemotherapy and discussed major examples of 'targeted' agents, that is, those with high specificity. We now briefly consider four examples to illustrate current strategies aimed at targeting key cancer drivers.

BRAF

The identification of oncogenic BRAF as a major driver in melanoma has led to the introduction of several small molecule inhibitors that preferentially block ATP binding to the mutant form (BRAF^{V600E}) of the protein. Although some of these have given remarkably high response rates in terms of initial tumour regression (Fig. 8.9), most patients develop resistance to the drugs within about a year. An astonishing range of molecular mechanisms has been unveiled by which resistance to these agents can arise. They fall into two broad categories: those that re-activate ERK signalling and those that promote proliferation by ERK-independent means. Stimulation of ERK can occur through *BRAF* amplification (Corcoran *et al.*, 2011), or by mutation of NRAS, RAF1 or the downstream MEK1 (Montagut *et al.*, 2008; Emery *et al.*, 2009; Nazarian *et al.*, 2010; Wagle *et al.*, 2011). In addition, a truncated form of BRAF^{V600E} (p61 BRAF^{V600E}) with a deleted exon causes BRAF to form dimers: this has the effect of blocking inhibition (Poulikakos *et al.*, 2011). MAPK signalling is activated independently of RAF by elevated expression of a distinct member of the MAP3K family (Johannessen *et al.*, 2010) or by PDGFRβ upregulation (Nazarian *et al.*, 2010) or IGF1R/PI3K signalling (Villanueva *et al.*, 2010).

One implication of these effects is that although BRAF inhibitors normally block MAPKK activation, in NRAS mutant cells they can *cause* ERK phosphorylation. This is

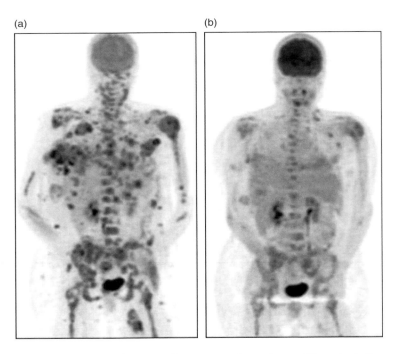

Figure 8.9 PET imaging of melanoma. Two scans of a patient with metastatic melanoma (a) before and (b) after treatment with the BRAF inhibitor PLX4032 (vemurafenib) showing a 70% reduction in metastatic burden (Bollag *et al.*, 2010).

Tailoring therapy on the basis of whole genome sequencing

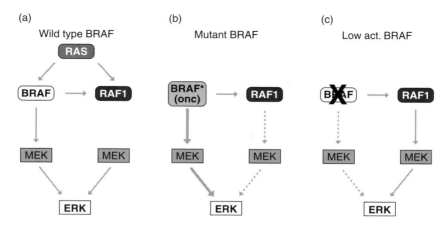

Figure 8.10 RAF signalling in normal and cancer cells. (a) Activation of BRAF-RAF1 binding and of the RAF-MAPK pathway by normal RAS; (b) mutant BRAF binds to RAF1 and activates the pathway independently of RAS; (c) complex formation between inhibited BRAF and RAF1, which can facilitate pathway activation by oncogenic RAS.

important because BRAF is mutated to a constitutively active kinase in ~50% of melanomas but in a further 20% the driving mutation is in NRAS. RAF inhibitors induce BRAF/RAF1 binding to mutant RAS, thus activating MAPKK – the inhibited BRAF acting as a scaffold. Thus, anti-BRAF drugs can, if secondary RAS mutations are acquired, become tumour promoters. In other words, depending on the cellular context, RAF inhibitors can be very effective anti-tumour agents or they can actually enhance tumour growth (Fig. 8.10). These extraordinary effects occur because RAF proteins themselves form dimers as well as binding to RAS proteins and, in melanomas, resistance to inhibition of mutant BRAF develops because the cells can switch to either of the other isoforms (ARAF or RAF1) to activate the MAPK pathway.

This diverse panoply of evasion strategies is a striking demonstration of genomic flexibility and a stark warning of the difficulties that can confront drug therapies for cancer. For our other examples we turn from the major melanoma oncoprotein to three drivers that play critical roles in many cancers, PI3K, MYC and p53.

PI3K

In the context of cancer it is more appropriate to think of PI3K signalling as a super-highway than a pathway. It is hyper-activated in many different types of human cancer and most of its components are vulnerable to mutation. The catalytic subunit of PI3K (p110α, PIK3CA) is mutated in 15% of human cancers, making it the most frequently altered kinase. PTEN, the phosphatase that reverses the effect of PI3K, is one of the most commonly inactivated tumour suppressor genes (see Fig. 6.5). Furthermore, PI3K signalling is anomalously activated by mutations in RTKs and RAS, even without mutation in the pathway itself.

A number of PI3K inhibitors are available but the pathway has diverse functions, contributed by there being four isoforms of the PI3K enzyme (α, β, δ and γ), and additional drugs are required. The information network that we discussed in Chapter 6 suggests that greater efficacy will come from combinations of therapeutics that target more than one pathway simultaneously. Some evidence for this has come by using PI3K inhibitors with the anti-ERBB2 (RTK) antibody trastuzumab or with BRAF inhibitors. Such strategies have been called 'horizontal' inhibition (i.e. of parallel pathways) in contrast to 'vertical' inhibition (i.e. of more than one protein in a single pathway).

MYC

Increased expression of MYC is a key 'driver' in most human cancers and for that reason an inhibitor of MYC is an attractive therapeutic concept for cancer. However, because MYC is essential for normal cells to divide, the notion of blocking its function in a whole animal carries obvious risks. Despite that reservation, preliminary evidence from mouse models gives grounds for optimism. Omomyc is a small, synthetic protein that blocks the interaction of MYC with its partner MAX. MYC–MAX dimers are the functional form in which MYC acts as a transcription factor to drive cell proliferation (Chapter 3). In a mouse model of lung cancer that closely resembles the human disease, Omomyc inhibition of MYC causes rapid regression of established lung tumours (Fig. 8.11). The same strategy of inhibiting the action of endogenous MYC also causes

(a)   (b)   (c)

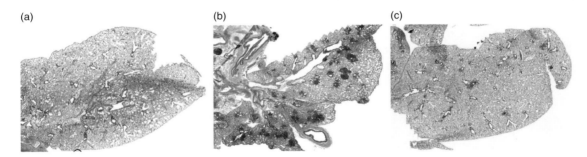

Figure 8.11 **Lung tumour regression induced by inhibition of MYC.** In this transgenic mouse model lung adenocarcinoma develops as a result of expression of oncogenic *Kras*. Histological staining of lung tissue shows (a) control lungs from transgenic mice in which *Kras* was never activated; (b) extensive tumour lesions after 18 weeks expression of oncogenic *Kras*; and (c) clearance of tumour lesions after systemic induction of the MYC inhibitor Omomyc for seven days, resulting from blockade of endogenous MYC activity. (Omomyc dimerises with MYC to prevent MYC-MAX interaction: MYC-Omomyc heterodimers cannot bind to E-box consensus recognition elements, so Omomyc blocks MYC-dependent transcriptional activation (Chapter 3).) Transgenic Omomyc expression is regulated by a tetracycline-responsive promoter element that is activated by a reverse tetracycline-controlled *trans*-activator when doxycycline is administered to the mice. These experiments indicate the feasibility of systemic expression of a dominant-interfering MYC mutant as a cancer therapy. (Images contributed by Gerard Evan and reproduced with permission from Soucek *et al.*, 2008.) (See plate section for colour version of this figure.)

regression of pancreatic tumours, demonstrating that the phenomenon is not confined to one type of cancer. In the latter model the specific expression in beta cells of simian virus 40 (SV40) large T and small t antigens ablates the action of p53 and the retinoblastoma protein and gives rise to highly angiogenic tumours. A prominent effect of MYC suppression is the rapid induction of endothelial cell death leading to collapse of the tumour vasculature. This dramatic perturbation of the microenvironment precedes tumour regression and reflects the role of MYC as a master transcription factor, regulating, *inter alia*, a range of cytokine genes together with *VEGF*, the most potent pro-angiogenic agent.

Other small molecule inhibitors that target *MYC* transcription have been shown to have significant anti-tumour activity against mouse xenografts of several human tumours. The most encouraging and unexpected finding in these mouse models, however, is that blockade of MYC activity over extended periods has no significant, adverse effects on the mice. The strategy of inhibiting this master regulator therefore remains open as a viable therapy.

p53

In Chapters 4 and 6 we saw that the tumour suppressor p53, together with INK4 and RB1, lies at the core of the protective anti-tumour network and when that system is overcome p53 activity is a frequent casualty. So there are two ways of looking at this: (1) most of us don't die from cancer and most who do are pretty old, so this randomly evolved protection system works rather well; or (2) because our protection system is the product of unintelligent design, we ought to be able to do better. We have mentioned one approach to 'p53 therapy' that attempts to target selectively cells in which it has been lost. Perhaps an alternative is to consider re-progamming the genome by, for example, boosting our INK4A or p53 stock? Studies of mice indicate that we should: transgenic mice engineered to express three normal copies of either the *P53* or the *Ink4a/Arf* genes (rather than the usual two) are significantly protected from cancer (Fig. 8.12). Given that p53 can drive senescence, you might wonder whether mice with an extra *P53* gene age prematurely but, so long as the extra *P53* is under normal control, they don't and seem otherwise fine. What's more, in mouse tumours generated by knocking out *P53*, the restoration of *P53* expression can cause tumours to regress and revert to a senescent form. The tumour cells actually get broken down and so the tumours disappear. Thus modest increases in the activity of *P53* or *Ink4a/Arf* confer a beneficial, cancer-resistant phenotype without affecting normal viability or ageing.

Other mouse models have shown that pre-malignant tumours contain senescent cells but malignant ones do not, and the restoration of p53 expression, again transgenically, can cause tumour regression and re-expression of senescence markers. These results suggest that, in principle, re-programming tumour suppressors may in the long term be a viable form of gene therapy. However, there is presently no therapeutic method corresponding to the introduction of transgenes that can be used in humans. Furthermore, even if this obstacle can be overcome, the complexity of *P53* in particular (e.g. the oncogenic activity of some p53 isoforms) may undermine the approach.

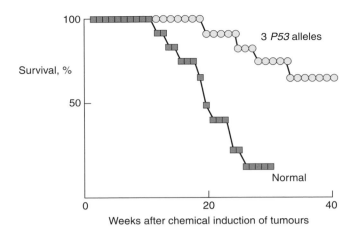

Figure 8.12 The protective effect of a third *P53* allele in transgenic mice treated with a chemical carcinogen. Similar protection is afforded by a third *Ink4a/Arf* allele (Garcia-Cao *et al.*, 2002; Matheu *et al.*, 2004).

Tumour biomarkers

Biomarkers are defined by the FDA as 'A characteristic that is objectively measured and evaluated as an indicator of normal biologic processes, pathogenic processes, or pharmacologic responses to a therapeutic intervention'. The leading candidates for biomarkers are membrane proteins, secreted proteins or extracellular matrix proteins that are more likely to be released into the circulation than intracellular proteins. Perhaps the most familiar tumour biomarker is prostate specific antigen (PSA) – a protease also called kallikrein 3. The detection of elevated blood levels of PSA would prompt the recommendation of a prostate biopsy. However, seven in ten men with raised PSA levels will not have cancer and more than one in five of those with prostate cancer will have normal circulating levels of PSA.

A second example is the glycoprotein mucin 16, otherwise known as cancer antigen 125 (CA125), a trans-membrane glycoprotein thought to mediate adhesion. It is normally expressed in a variety of cell types but is up-regulated in ovarian cancer cells. Blood levels of CA125 are raised in about 50% of these cancers at an early stage and in about 90% of advanced cases. Decrease in the circulating level of CA125 has been associated with the response of ovarian cancers to treatment.

These examples illustrate the current cancer predicament over biomarkers. The PSA test has FDA approval but the high false-positive and false-negative rates indicate its limitations. Similarly, CA125 has approval for monitoring the response to treatment of ovarian cancers but is severely limited as a diagnostic assay. In pursuit of potential markers for the major cancers hundreds of compounds have been investigated and occasionally something of promise emerges. Thus, for example, the FDA has recently approved epididymal protein 4 (encoded by *WFDC2* and also known as HE4) both for detecting stage 1 ovarian cancer and for monitoring treatment response. A broader approach is to identify combinations of genes (say up to five showing coherent responses) that collectively provide a more sensitive indicator than single biomarkers. Nevertheless, there are currently no serum biomarkers for cancer that approach an ideal specification and optimism that this position will change rests heavily on the development of metabolomic analyses (Chapter 7).

An alternative approach to tumour biomarkers is through genomic analyses that, in the first instance, have been used to quantify tumour burden. Thus, for example, the levels of BCR-ABL1 mRNA in blood from CML patients can be used to follow responses to imatinib treatment. A decrease of 1,000-fold in response to therapy is associated with prolonged progression-free survival and these findings indicate the potential for this type of approach both to monitor response and to provide an early indication of relapse.

The detection of transcripts of the *BCR-ABL1* translocation is, of course, a signature of some forms of leukaemia. In addition to leukaemic cells, circulating tumour cells (CTCs) – cells that have detached from a primary tumour and entered the bloodstream – can also be detected in blood, as can free, tumour-derived nucleic acids from a range of solid cancers. These have revealed a variety of genetic alterations including point mutations (e.g. in KRAS, MYC and EGFR) and epigenetic changes. The level of tumour-derived DNA may be as little as 0.01% of the total of normal fragments in the circulation but, if mutations specific to a primary tumour are known, mutant DNA in plasma samples may be used to follow response to therapy. Because most solid tumours carry gene rearrangements, resembling those driving leukaemias, that are specific to each tumour, they can be identified by whole genome sequencing. The use of the polymerase chain reaction with primers spanning the breakpoints provides an exquisitely sensitive method for the detection of these 'personalised biomarkers' that reflect tumour burden and hence patient response to therapy. These methods are currently unable to provide initial tumour detection screens but this limitation may be relieved as mutation signatures for at least the major sub-types of cancers are defined and become utilised for individual screening.

Microfluidics and the isolation of circulating tumour cells

One of the most exciting recent developments has been the application of silicon chip technology to the detection of CTCs (Fig. 8.13). The chips, which are the size of a microscope slide, have about 80,000 microscopic columns etched on their surface that are coated with an array of antibodies that recognise antigens expressed on the surface of epithelial cells. CTCs are present in tiny amounts in circulating blood – one in a billion normal cells – but the combination of antibody specificity and the flow properties of the cell are such that about 100 CTCs can be captured from a teaspoon (5 ml) of blood.

This remarkable technology may offer both the most promising way to early tumour detection and of determining responses to chemotherapy. It also provides a bridge between proteomic and genomic technologies because DNA, extracted from the

→ Identification of CTCs

→ Genomic DNA extraction and sequencing

Figure 8.13 Tumour cell isolation from whole blood by a CTC-chip. Whole blood is circulated through a flow cell containing the capture columns (Sequist *et al.*, 2009).

captured cells, can be used for whole genome sequencing. Thus, for example, specific EGFR mutations, confirmed by sequencing tumour biopsies, have been detected in DNA isolated from CTC-chips in over 90% of one set of samples. If this system is able to capture cells from most major types of tumour it provides a rapid route from early detection through genomic analysis to tailored chemotherapy without the requirement for tumour biopsies.

Drug development

The upshot of the sequencing revolution is that we will soon have a complete view of the cancer scene in that essentially all mutations will be known. This will shift the focus to the problem of developing treatments: how to produce agents that can specifically target dominant oncoproteins and how to re-activate lost tumour suppressors. For the first category another revolution is underway. It is now possible to generate huge chemical libraries of hundreds of thousands of related compounds. In addition, high throughput screening systems now mean that the time from making a compound to showing that it has potency has been reduced.

Hitherto a major problem in identifying potential anti-cancer agents has been the absence of effective methods for their identification without carrying out expensive and time-consuming tests in animals. One solution is emerging in a collaboration between the Cancer Genome Project at the Wellcome Trust Sanger Institute (UK) and the Center for Molecular Therapeutics, Massachusetts General Hospital Cancer Center (USA). This has established over 1,000 genetically characterised human cancer cell lines that can be used to screen new agents for toxicity and anti-proliferative effects that can be correlated with the genetic profile of individual cell lines. The rationale behind this approach is that, although a cell line will acquire further mutations as it is grown in the lab – so it is not identical to the human tumour of origin – in general a closely similar mutational pattern is retained. This represents a step towards tailoring therapy programmes to genetically defined sub-sets of cancers. Agents that pass this test are still a long way from becoming clinically useful but these technical developments are greatly accelerating the rate at which new drugs are becoming available. This programme has already generated substantial amounts of data that can be viewed at the Catalogue of Somatic Mutations in Cancer (COSMIC) website (www.sanger.ac.uk/genetics/CGP/cosmic/).

Other initiatives have also been established to integrate genomics with drug production and screening, for example, the Cancer Target Discovery and Development Network (CTD2N) of the National Cancer Institute. Progress is illustrated by the relative rapidity with which a specific inhibitor of BRAF became clinically available following the identification of the activating mutation – a period of about eight years by comparison with the decades for drugs targeting BCR-ABL1 or mutant forms of the EGFR. This in turn has been eclipsed by the three years taken for the ALK kinase inhibitor crizotinib to be used for the treatment of non-small cell lung cancer from the discovery of the activating translocation. It is evident therefore that these developments designed to expedite translational research are already beginning to yield results.

At this point we should remind ourselves of a feature of cancer that we have encountered several times in the story: namely that tumours are heterogeneous. That is, not merely is the complete mutational pattern revealed by the whole genome sequence unique to each cancer, but that it represents an average from a mixed population of cells – a reflection of all the clones that comprise that tumour. The implications of this picture are immense because it means that any tumour is made up of groups of genetically distinct cells that are moreover continually evolving, i.e. mutating. Therapeutic strategies are therefore confronting not only mixed targets but moving targets. At first sight the concept of a moving therapeutic target comprising multiple cancer propagating cells is somewhat daunting because, intuitively, it would seem easier to have to eliminate just one set of identical cells. However, all may not be lost if the shifting mutational pattern mainly reflects different components in key signal pathways. If that were true then treatment that neutralised the oncogenicity of a given pathway would work, provided its target was downstream of any activating mutation. In other words, if drug cocktails can be produced that target the critical 'driver' mutations detected in a tumour they may still be sufficiently effective at controlling the disease even though major components of the mutational signature come from different cells – i.e. different clones that comprise the bulk of the tumour.

The expanding field of cancer treatment

We made the observation in Chapter 6 that, despite the massive incidence of cancer cases and resultant deaths, at a cellular level the emergence of a tumour is a very rare event. In our discussion thus far we have focused almost exclusively on the nature of the molecular changes – mutations – that combine, albeit infrequently, to drive cells from normality into cancer. However, an underlying question is also of great interest, namely, in the genetic upheaval associated with producing a tumour cell, what is the critical core of molecular normality that must be retained for the cell to remain viable? That is, in addition to the several hundred genes that can cause tumours by their aberrant action as

Table 8.1 Drug targets that may reflect oncogene addiction or non-oncogene addiction.

Oncogene	Drug
EGFR	erlotinib, gefitinib, cetuximab
ERBB2	trastuzumab
BCR-ABL1	imatinib, nilotinib
BCR-ABL1, SRC family, KIT, PDGFR	dasatinib
BRAF	GDC-0879 and PLX4720
PI3K	BEZ235
RAR, RXR	retinoic acid
Tumour suppressor	
BRCA1	Olaparib, KU0058684
Non-oncogenes	
DNA	5-fluorouracil cisplatin, oxaliplatin
VEGF	bevacizumab (Avastin®)
VEGFR, RAF1, KIT, PDGFRB	sorafenib, sunitinib, SU10944
Topoisomerase 1	irinotecan
DHFR	methotrexate
Mitotic spindle	paclitaxel, vinblastine, vincristine
mTOR	temsirolimus
CD20	rituximab
Immune system	prednisone

oncogenes and tumour suppressors, it is self-evident that there are many normal genes whose expression is essential to support the growth and survival of cancer cells. By definition, therefore, these normal, supporting players form a separate category of potential therapeutic targets. In an extension of the concept of oncogene addiction (Chapter 6), Stephen Elledge and his colleagues have termed this dependence of tumour cells on normal genes that do not undergo mutation 'non-oncogene addiction'.

The notion that, notwithstanding the characteristic genomic instability of most tumour cells, many basic molecular pathways remain functional has been implicit in our discussion of cancer biology and we have already encountered several features of tumour cells that encompass non-oncogene addiction. These include perturbed metabolism, the escape from immune surveillance and adaptation to an hypoxic environment. It also seems probable that, although defective DNA repair systems directly promote genetic instability, there is a pressure on cancer cells to retain a level of genomic maintenance consistent with survival. Table 8.1 lists some of the drug targets mentioned earlier (Chapters 5 and 7), categorised as oncogenes/ tumour suppressors or non-oncogenes. It may be recalled that some cytokines are not readily classifiable in this way, for example IL6 and TGFβ that can promote or inhibit tumour progression, depending on the context. This functional ambiguity may be reflected in the fact that some immunosuppressant drugs are associated with increased cancer risk, including tocilizumab, an inhibitor of IL6 signalling with FDA approval for the treatment of rheumatoid arthritis.

This concept of a trade-off between rampant genetic instability and the retention of intrinsic proliferative capacity is consistent with the finding in a number of major cancer types (including breast, ovarian, stomach and non-small cell lung) that extreme chromosomal instability correlates with a better clinical outcome. In other words, extreme genomic disruption compromises cell viability.

The identification of specific genes upon which survival of a cancer cell depends has only recently become feasible with the development of methods for gene knock-down using small interfering RNA (siRNA) that now permit the individual interrogation of the majority of human genes. That is, the construction of short hairpin RNA (shRNA) libraries has made it possible to screen cell lines to determine the phenotypic effect of loss of function of each gene in turn. A specific application of this approach has focused on the RAS signalling pathway, initially by using human cells carrying an activating KRAS mutation (G13D: Chapter 5). Targeting over 32,000 unique mRNAs, shRNAs were identified that inhibited the activity of the MAPK pathway and displayed synthetic lethality with mutant but not wild-type RAS. From these screens a functionally diverse group of about 100 genes emerged, encoding proteins involved not only in signal transduction, cell proliferation and apoptosis but also in protein and nucleic acid metabolism and intracellular transport. Surprisingly few of these candidate RAS synthetic lethal genes were in the MAPK or PI3K pathways. Genes that were particularly prominent encoded components of the machinery of mitosis and it appears that RAS mutation confers increased mitotic stress. That is, cells with mutant RAS are much more sensitive to inhibition of mitotic spindle function (e.g. by paclitaxel) than are normal cells and, in general, these cells are hypersensitive to mitotic inhibitors (e.g. of the anaphase-promoting complex/cyclosome (APC/C) complex, the E3 ubiquitin ligase that labels cell cycle proteins for degradation by the 26S proteasome, or of the mitotic regulator Polo-like kinase 1 (PLK1)). The effect of inhibition is to cause prometaphase accumulation leading to cell death. A second serine/threonine-protein kinase, STK33, emerged from these RNA silencing screens as being essential for the survival of KRAS-mutant cells. Although STK33 can regulate ribosomal protein S6 kinase (RPS6KB1), an effector of mTOR that can integrate growth factor and nutrient signals and also modulate apoptosis via phosphorylation of BAD, it has not hitherto been associated with RAS signalling. Even more surprisingly, tumour cells with mutant forms of HRAS or NRAS do not show STK33 dependence.

In common with all experimental methods, interference RNA is not without its problems one of which is the variability between shRNAs in the efficiency with which they reduce protein expression. Even so, the power of the approach is indicated by the detection of novel therapeutic targets in RAS mutant cells. A further striking finding is that the expression pattern of genes encoding subunits of APC/C correlates with prognosis in human KRAS-mutant lung tumours while there is no correlation in tumours with normal RAS. Several proteasome inhibitors are under development and, bortezomib (Velcade®) was approved by the FDA in May 2003 for the treatment of multiple myeloma.

Balancing priorities

We have referred to some of the seminal contributions to science by the polymath J. B. S. Haldane, who was a victim of colorectal cancer at the relatively early age of 72. In characteristic fashion, when Haldane knew he was dying, he penned a witty poem on the subject of his nemesis in which he noted that 'Yet, thanks to modern surgeons' skills, It can be killed before it kills, Upon a scientific basis, In nineteen out of twenty cases.' His main point was to encourage people to consult a doctor early at the first sign of cancer.

In the 50 years since Haldane's death, the field of cancer has been transformed. Nevertheless, his advice is as valid as ever because early detection and surgery remains the first and most effective treatment for many cancers. This in turn suggests that mass screening programmes for major cancers can only be beneficial. It has transpired though that screening programmes have inherent deficiencies that present problems when it comes to making an overall cost versus benefit assessment. These have become particularly prominent in the context of breast cancer. The plus side of the balance sheet is that screening saves lives when early detection permits appropriate treatment. On the negative side, four factors make a contribution. (1) Any screening system will miss some tumours that should be treated, that is, it will give false negatives. For breast screening about 10% of tumours are missed. (2) Conversely, there will be a false positive rate, typically 7%, most of which (>80%) have a benign cause. (3) Tumours are identified and treated, even though they would not have developed to become life-threatening during the normal lifespan of the patient – an occurrence referred to as over-diagnosis. (4) Finally, although the radiation exposure required for an X-ray or CT scan is low it is nevertheless weakly carcinogenic. Collectively these factors cause patient distress, increase the cost and lead to unnecessary treatment.

The problem is one of risk assessment and judgement – the probability of saving a life compared with the overall risk of doing harm. The US Preventive Services Task Force has reviewed eight large trials and concluded that the benefits of breast cancer screening were so marginal that it recommended reducing the US screening pro-gramme from annually for women over the age of 40 to biannually commencing at age 50. In other words concluding that benefit is negligible for women under 50. A corresponding analysis concluded that the Danish screening programme had con-ferred no significant benefit. The suggestion of the US Task Force has not been implemented but it brings into sharp focus the question of how best to use resources that will inevitably be limited.

To illustrate this problem, consider the Cochrane Collaboration report of 2009 concluding that mammography brought about an overall reduction in breast cancer deaths of 15%. This translates to one life prolonged from every 2,000 women screened over a ten-year period but in that time there will be ten healthy women who have undergone unnecessary treatment as a result of over-diagnosis. These findings have been predictably controversial because they rely on the estimation of risk ratios that are small and vary between different studies. Overall, however, it seems reasonable to conclude that, in the UK, breast cancer screening has played a similar role to the various chemotherapy advances summarised earlier. Each has made small but significant contributions to the national trend of survival rate. Similar concerns have also been raised over other screening programmes, notably that for prostate cancer, and we have already mentioned the shortcomings of the PSA test for this disease. Much less controversial has been the highly effective screening method for cervical carcinoma. Because this detects *pre*-malignant lesions that can be treated, it has led to a progressive decline in UK deaths from this disease over the last 40 years to the current level of about 1,000.

Despite the statistical variation, the evidence indicates that, although some individuals will undoubtedly benefit from screening programmes, there are substantial drawbacks and that these might be made clearer to the general public. In the end the question revolves around the best use of financial resources in circumstances that are in continuous flux as technologies develop. For example, over the last ten years the number of metabolites identifiable by the metabolomics methods summarised in the previous chapter has risen from 30 to over 1,500. While very few of these are likely to emerge as useful in cancer diagnosis, due to the broad range of natural variation, it nevertheless seems reasonable to believe that an extended repertoire of tumour biomarkers, particularly those provided by genomic analysis, will progressively improve screening quality, so that we might anticipate that the shortcomings in such programmes will diminish as the technology is refined.

In this review of cancer biology we have encountered remarkable technical developments that cover improvements in screening methods, a progressively increasing range of biomarkers with enhanced sensitivity of detection, the complete identification of the mutational profile of individual tumours and the identification of the pattern of functionally critical non-oncogenes that characterise specific types of tumours. To these advances is coupled the promise of being able to monitor tumour response to treatment from the earliest stages by the essentially non-invasive techniques of imaging and/or genomics. Together with developments in drug design and testing that are already accelerating the rate at which treatments progress to clinical application, these prospects represent the spectacular scale upon which cancer can now be tackled at the molecular level – a breadth of strategies of immense potential that would have been almost inconceivable just a few years ago. As we have noted, the success of these approaches depends critically on the capacity to develop drugs and therapeutic methods to exploit the targets. The fact that resources are limited means that the cost of screening will have to be balanced against the expansion of drug development and testing as one of the anticipated benefits from WGS. This implies that diagnosis will become a more precise science and offers the hope that specific chemotherapy could become the first line of defence, assuming an adequate supply of new drugs of appropriate specificity and with acceptable side effects. That desirable goal will not, however, remove the need for continued public debate on how to apportion cancer funding, a discussion in which clear and informed input from the scientific community will be an essential contribution.

Conclusions

The completion of the sequence of the human genome in 2003 was one of the great milestones in scientific history. It made an almost immediate impact on cancer with the discovery of a major mutation in melanoma that rapidly led to a new drug treatment. Equally dramatic has been the advances in technology that now mean complete genomes can be sequenced at low cost in a day or so. This is generating an avalanche of data from which has already emerged not only new cancer genes but the capacity to sub-classify cancers using molecular signatures. We are still at a very early stage in the genomic revolution but it has already had direct effects on patient treatment. As the application of genomics and associated high-throughput screening methods gathers pace breathtaking horizons are opening before us. The possibility of using chips to isolate single tumour cells from the circulation that can then be completely sequenced offers the possibility of biomarker detection at the earliest stages of cancer progression. If the massive screening programmes being applied to drug identification can fulfil their promise, the dream that chemotherapy might supplant surgery as the most effective first-line treatment for cancer may yet be realised.

Key points

- Since the human genome was sequenced in 2003 advances in sequencing technology have initiated a revolution in medicine.
- The advances have reduced the time required for complete sequencing of individual genomes to about a day and selective sequencing of specific regions, known as 'genomic partitioning', can reduce the cost and increase yet further the speed with which diagnostic data can be obtained.
- Coupled with gene expression profiling and proteomic analysis, comprehensive pictures of individual cancers can now be obtained.
- These are revealing both novel cancer genes and signalling pathways that are consistently aberrant in specific types of tumour even though the mutated genes differ.
- These methods are also driving the application of silicon chip technology to the detection of circulating tumour cells, offering the possibility that the mutational signature of a tumour could be obtained by an essentially non-invasive method long before any overt manifestation of cancer.
- The ultimate aim of these methods is to identify novel drug targets and to use the molecular profile of individual cancers to prescribe the optimal treatment.
- Implicit in these developments is the possibility that early identification and profiling of incipient cancers may eventually replace surgery as the first and most effective defence.

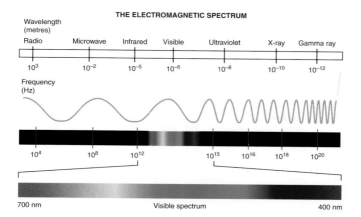

Figure 2.1 The electromagnetic spectrum. The wavy line shows the relationship between frequency (v) and wavelength (λ). One complete oscillation of the field (called the period) is the distance between two adjacent peaks. The wavelength of a complete cycle is related to the frequency by the speed of light, c ($\lambda = c/v$). Ultraviolet (UV) radiation (i.e. 'beyond' violet light in the sense of being of shorter wavelength and hence higher energy) occupies the wavelength bands from 400 to 315 nm (UVA), 315 to 280 nm (UVB) and 280 to 100 nm (UVC). 1 nm (nanometre) $= 10^{-9}$ m.

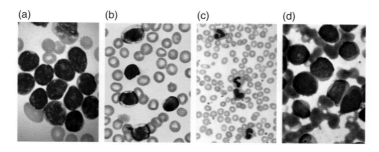

Figure 2.2 Leukaemic cells. Leukaemias arise in the blood-forming tissues, the bone marrow and lymphatic system, giving rise to abnormally high levels of circulating cells. These are usually leucocytes (white blood cells). (a) Acute lymphocytic leukaemia (ALL). The cells stained purple are white blood cells (lymphocytes) at a very immature stage (called blasts): they have very large nuclei and little cytoplasm. The majority of the cells are normal red blood cells. ALL is the most common type of childhood cancer. (b) Chronic lymphocytic leukaemia (CLL). The slide shows relatively high numbers of mature lymphocytes (purple nuclei). CLL most commonly occurs in older adults. (c) Chronic myeloid leukaemia (CML): marked leucocytosis (i.e. raised white blood cell count) with increased numbers of precursor neutrophils. (d) Acute myeloblastic leukaemia (AML): a bone marrow smear showing a high proportion of blast cells (immature precursors of granulocytes) of large size and with visible nucleoli. Acute erythroblastic leukaemia is a variant of AML in which high levels of immature red blood cells are produced. (Source: http://commons.wikimedia.org/wiki/)

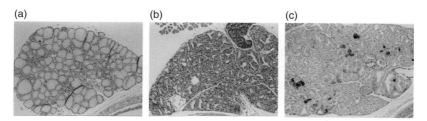

Figure 2.4 A tissue section of a normal and an enlarged thyroid gland (a mouse goitre). Goitres can be caused by low iodine levels and they increase the chance of developing thyroid cancer. (a) and (b) have been stained with haematoxylin and eosin. (a) Normal thyroid architecture with a portion of the trachea. (b) A chemically induced goitre, showing diffuse thyroid hyperplasia with part of the parathyroid. (c) A section of a goitre that has been treated with an anti-vascular drug, which has caused localised damage manifested as multiple thrombi that stain positive for von Willebrand factor (brown). (Griggs *et al.*, 2001.)

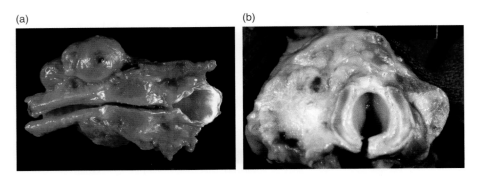

Figure 2.3 The thyroid gland. The normal thyroid gland has two small lobes that wrap around the trachea. (a) Shows an opened trachea with a goitre, an enlargement of the thyroid most frequently caused by iodine deficiency. (b) Shows a cross-section through the thyroid and trachea of a metastatic **carcinoma**. (© Dr Peter Anderson, University of Alabama at Birmingham, Department of Pathology.)

(c)

(d)

(e)

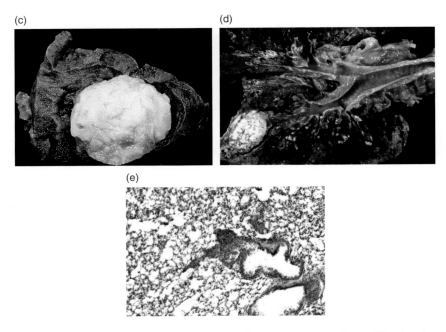

Figure 2.5 Lung tumours. (c) A pulmonary hamartoma, the most common form of **benign** lung neoplasm. The yellow and white tissue of the well-circumscribed hamartoma is fat and cartilage, respectively. (d) Lung primary carcinoma (at base of left lower lobe). Many tumours metastasise to the lung and secondary growths are more common than primary lung tumours. (e) Lung section stained with haematoxylin and eosin (H & E) showing numerous alveoli and two bronchioles. In the centre is a nest of tumour cells that have metastasised from a blood vessel. These have the characteristic dark purple colour of hyperchromatic cancer cells stained with H & E. (© University of Alabama at Birmingham, Department of Pathology.)

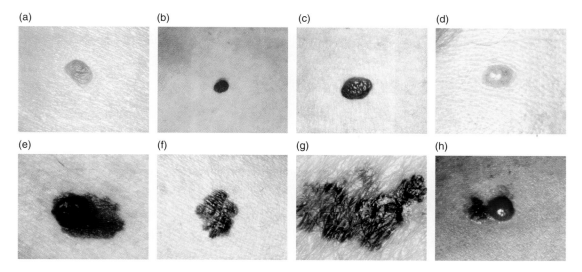

Figure 2.6 Moles and melanomas. (a) to (d) Four normal moles. These are benign birthmarks, a category that also includes café au lait spots, light brown areas that may be associated with the genetic disorder neurofibromatosis type I. (e) to (h) Melanomas ((e) to (g)) and an advanced malignant melanoma (h). Melanoma is a malignant tumour of melanocytes and usually arises in moles or in skin of normal appearance. The most common type of skin cancers are basal cell carcinomas, one of the two major groups of non-melanoma skin cancers (the other being squamous cell carcinoma). (Photographs courtesy of the Skin Cancer Foundation with the permission of the National Cancer Institute.)

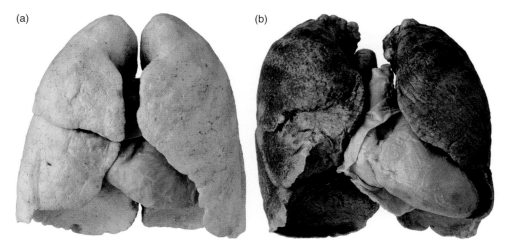

Figure 2.8 The effect of smoking on the lungs. (a) Non-smoker's lungs. (b) Smoker's lungs. (Gunther von Hagens' Bodyworlds, Institute for Plastination, Heidelberg, Germany, www.bodyworlds.com.)

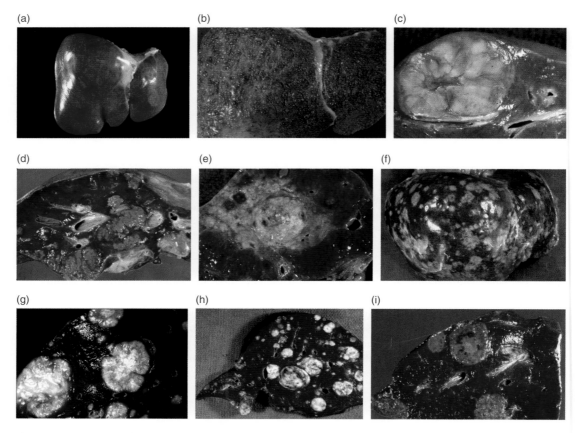

Figure 2.9 Human liver. (a) A normal liver. There are four lobes of unequal size and shape. The gallbladder can also be seen. (www.aafp.org/afp/2006/0901/p756.html) (b) Cirrhosis: external view of micronodular cirrhosis. The result of damage, this condition causes hardening of the organ due to the formation of non-functional scar tissue and surface nodules, here ~3 mm in diameter. The most common causes of cirrhosis are alcoholism and infection with hepatitis viruses. The condition can lead to liver cancer (hepatocellular carcinoma). (c) Liver adenoma showing the natural colour in a close-up view of a well demarcated lesion. (d) Hepatitis: inflammation of the liver characterised by the presence of inflammatory cells. Regions of atrophy can be seen caused by chronic blood vessel damage. (e) Hepatocellular carcinoma (malignant hepatoma): a primary malignancy of the liver. (f) Metastatic carcinoma from a primary stomach tumour on the liver surface. (g) Metastatic carcinoma showing necrosis in the centre of tumour masses. (h) Metastatic lesions in the liver from a primary stomach tumour ranging in diameter from <1 mm to several cm. (i) Magnified view of metastatic lesions in the liver from a primary tumour. (© University of Alabama at Birmingham, Department of Pathology.)

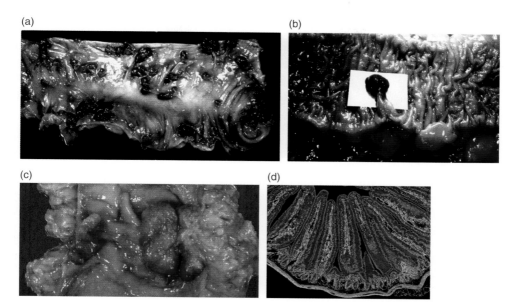

Figure 2.10 Colon cancer. (a) A section of colon with multiple polyps (typically 5 mm in diameter); (b) a colon polyp showing also the convoluted structure of adjacent normal colon; (c) malignant colon carcinoma (~5 × 7 cm). (© University of Alabama at Birmingham, Department of Pathology.) (d) Mouse intestine section imaged with wide-field multi-photon microscopy, showing the villi that provide a large area for digestion and absorption. Actin (green), lamin (red), nuclei (blue). (Thomas Deerinck, NCMIR, UCSD.)

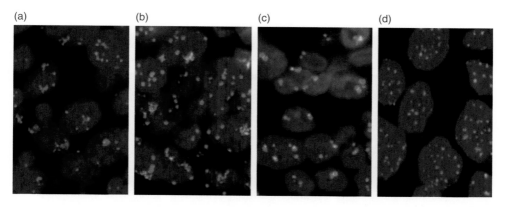

Figure 4.9 EGFR gene amplification in non-small cell lung carcinoma. The EGFR gene copy number was determined by fluorescence *in situ* hybridisation (FISH) assay using two fluorescent probes one targeting the centromere of the chromosome (green), the other the EGFR gene (red). The pictures show four patterns of EGFR gene amplification. (a) Large EGFR gene clusters; (b) co-localised clusters of EGFR and centromere signals; (c) large and bright EGFR signal, larger than the centromere signals in tumour cells; (d) high frequency of balanced EGFR and centromere signals (Varella-Garcia, 2006).

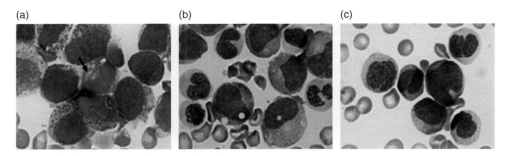

Figure 4.13 Acute promyelocytic leukaemia (APL): induction of differentiation in bone marrow blast cells. (a) No treatment: showing immature promyelocytes with characteristic granularity and Auer rods, elongated needles found in the cytosol of leukaemic blasts (see arrow). (b) Cells from patients treated for three weeks with retinoic acid. (c) Cells from patients treated for three weeks with As_2O_3. In (b) and (c) the granulocytes have started to differentiate and have large areas of clear cytoplasm. Nuclear remodelling is also observed, ranging from a simple indentation to the polylobular nuclei of terminally differentiated granulocytes (Zhu et al., 2002).

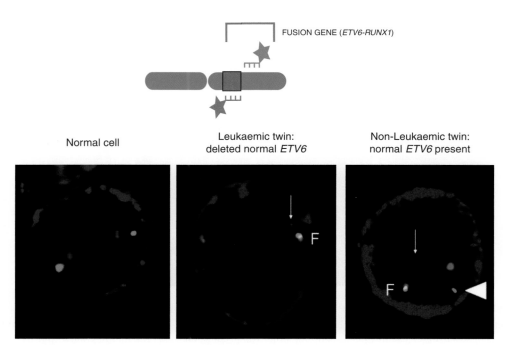

Figure 4.14 The ETV6-RUNX1 translocation in childhood ALL. Fluorescent probes identifying the two genes show them on distinct chromosomes in a normal cell (left) and juxtaposed in cells from the leukaemic twin (centre) and healthy twin (right). ETV6 is green, RUNX1 red and the fusion gene ETV6-RUNX1 is green-red (F). The weak red signals (arrows) are the remnants of the RUNX1 locus left by the translocation. The green signal (open arrowhead) is the intact second ETV6 allele. CD19 (blue) is a surface antigen expressed by these leukaemia cells. (Images contributed by Mel Greaves (Institute of Cancer Research) from Hong et al. (2008).)

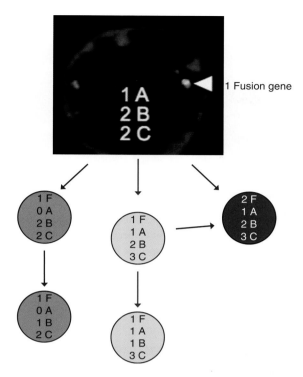

Figure 4.15 **Differing proportions of genetically distinct cells from a patient with leukaemia caused by the** *ETV6-RUNX1* **translocation.** The numbers refer to the copy numbers of the fusion gene *ETV6-RUNX1* (F), *ETV6* (A), *PAX5* (B) and *RUNX1* (C). Image contributed by Mel Greaves (Institute of Cancer Research); figure redrawn from Greaves, 2009.

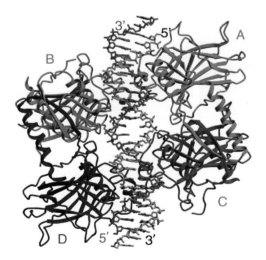

Figure 4.21 **The protein p53 is a transcription factor, binding to DNA.** The figure shows the three-dimensional structure of two p53 dimers bound to a p53-response element. The four p53 molecules (green, yellow, red and blue ribbons) are labelled A to D: in essence they form a clamp around DNA (cyan). The individual p53 molecules bind to grooves in DNA (blue): amino acids in mutational hot spots can directly contact DNA. (From Emamzadah *et al.*, 2011.)

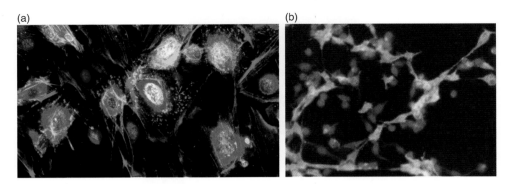

Figure 5.10 Endothelial cells in culture: effect of reduced oxygen. (a) Shows human endothelial cells grown under normoxic conditions (21% oxygen). The cells have been stained with three fluorescent markers (antibodies detecting actin (green) and the endothelial cell-specific von Willibrand factor (red) and the nuclear stain DAPI). (b) The same cells after nine hours under hypoxic conditions (oxygen <2%). These cells have started to form tubular structures as they would in vivo when initiating the growth of new blood vessels. (Photographs by Emily Hayes.)

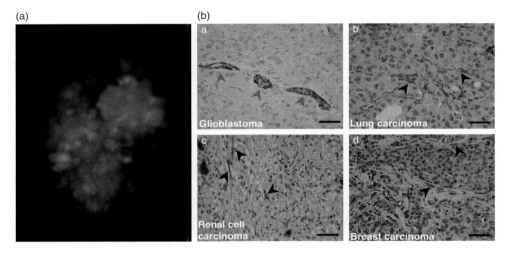

Figure 5.18 Metastatic seeding of tumour cells associated with stromal fibroblasts. Left: clump of cells shed from a kidney tumour isolated from circulating blood. The tumour cells (red) are associated with stromal cells (green). **Right: Clinical evidence for carryover of primary tumour stromal cells in human metastases.** Representative microscopy images: (a) glioblastoma (brain tumour), (b–d) human brain metastases originating from lung carcinoma (b), renal cell carcinoma (c) and breast carcinoma (d). Red arrowheads indicate tumour vessels after α-smooth muscle actin (αSMA)/CD31 double staining. In normal human brain and primary brain tumours only vessel associated pericytes and vascular smooth muscle cells are αSMA-positive. Vascular endothelial cells are CD31-positive. In the metastases, in addition to blood vessel staining, tumour-associated fibroblasts (αSMA-positive) are also present (black arrowheads). Fibroblasts are not present in normal brain tissue and must therefore have come from the original tumour. Scale bars: 50 μm. (Photos contributed by Rakesh Jain and Dan Duda (Duda *et al.*, 2010).)

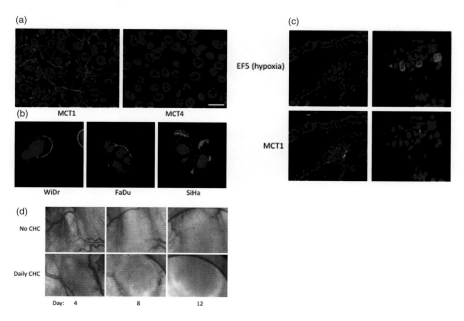

Figure 5.12 Expression and activity of the lactate transporter MCT1. (a) Fluorescent labelling of MCT1 (red) and MCT4 (red) and nuclei (blue) showing that oxidative tumour cells express high levels at the plasma membrane of the protein that carries lactate in (MCT1) and very little that carries it out (MCT4). Scale bar: 20 μm. (b) A variety of oxidative tumour cells express high levels of the protein that carries lactate in (MCT1) and very little that carries it out (MCT4).Fluorescent labelling of three cell lines (WiDr, FaDu and SiHa) shows the plasma membrane location of the lactate transporter MCT1 (red) and nuclei (blue). (c) MCT1 and hypoxia detected in cryoslices of a primary human lung cancer by confocal microscopy. MCT1 staining (green) and the red hypoxia marker hypoxia (2-[2-nitro-1H-imidazol-1-yl]-N-[2,2,3,3,3-pentafluoropropyl] acetamide [EF5] do not overlap. (d) Intravital microscopy shows that sustained MCT1 inhibition completely prevented angiogenesis of Lewis lung carcinoma tumours in mice. The MCT1 inhibitor was α-cyano-4-hydroxycinnamate (CHC). (Images kindly contributed by Dr Pierre Sonveaux, University of Louvain Medical School, reproduced by permission from Sonveaux *et al.*, 2008, 2012.)

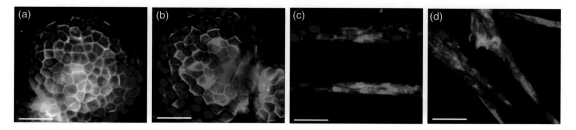

Figure 5.15 Images of the epithelial–mesenchymal transition. RAS-transformed mammary epithelial cells in collagen gel cultures form three-dimensional hollow, alveolar structures of polarised epithelial cells (a and b) expressing E-cadherin and a cortical actin ring but no mesenchymal markers (e.g. vimentin). Addition of TGFβ (6 days) induces these cells to undergo epithelial–mesenchymal transition (EMT), characterised by unorganised structures of fibroblastoid cells expressing cytoplasmic actin and vimentin (c and d) but not E-cadherin. (Wiedemann, IMP)

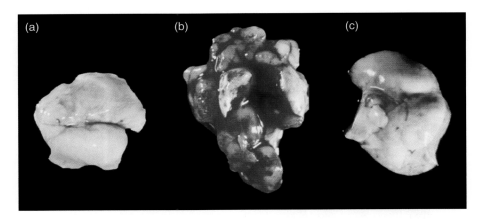

Figure 5.19 Metastatic tumour in mouse lung and the protective effect of combretastatin. (a) Normal mouse lung. (b) Extensive metastases as a result of which the weight of the lung has doubled, resulting in complete respiratory failure. (c) Lung of normal appearance in which metastases have been blocked by the administration of the drug combretastatin to the mouse.

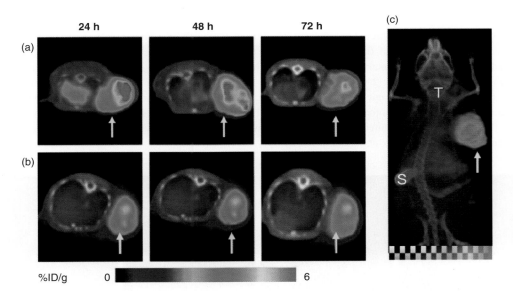

Figure 7.2 SPECT/CT of chemokine receptor 4 (CXCR4) expression in xenografts of a human glioblastoma cell line detected using ^{125}I-labelled antibodies. Severe combined immunodeficient mice bearing subcutaneous tumours were injected with: (a) ^{125}I-anti-CXCR4 antibody or (b) ^{125}I-IgG$_{2A}$: sections through the tumour are shown. (c) Whole mouse imaged 48 h after injection. The images show radioactivity accumulating in the tumours over 48 h and beginning to clear by 72 h. IgG$_{2A}$ (b) is a control that also shows some accumulation. %ID/g = percentage of injected dose per gram of tissue; arrows indicate tumour; S = spleen; T = thyroid. (Nimmagadda *et al.*, 2009.)

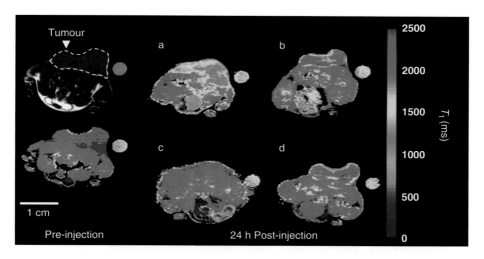

Figure 7.6 Magnetic resonance imaging of the accumulation of a phosphatidylserine-binding contrast agent in the tumours of drug-treated animals 24 hours after injection. The smart contrast agent used was gadolinium conjugated to a protein that binds to phosphatidylserine. The position of the tumour is indicated on the grey-scale image (left). Colour scale indicates T_1 values for image voxels. Drug-treated and untreated tumours are shown in (a) and (c), respectively. (b) and (d) show controls using a non-phosphatidylserine binding reagent, drug-treated and untreated, respectively. The treated tumour (a) shows marked accumulation of the PS-active contrast agent, illustrating the potential of this approach for monitoring responses to therapy (Krishnan *et al.*, 2008).

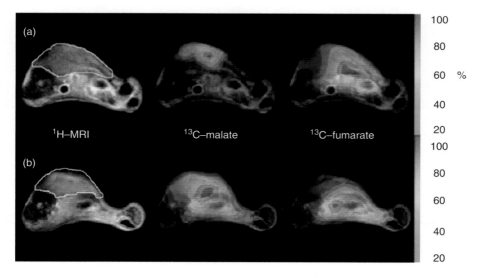

Figure 7.8 Representative transverse images from (a) untreated and (b) etoposide-treated mice with implanted lymphoma tumours. The four false colour images on the right represent hyperpolarised [1,4-^{13}C]malate and [1,4-^{13}C]fumarate following intravenous injection of hyperpolarised fumarate. The increased malate signal in the drug-treated tumour is thought to be due to tumour cell necrosis. Image reproduced by permission of Dr Ferdia Gallagher, Prof. Kevin Brindle and colleagues (Gallagher *et al.*, 2009).

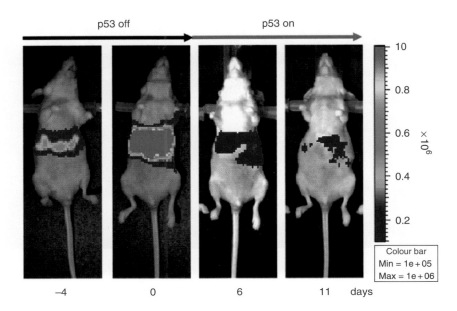

Figure 7.10 Reactivation of p53 in p53-deficient tumours causes regression. In this transgenic mouse experiment bioluminescence imaging monitors the size of the lung tumour that arises in *P53*-negative mice (days −4 to 0) but regresses (days 0 to 11) when p53 expression is activated. (Images contributed by Dr Scott Lowe, Sloan-Kettering Institute (Xue *et al.*, 2007).)

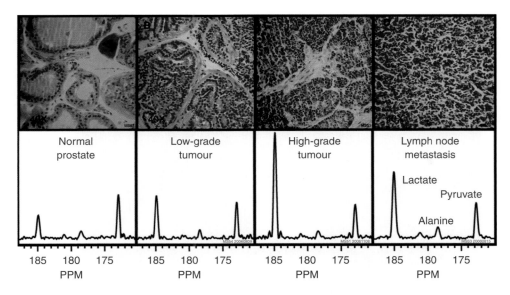

Figure 7.7 Stained sections and hyperpolarised ^{13}C spectra for histologically defined groups of a transgenic mouse model for prostate adenocarcinoma. The hyperpolarised ^{13}C spectra illustrate the strong correlation that exists between the amount of hyperpolarised ^{13}C lactate and the progression of the disease from normal prostate through low-grade primary tumours to high-grade primary tumours (Albers *et al.*, 2008).

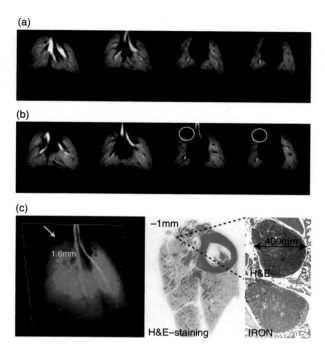

Figure 7.9 Lymph node metastasis of human prostate carcinoma cells in a mouse detected by MRI. This study used luteinising hormone-releasing hormone (LHRH) as the ligand to target tumour cells in lung metastases followed by high-resolution MRI of hyperpolarised helium (^{3}H). (a) No LHRH-SPION injection. (b) LHRH-SPION injection showing a lymph node metastasis (circled). (c) Three-dimensional representation of the lungs showing the site of the metastasis and a magnified view of the excised tissue after staining for iron and with haematoxylin and eosin, showing the high density of SPIONs in the metastasis. Reproduced with permission (Branca *et al.*, 2010).

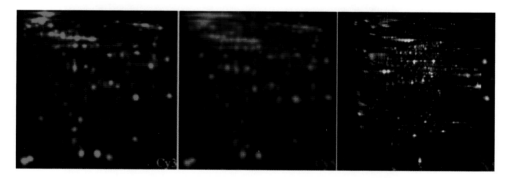

Figure 7.12 An overlay image from a two-dimensional DiGE gel of proteins labelled with two fluorescent cyanine dyes. In this experiment 50 µg of wild-type *Xenopus* embryo lysate was labelled with Cy3, which minimally labels lysine residues. The same amount of protein from a corresponding sample from a mutant strain was labelled with Cy5. Both samples were pooled prior to two-dimensional electrophoresis. Fluorescent images were captured using excitation and emission parameters unique to Cy3 and Cy5. The two resulting images were overlayed and false coloured (Cy3 green, Cy5 red). Proteins present in both samples at similar abundances will directly overlay and appear yellow on this false colour image. The ratio of Cy3:Cy5 intensity for any spot reflects the relative abundance of the protein(s) present in that spot between the two samples. (Image kindly contributed by Kathryn Lilley, Cambridge Centre for Proteomics.)

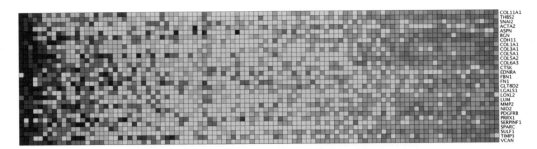

Figure 7.13 Gene expression profiling. This method permits the simultaneous determination of the levels of expression of thousands of genes in a cell sample. Heat maps show the variation in the relative amounts of mRNA between samples for individual genes. Each row represents the expression level of an individual gene: each column is a different sample. In the above example red indicates high and blue reduced expression. One of the first studies to use oligonucleotide microarrays for expression profiling screened 6,827 mRNAs from 72 patients and showed that the patterns of 50 genes permitted differentiation between classes of acute leukaemia – ALL and AML (Golub *et al.*, 1999). Numerous subsequent studies have utilised the power of this technique to classify sub-types of cancer and to identify gene expression 'signatures' that, for example, are prognostic indicators for metastatic development. The heat map shown above is from a recent study in which RNA was profiled from xenografts of individual human cancer cell lines. This identified a signature that is significantly triggered only when a cancer has reached a particular stage of invasiveness. The heat map shown is for colon cancers but the invasion-associated signature occurs in many other human cancers (e.g. breast, lung, ovarian). It includes many genes associated with the epithelial to mesenchymal transition (Chapter 5) including COL11A1 (collagen type XI), SNAI2 (slug), ACTA2 (α-smooth muscle actin), CDH11 (cadherin 11), FN1 (fibronectin 1), PDGFRB (platelet-derived growth factor receptor β). (Image kindly contributed by Dimitris Anastassiou, Columbia University; Anastassiou *et al.*, 2011.)

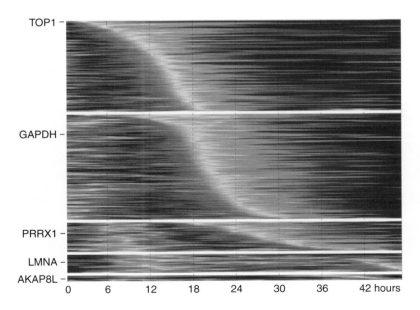

Figure 7.14 Temporal patterns of protein fluorescence intensity in response to a drug. Each row corresponds to one protein averaged over all cells in the movie at each time point. Proteins were clustered according to their dynamics and show waves of accumulation and fall in intensity (Cohen *et al.*, 2008). Among the most rapid responders of the 1,000 tagged proteins was the target of the drug (camptothecin) – topoisomerase-1 (TOP1) that is degraded within a few minutes. Other classes of dynamic behaviour are represented by GAPDH (level increases over 12 h before declining), PRRX1, LMNA and AKAP8L (increases after 30 h).

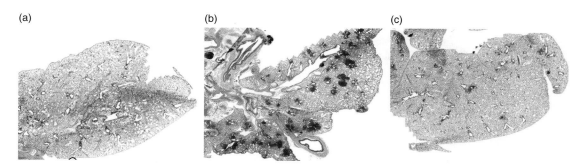

Figure 8.11 Lung tumour regression induced by inhibition of MYC. In this transgenic mouse model lung adenocarcinoma develops as a result of expression of oncogenic *Kras*. Histological staining of lung tissue shows (a) control lungs from transgenic mice in which *Kras* was never activated; (b) extensive tumour lesions after 18 weeks expression of oncogenic *Kras*; and (c) clearance of tumour lesions after systemic induction of the MYC inhibitor Omomyc for seven days, resulting from blockade of endogenous MYC activity. (Omomyc dimerises with MYC to prevent MYC-MAX interaction: MYC-Omomyc heterodimers cannot bind to E-box consensus recognition elements, so Omomyc blocks MYC-dependent transcriptional activation (Chapter 3).) Transgenic Omomyc expression is regulated by a tetracycline-responsive promoter element that is activated by a reverse tetracycline-controlled *trans*-activator when doxycycline is administered to the mice. These experiments indicate the feasibility of systemic expression of a dominant-interfering MYC mutant as a cancer therapy. (Images contributed by Gerard Evan and reproduced with permission from Soucek *et al.*, 2008.)

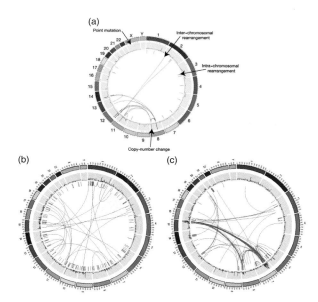

Figure 8.2 Somatic mutations in breast cancer represented as genome-wide Circos plots. (a) The mutational information included in the plots. An ideogram of a normal **karyotype** is shown in the outer ring. The blue lines show copy number variation, green denotes an intra-chromosomal rearrangement and each purple line an inter-chromosomal rearrangement. (b) and (c) show somatic rearrangements in two primary breast cancers. (b) A triple negative tumour (oestrogen receptor (ER^-), progesterone receptor (PR^-) and ERBB2 negative) in which hardly any gene amplification has occurred but there are about 150 tandem duplications.(c) An ER^+, PR^+, $ERBB2^-$ tumour with nearly 200 amplified genes but very few tandem duplications (see Appendix E: Breast cancer). (Reproduced with permission from Stratton *et al.*, 2009 and Stephens *et al.*, 2009.)

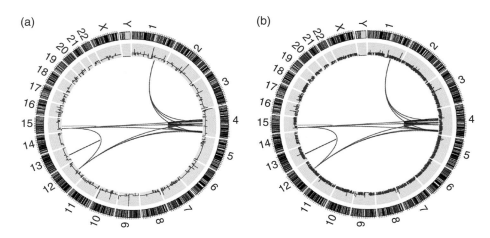

Figure 8.4 Genomic rearrangements localised to chromosome 4q in a chronic lymphocytic leukaemia. The two plots represent the genome-wide profile of rearrangements before (a) and at relapse 31 months after chemotherapy (b). The patterns are not significantly different, indicating that in this cancer no further evolution had occurred (Stephens *et al.*, 2011: Fig. 1c,d).

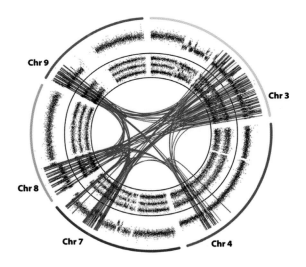

Figure 8.5 Chromothripsis involving five chromosomes. This plot represents 147 rearrangements between chromosomes 3q, 4q, 7q, 8p and 9p that have occurred in a chordoma, a rare type of bone tumour (Stephens *et al.*, 2011: Fig. 4a).

Future directions

- Whole genome sequencing is currently generating a complete picture of mutations in all major cancers. Although there are many drugs in chemotherapeutic use, fewer than 20 target specific proteins. A major requirement is therefore to expedite the flow of novel agents through clinical trials and into general use with the aim of generating cocktails that are tailored to individual tumours on the basis of their mutational profile.

- In the meantime, for clinical purposes sequencing that focuses on specific gene groups (e.g. kinases) or signalling pathways is a more practical use of the technology.

- Despite the present limitations, the range of techniques that is unfolding for the detection, monitoring and treatment of cancers is astonishing and offers the possibility that, within the not too distant future, chemotherapy might replace surgery as the first method of treatment.

Further reading: reviews

HUMAN GENOME SEQUENCING

Branton, D., Deamer, D. W., Marziali, A. *et al.* (2008). The potential and challenges of nanopore sequencing. *Nature Biotechnology* **26**, 1146–53.

England, R. and Pettersson, M. (2005). Pyro Q-CpG™: quantitative analysis of methylation in multiple CpG sites by Pyrosequencing®. *Nature Methods* **2** doi:10.1038/nmeth800.

Pareek, C. S., Rafal, S. and Andrzej, T. (2011). Sequencing technologies and genome sequencing. *Journal of Applied Genetics* **52**, 413–35.

Ryan, D., Rahimi, M., Lund, J., Mehta, R., Parviz, B. A. *et al.* (2007). Toward nanoscale genome sequencing. *Trends in Biotechnology* **25**, 385–9.

Shendure, J. and Ji, H. (2008). Next-generation DNA sequencing. *Nature Biotechnology* **26**, 1135–45.

Turner, E. H., Ng, S. B., Nickerson, D. A. and Shendure, J. (2009). Methods for genomic partitioning. *Annual Review of Genomics and Human Genetics* **10**, 263–84.

WHOLE GENOME SEQUENCING (WGS) AND CANCER

McDermott, U., Downing, J. R. and Stratton, M. R. (2011). Genomics and the continuum of cancer care. *New England Journal of Medicine* **364**, 340–50.

Stratton, M. R., Campbell, P. J. and Futreal, P. A. (2009). The cancer genome. *Nature* **458**, 719–24.

CHROMOTHRIPSIS

Maher, C. A. and Wilson, R. K. (2012). Chromothripsis and human disease: piecing together the shattering process. *Cell* **148**, 29–32.

GENOME-WIDE ASSOCIATION (GWA) STUDIES AND INHERITED CANCER PREDISPOSITION

Freedman, M.L., Monteiro, A.N.A., Gayther, S.A. *et al.* (2011). Principles for the post-GWAS functional characterization of cancer risk loci. *Nature Genetics* 43, 513–18.

BRAF

Downward, J. (2011). Targeting RAF: trials and tribulations. *Nature Medicine* 17, 286–8.

MYC

Sawyers, C.L. (2009). Finding and drugging the vulnerabilities of RAS-dependent cancers. *Cell* 137, 796–8.

Soucek, L., Whitfield, J., Martins, C.P. *et al.* (2008). Modelling Myc inhibition as a cancer therapy. *Nature* 455, 679–83.

Soucek, L. and Evan, G.I. (2010). The ups and downs of Myc biology. *Current Opinion in Genetics & Development* 20, 91–5.

TUMOUR BIOMARKERS

Fleischhacker, M. and Schmidt, B. (2007). Circulating nucleic acids (CNAs) and cancer: a survey. *Biochim Biophys Acta* 1775, 181–232.

Fleischhacker, M. and Schmidt, B. (2010). Free circulating nucleic acids in plasma and serum (CNAPS): useful for the detection of lung cancer patients? *Cancer Biomark* 6, 211–19.

SYSTEMS BIOLOGY

Akavia, U.D., Litvin, O., Kim, J. *et al.* (2010). An integrated approach to uncover drivers of cancer. *Cell* 143, 1005–17.

Gonzalez-Angulo, A.M., Hennessy, B.T.J. and Mills, G.B. (2010). Future of personalized medicine in oncology: a systems biology approach. *Journal of Clinical Oncology* 28, 2777–83.

Leary, R.J., Lin, J.C., Cummins, J. *et al.* (2008). Integrated analysis of homozygous deletions, focal amplifications, and sequence alterations in breast and colorectal cancers. *Proceedings of the National Academy of Sciences of the USA* 105, 16224–9.

THE EXPANDING FIELD OF CANCER TREATMENT

Dancey, J.E., Bedard, P.L., Onetto, N. and Hudson, T.J. (2012). The genetic basis for cancer treatment decisions. *Cell* 148, 409–20.

Downward, J. (2009). Finding the weakness in cancer. *New England Journal of Medicine* 361, 922–4.

International Cancer Genome Consortium. (2010). International network of cancer genome projects. *Nature* 464, 993–8.

BREAST CANCER SCREENING

Raftery, J. and Chorozoglou, M. (2011). Possible net harms of breast cancer screening: updated modelling of Forrest report. *British Medical Journal* 343: d7627

WEBSITES

Cancer Genome Atlas: http://cancergenome.nih.gov/
cBio Cancer Genomics Portal: www.cbioportal.org
Pathway Commons: www.pathwaycommons.org

Tumour grading and staging

Tumour grading is a way of classifying cancer cells on the basis of their appearance when viewed using a microscope. The grade of a tumour is important in considering treatment because it may reflect the way the tumour will behave. There are four tumour grades (1–4) that, in essence, reflect differing degrees of severity. The grade is a measure of the difference between the cancer cell and the type of normal cell from which it is thought to have derived. Thus Grade 1 tumour cells closely resemble their normal counterparts: they grow and divide slowly, are generally the least aggressive (slowest growing) and their cells are often well differentiated. Grade 2, 3 and 4 tumour cells are, increasingly to Grade 4, not readily identifiable in terms of their precursor: they tend to proliferate more rapidly than lower grade tumour cells, are the most aggressive and their cells are usually poorly differentiated or undifferentiated.

The term 'differentiated' crops up quite often in the context of cancer and its meaning takes us back to the very beginning, that is, the fact that animals begin life as a single cell – a fertilised egg. However, those develop into an extraordinary mixture of different types of cell, which means, in effect, that cells do two things during the making of an organism: they grow and divide to make more of the same type of cell and they change into different types of cell. This change of type to make specialised cells is called 'differentiation'. If cancer cells look rather like a type of normal cell they are said to be 'well differentiated'. If it's hard to make out any resemblance to a normal cell they are 'poorly differentiated'. This abnormality of cancer cells arises from the large number of essentially random genetic changes that make up their genomes – and it emphasises the point that in molecular terms cancer is a pretty rare event. Put another way, it's amazing that the chaotic cancer genome can support a functional cell at all. After all, in any kind of engineering context if you jumbled up the blueprint the result would almost inevitably be that your bridge would fall down or your power station would blow up. The important point is that generally:

- low grade = well differentiated = less aggressive, i.e. less malignant and less dangerous
- high grade = poorly differentiated = aggressive, i.e. malignant and dangerous.

Lymphomas – cancers of the cells of the immune system – are an exception in that low grade tumours tend to grow slowly and, because there are no effective treatments, are ultimately fatal, although they may not significantly affect life span. High grade lymphomas grow rapidly and they're aggressive, but are generally curable.

The term anaplasia (meaning 'to form backward') is used to describe de-differentiation. That is, when the structural and functional features of a normal cell are lost and the cell accomplishes a reversion that is characteristic of malignant tumour cells. However, cancers may also arise from stem cells when the mechanism is failure to differentiate rather than reversal of differentiation. Anaplastic cells stain more darkly (they're 'hyperchromatic') and have larger nuclei than normal cells. These features, together with the rate of division, are used to assess the grade of tumour cells.

Tumour staging

It seems fairly obvious that it would be useful to have a way of categorising cancers so that doctors and scientists from all over the world would immediately have some idea what kind of condition you were talking about – even, perhaps especially, for diseases as complex as the cancers. Obvious this now may be, but the fact that we have such a system owes much to a Frenchman who was not only a gifted clinician (he was Professor of Clinical Oncology in the University of Paris and Director of the Institut de cancérologie Gustave Roussy, the first designated anti-cancer centre in Europe) and far-sighted enough to promote the application of statistics to medical problems in the 1950s, but he was also a pretty determined chap (he'd won a Croix de Guerre in the war and was a Commander of the Légion d'honneur). He needed to be because when Pierre Denoix came up with the idea that cancers could be defined by numbers, the TNM staging system (Denoix, 1946; see TNM Online), he came in for a fair bit of criticism on the general ground that they were too complicated to be readily classified. However, he was right and his vision led to an international language for cancer. In fact there are now two broad ways in which cancers are defined: by grade and by stage.

The idea behind the system evolved by Denoix could not have been simpler: there are three main stages to cancer development, growth of the primary tumour (T), spread to lymph nodes (N) and metastasis (M). Denoix's TNM system assigns a number to each stage: the bigger the number the further the tumour has progressed. This simple concept is now applied to most cancers except for those of the blood (leukaemias) and brain and it has largely replaced earlier scoring methods (e.g. Dukes for colorectal and Gleason for prostate cancer). Needless to say, given the complexity of cancers, there's a bit more to it than that and some have sub-groups and individual criteria.

The general rules for the TNM classification are as follows:

T (tumour): 1–4 indicates increasing tumour size.
    T0: localised tumour (no invasion of local tissues).
    Ta: non-invasive (benign) tumour.
    Tis: carcinoma *in situ* (abnormal cell proliferation in the normal cellular location).
    T1: tumour not more than 2 cm in diameter:
        T1a: between 0.1 cm and 0.5 cm
        T1b: between 0.5 cm and 1.0 cm
        T1c: between 1 cm and 2 cm.
    T2: tumour more than 2 cm not more than 5 cm in diameter.
    T3: tumour more than 5 cm in diameter.
    T4: tumour invasion:

T4a, T4b and **T4c** refer to the extent of local invasion into surrounding structures by the tumour.

N (lymph node): 1–4 indicates increasing spread to lymph nodes.

NX: lymph nodes cannot be assessed (e.g. if they have been removed).

N0: no spread to lymph nodes.

N1: spread to one local lymph node.

N2: spread to multiple local lymph nodes that are mobile (tender and can be moved from side to side as is usual when they enlarge due to infection).

N3: spread to multiple local lymph nodes (malignant lymph nodes tend to be hard and immobile).

N4: tumour cells in more distant lymph nodes.

M (metastasis):

M0: no evidence for metastases.

M1: distant metastases, e.g. rectal cancer with metastases in lung or liver.

Examples

Breast cancer: T3 N2 M0: a 7 cm tumour that has metastasised to three lymph nodes but not elsewhere.

Prostate cancer T2 N0 M0: 3 cm tumour confined to the prostate with no dissemination.

Lung cancer: T1 N1 M1: 0.5 cm tumour confined to left lung; one mediastinal lymph node; metastases in liver.

Although TNM combinations give a detailed description of the progression of cancer, in practice a full cancer staging can be irrelevant (if the degree of spread doesn't affect treatment choices), time consuming and costly for health services. Therefore a simplified staging system is sometimes used, and TNM roughly correspond to one of its five stages:

Stage 0: carcinoma *in situ.*

Stages 1, 2 and 3: increasing number indicates tumour growth to the point of local invasion.

Stage 4: metastasis.

The TNM system is applicable to most cancers although there are some variants (the International Federation of Gynecology and Obstetrics (FIGO) system for cancers of the cervix, uterus, ovary, vagina and vulva; the Children's Oncology Group system for childhood cancers).

Brain and spinal cord cancers are staged according to cell type and grade. The Ann Arbor staging classification is used for lymphomas. Different systems are used for the leukaemias (see Leukaemia tables in Appendix E).

These methods for staging cancers therefore provide a common language for describing the extent to which tumours have developed. Their value lies in providing a basis for treatment decisions and for predicting the course that the disease will most probably take (prognosis). The evidence on which staging is based comes essentially from examining tumour samples (biopsies) and from imaging methods (PET/CT or PET/MRI), discussed in Chapter 7. As a broad generalisation, early stage tumours can be effectively treated by a combination of surgery, chemotherapy and radiotherapy. As the stage advances, the efficacy of available therapies declines so that, once tumour cells have spread to distant sites, the prognosis is poor and treatment is largely palliative. The greatest challenge currently facing cancer medicine is to find ways of arresting the growth of metastases. As ever with cancers, however, there are no absolutes and some individuals make full recoveries even from these extreme forms of the disease.

Targets of specific anti-cancer drugs

Type	Mechanism	Drug	Target
A	Tyrosine kinase inhibition	Gefitinib	Small molecule EGFR tyrosine kinase inhibitor: competitive inhibitor of ATP binding
		Lapatinib	Inhibits HER2 tyrosine kinase
		PF002998094, HKI-272, BIBW-2992	Inhibits EGFR, HER2, tyrosine kinases irreversibly
		ARRY-3334543	Inhibits EGFR, HER2 and HER4 reversibly
		Panitumumab (Vectibix®)	Inhibits EGFR tyrosine kinase
		Imatinib	Inhibits BCR-ABL tyrosine kinase by blocking ATP binding
		Dasatinib	Inhibits SRC
		Erlotinib	Small molecule EGFR tyrosine kinase inhibitor: competitive inhibitor of ATP binding
		Sorafenib	Small molecule VEGFR1, VEGFR2, VEGFR3, PDGFRB inhibitor; also inhibits RAF1 and BRAF
		MK-0646	Humanised monoclonal antibody: insulin-like growth factor receptor type 1 (IGF1R) inhibitor
		Sunitinib	Small molecule VEGFR1, VEGFR2, VEGFR3, PDGFRB and RET inhibitor
		Pazopanib (Votrient®)	Small molecule VEGFR, PDGFR and RET inhibitor
		Nilotinib	Inhibits BCR-ABL, KIT and other tyrosine kinases
		PHA665752	MET
		BIBF 1120	Small molecule VEGFR, PDGFR, FGFR inhibitor

Type	Mechanism	Drug	Target
		Crizotinib	EML4-ALK inhibitor
		XL184	MET and VEGFR2 small molecule inhibitor
		SU11274	MET small molecule inhibitor
		Lestaurtinib	FLT3, JAK2, TRKA, TRKB, TRKC small molecule inhibitor
		XL999	VEGFR2/KDR, FGFR1/3, PDGFR-β, FLT3, RET, KIT and SRC inhibitor
B	Heat shock protein inhibition	Tanespimycin/17-AAG	HSP90
		IPI-504	HSP90
C	Receptor inhibition/ modulation	Trastuzumab	Monoclonal antibody: binds to HER2; modulates signalling; marks cells for immunological attack
		Pertuzumab	Inhibits dimerisation, marks cells for immunological attack
		Ertumaxomab	Bispecific affinity recruits T cells
		AMG 888 or U3-1287	HER3
		Cetuximab	Monoclonal antibody: blocks EGFR
		Atrasentan	Inhibits endothelin (subtype A) receptor
		Bexarotene	Inhibits retinoid X receptor
		Testolactone	Derivative of progesterone: inhibitor of aromatase (steroid), reducing estrone synthesis
		Ofatumumab (Arzerra®)	Fully human MAb to CD20
		Ibritumomab tiuxetan	Ibritumomab is a mouse MAb used in conjunction with the chelator tiuxetan, to which a radioactive isotope (either yttrium-90 or indium-111) is attached
		Tositumomab-I^{131} (Bexxar®)	Mouse MAb to CD20: tositumomab is infused followed by tositumomab-I^{131}

Type	Mechanism	Drug	Target
		Alemtuzumab (Campath®)	MAb to CD52 on mature lymphocytes
		Denileukin diftitox (Ontak®)	Fusion protein (fragment of diphtheria toxin fused to interleukin 2): targets IL 2 receptors on the surface of malignant cells and some normal lymphocytes
		Ipilimumab	MAb to cytotoxic T-lymphocyte antigen 4 (CTLA-4)
D	Cytotoxic drug delivery	Gemtuzumab ozogamicin	MAb to CD33: directs cytotoxic calicheamicin (antibiotic) to most leukaemic blast cells and to normal hematopoietic cells
		Trastuzumab-DM1	MAb binds HER2: directs anti-microtubule agent DM1 (a maytansine derivative) to HER2 +ve cells
E	Signalling pathway inhibition	Sirolimus	Immunosuppressant: binds FKBP12 and inhibits mTOR
		Temsirolimus, Everolimus (Rapamycin derivatives)	Small molecule inhibitors of mTOR
		Tamoxifen	Blocks (partially) oestrogen receptor
		Vemurafenib (PLX4032), PLX4720, GDC-0879, SB590885	BRAF
		AZD6244 (ARRY-142886)	MEK1/2 inhibitor
		XL281 (BMS-908662)	RAF, BRAF, CRAF inhibitor
		Salirasib	Inhibitor of RAS activation
		Robotnikinin	Sonic hedgehog protein inhibitor
		BMS 833923, CUR-61414, IPI-926, LDE225, GDC-0449	Inhibit Smoothened (SMO) the receptor that is activated by Hedgehog ligands binding to PTCH
		Gant61, Physalin F	Inhibit GLI transcription activation and hence Hedgehog signalling
		BEZ235, XL765, Torin1, PP242	mTORC1, mTORC2, PI3K inhibitors

Type	Mechanism	Drug	Target
		GDC-0941, BKM120	PI3K inhibitors
		AR-A014418	Glycogen synthase kinase 3-β inhibitor
		Perifosine	AKT inhibitor
		PKC 412	Small molecule PKC (α, β and γ), SYK, FLK1, AKT, PKA, KIT, FGR, SRC, FLT3, PDGFRβ, VEGFR1 and VEGFR2 inhibitor
F	Apoptosis activation	TRAIL, Tretinoin, Trabectedin	Activates death receptors and apoptosis. Superoxide synthesis causes DNA strand cleavage that promotes apoptosis
G	DNA alkylation	*Nitrogen mustards:* Mechlorethamine, Cyclophosphamide (Ifosfamide®, Trofosfamide®), Chlorambucil (Melphalan®, Prednimustine®), Bendamustine, Uramustine, Estramustine	Alkylates DNA
		Nitrosoureas: Carmustine, Lomustine (Semustine®), Fotemustine, Nimustine, Ranimustine, Streptozocin	
		Alkyl sulfonates: Busulfan, Mannosulfan, Treosulfan	
		Aziridines: Carboquone, ThioTEPA, Triaziquone, Triethylenemelamine	
H	DNA cross-linking	Carboplatin, Cisplatin, Nedaplatin, Oxaliplatin, Triplatin tetranitrate (BBR3464), Satraplatin	By cross-linking DNA these agents activate the damage repair pathways that lead to apoptosis when repair cannot be effected
		Hydrazines: Procarbazine	
		Triazenes: Dacarbazine, Temozolomide (Temodar® and Temodal®), Altretamine, Mitobronitol	

Type	Mechanism	Drug	Target
I	DNA intercalation	Actinomycin D (Dactinomycin), Bleomycin, Mitomycin, Plicamycin	Inhibit DNA replication
J	Photosensitisers/ photodynamic therapy	Aminolevulinic acid, Methyl aminolevulinate, Efaproxiral *Porphyrin derivatives:* Porfimer sodium, Talaporfin, Temoporfin, Verteporfin	Allosteric activator of haemoglobin: increases oxygen capacity; used to improve chemotherapy in hypoxic tumours
K	Cell cycle arrest	*Vinca alkaloids:* Vinblastine, Vincristine, Vinflunine, Vindesine, Vinorelbine	Block microtubule disassembly and hence mitosis (between metaphase and anaphase)
		Taxanes: Docetaxel, Larotaxel, Ortataxel, Paclitaxel, Tesetaxel	
		Epothilones: Ixabepilone Alvocidib *R*-roscovitine	Cyclin-dependent kinase inhibitor Cyclin-dependent kinase inhibitor: inhibits CDK2, CDK7 and CDK9 and hence cell cycle progression
		Nutlin 3	MDM2 inhibitor
L	Anti-metabolites: nucleic acid synthesis	Pentostatin, Cladribine, Clofarabine, Fludarabine, Thioguanine, Mercaptopurine	Adenosine deaminase inhibitor. Ribonucleotide reductase inhibitors. Thiopurines
	Inhibition: purine analogues Nucleic acid synthesis inhibition: pyrimidine analogues	5-Fluorouracil, Capecitabine Tegafur (with uracil), Carmofur, Floxuridine Cytosine arabinoside (Cytarabine)	Thymidylate synthase inhibitors Prodrug: converted to 5-fluorouracil Converted to cytosine arabinoside triphosphate: DNA polymerase inhibitor
		Gemcitabine, Azacitidine, Decitabine	Ribonucleotide reductase inhibitor (cytidine analogue) Hypo-methylating agents
	Deoxyribonucleotide	Hydroxyurea	Ribonucleotide reductase inhibitor
	Folic acid	Aminopterin, Pemetrexed, Methotrexate	Dihydrofolate reductase inhibitors
		Raltitrexed, Pemetrexed	Thymidylate synthase inhibitors

Type	Mechanism	Drug	Target
	Topoisomerase inhibition	Camptothecin, Topotecan, Irinotecan hydrochloride, Rubitecan, Belotecan	Topoisomerase I
		Etoposide, Teniposide	Topoisomerase II
	Topoisomerase II and intercalation into DNA	Aclarubicin, Amsacrine, Daunorubicin, Doxorubicin, Epirubicin, Idarubicin, Amrubicin, Pirarubicin, Valrubicin, Zorubicin, Mitoxantrone, Pixantrone	Intercalation into DNA stops replication
	Metabolic modulation	Asparaginase, Pegaspargase	Enzyme that converts asparagine to aspartic acid. Used to target acute lymphoblastic leukaemia cells that cannot make the non-essential amino acid asparagine
		Metformin	Activates AMP-activated protein kinase
		Methazolamide, Indisulam	Carbonic anhydrase inhibitors
		Phloretin	Competitive inhibitor of glucose transport
		Phenylacetate	Depletes serum levels of glutamine
		Cariporide	Na^+/H^+ exchanger 1 inhibitor
		Lonidamine, 2-deoxyglucose	Hexokinase 2 inhibitors
		TLN-232	Pyruvate kinase M2 (M2PK) inhibitor
		AR-C117977, α-cyano-4-hydroxycinnamate (CINN)	Monocarboxylate transporter inhibitors
		Dichloroacetate	Pyruvate dehydrogenase kinase inhibitor stimulating pyruvate dehydrogenase
		SB-204990: prodrug	ATP citrate lyase inhibitor
		C75, GSK837149A	Fatty acid synthase inhibitors
		Orlistat	Lipase inhibitor
	Other anti-neoplastics	Arsenic trioxide, Celecoxib, Demecolcine, Elesclomol, Elsamitrucin, Etoglucid, Lonidamine, Lucanthone, Mitoguazone, Mitotane, Oblimersen, Crisantaspase, Hydroxycarbamide	

Targets of specific anti-cancer drugs

Type	Mechanism	Drug	Target
	Histone deacetylase inhibitor	Vorinostat	
	Oestrogens	Diethylstilbestrol, Ethinylestradiol	
	Progestogens	Medroxyprogesterone acetate, Megestrol acetate, Norethisterone	
		Anastrozole, Exemestane, Letrozole	Aromatase inhibitors: block oestrogen synthesis
	Androgens	Fulvestrant, Tamoxifen, Toremifene, Raloxifene	Selective oestrogen receptor modulators (SERMs)
	Prostate cancer and gonadorelin analogues	Bicalutamide, Buserelin	
	Anti-androgens	Abiraterone acetate	Inhibits CYP27A1 & testosterone synthesis
		Cyproterone acetate, Flutamide, Goserelin	Inhibits testosterone-receptor binding
		Leuprorelin acetate Triptorelin	Suppresses testosterone and oestrogen synthesis
	Somatomedin analogues	Lanreotide, Octreotide	
M	Protein breakdown inhibition	Bortezomib	The first proteasome inhibitor used in humans. It is a tripeptide (with a leucine modified to include a boron atom) that blocks the catalytic activity of the 28S proteasome
N	Ion channel modulation	Tipifarnib	Farnesyltransferase inhibitor
		Anagrelide	Phosphodiesterase inhibitor that blocks platelet maturation
		Tiazofurin	IMP dehydrogenase inhibitor
O	Poly ADP ribose polymerase	Olaparib, MK-4827, KU0058684, KU0058948	Poly ADP ribose polymerase (PARP) inhibitors
		Masoprocol	Lipoxygenase inhibitor

Classes of major oncoproteins

Class 1 Growth factors

HSTF1/HST-1, INT-2, PDGFB/SIS, WNT1, WNT2, WNT3.

Class 2 Tyrosine kinases

Receptor-like tyrosine kinases:
   EGFR/ERBB, CSF1R/FMS, KIT, MET, HER2/NEU, RET, TRK.
Non-receptor tyrosine kinases:
   ABL1, FPS/FES.
Membrane associated non-receptor tyrosine kinases:
   SRC and SRC-related kinases.

Class 3 Cytoplasmic protein serine kinases

BCR, MOS, RAF/MIL.

Class 4 Membrane-associated G proteins

HRAS, KRAS, NRAS.

Class 5 Protein serine-, threonine- (and tyrosine) kinase

AKT1, AKT2.

Class 6 Cell cycle regulators

Cyclin D1, CDC25A, CDC25B.

Class 7 Transcription factors

E2F1, ERBA, ETS, FOS, JUN, MYB, MYC, REL, TAL1, SKI.

Class 8 Intracellular membrane factor

BCL2.

Major tumour suppressor genes

Gene	Function	Principal cancer associations
RB1	Cell cycle regulator Protects from apoptosis	Retinoblastoma, lung, bladder, breast, pancreatic carcinomas
P53	Transcription factor Promotes growth arrest and apoptosis	Most: sarcomas, breast carcinoma, leukaemias
E2F1	Transcription factor	Breast and others
INK4A	Cyclin-dependent kinase inhibitor	Most: melanoma, acute lymphoblastic leukaemia, pancreatic carcinomas
APC	Binds α- and β-catenin: mediates adhesion, cell cycle progression	Colon carcinoma
BRCA1	Transcription factor, DNA repair	Breast, ovarian
CDH1/E-cadherin	Calcium-dependent intercellular adhesion	Many: breast, ovarian
NF1	RAS GTPase activating protein	Neurofibromas, chronic myelogenous leukaemia
WT1	Transcription factor	Renal cell carcinoma
VHL	Inhibits RNA polymerase II via elongin	Renal cell carcinoma
MSH2, MLH1	DNA mismatch repair	Hereditary non-polyposis colon cancer

Ten major cancers at a glance

The following sections summarise the main facts about the ten leading cancers worldwide. Though not repeated in each table, a symptom common to most cancer is cachexia (also called wasting syndrome): loss of weight, muscle atrophy, tiredness and loss of appetite. For additional information there are numerous websites, some of which are included in the individual sections. General information websites include:

National Cancer Institute: www.cancer.gov/
American Cancer Society: www.cancer.org/
Cancer Research UK: http://info.cancerresearchuk.org/cancerstats/
National Institute for Health and Clinical Excellence (NICE): www.nice.org.uk/

Lung cancer

Lung cancer is the most common cause of cancer-related mortality in the world. It is responsible for the deaths of one in every ten people who die in their 60s in the industrialised countries. Ninety per cent of lung cancer cases are directly caused by smoking and lung cancer is the reason why smokers live, on average, ten years less than non-smokers. Main types of lung cancer are shown in the figure.

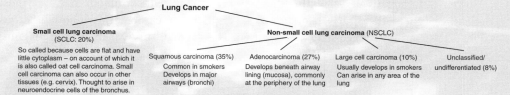

Lung Cancer

Small cell lung carcinoma (SCLC: 20%)

So called because cells are flat and have little cytoplasm – on account of which it is also called oat cell carcinoma. Small cell carcinoma can also occur in other tissues (e.g. cervix). Thought to arise in neuroendocrine cells of the bronchus.

Non-small cell lung carcinoma (NSCLC)

Squamous carcinoma (35%)
Common in smokers
Develops in major airways (bronchi)

Adenocarcinoma (27%)
Develops beneath airway lining (mucosa), commonly at the periphery of the lung

Large cell carcinoma (10%)
Usually develops in smokers
Can arise in any area of the lung

Unclassified/undifferentiated (8%)

Primary lung cancers develop in the epithelial cells that form the lining of the lung and there are two main types: small cell lung carcinoma (SCLC) and non-SCLC (NSCLC). There are three types of NSCLC (squamous cell carcinoma, adenocarcinoma and large cell carcinoma). Squamous cell cancer is the most common but all three behave similarly including in their response to treatment.

Small cell lung carcinoma usually develops in the bronchi at the centre of the lung. Small cell lung carcinoma accounts for 20% of all lung tumours and is almost always caused by smoking. The cells involved are small and mainly localised to midlevel bronchioles. Despite a generally good initial response to chemotherapy, it has a particularly poor prognosis, because of early extrathoracic dissemination and frequent disease relapse.

The major form of NSCLC, squamous carcinoma, begins when ciliated epithelial cells (cilia are tentacles used to waft detritus out of the airway) are lost from the lining of the airways and some of the underlying columnar cells begin to grow abnormally and change to a flatter shape (squamous cells). This leads to 'carcinoma *in situ*', an abnormal growth that has not invaded adjacent tissues. When this acquires invasive capacity and spreads within the lung and then to other organs it has become a malignant lung cancer. Squamous cell carcinoma is highly associated with tobacco smoking and is the most common type of lung cancer in men.

Adenocarcinoma is the most common form of NSCLC in both women and non-smokers. It develops mostly from the junction between the terminal bronchiole and the alveolus, termed 'bronchoalveolar duct junction', the source of mucus.

Large cell carcinoma, less common than the other two forms of NSCLC, forms near the surface of the lungs and comprises a class of rather poorly differentiated and less aggressive tumours. The most frequent sub-type is large cell neuroendocrine carcinoma.

In addition to SCLC and NSCLC there are several much rarer forms of cancer that can affect the lung:

Mesothelioma

Mesothelioma is a rare cancer of mesothelial cells that cover most internal organs. Approximately 70% arise in the pleura (the membrane that surrounds the lung), the majority as a result of chronic inflammation caused by exposure to asbestos, the risk being 500 times higher from blue asbestos than that from white asbestos. The average time for this disease to develop is between 30 and 40 years after exposure. With the cessation of the use of asbestos the incidence of mesothelioma is expected to decline from the estimate of 3,000 UK cases in 2015. Smoking alone does not appear to cause mesothelioma but it greatly increases the risk due to asbestos. The main symptoms are breathlessness and chest pain. Surgery is an option only in rare cases where the tumour is very localised. Radiotherapy may reduce tumour size. The most effective chemotherapy is pemetrexed + cisplatin or carboplatin; vinorelbine, gemcitabine and raltitrexed may also be used. The average survival time from diagnosis is 11 to 14 months.

Carcinoid

Carcinoid is a rare (<2.5% of all lung tumours) and less malignant form of small cell carcinoma that arises most frequently in the gastrointestinal tract.

Classic or *typical bronchial carcinoid* tumours develop in the relatively young and women are ten times more likely to be affected than men. They are the least aggressive of lung tumours: fewer than 3% metastasise beyond the regional lymph nodes and they are treatable by surgery with a five-year survival rate of 94%.

Atypical bronchial carcinoid tumours are more aggressive than typical carcinoids and mainly occur in older men. About one third metastasise to distant sites, a capacity reflected in a five-year survival rate of 57%.

Large cell neuroendocrine carcinoma

Large cell neuroendocrine carcinoma (LCNEC) of the lung is so named because the cells of the tumour are at least three times the size of SCLC. Large cell neuroendocrine carcinoma patients may be treated by surgery but the five-year survival rate is similar to that of SCLC.

Tracheal cancer

Tracheal cancer (i.e. cancer of the windpipe) is rare (0.1% of all cancers, <1% of lung cancers). There are two main forms, squamous cell carcinoma (developing from cells lining the airway) and adenoid cystic carcinoma (developing from glandular tissue). Treatment is by surgery followed by radiotherapy and chemotherapy (cisplatin or carboplatin). For further information on lung cancer consult the British Thoracic Society: www.brit-thoracic.org.uk/.

New cases/year	World 2008: 1,607,000 (males: 1,092,000, females: 515,000); USA 2011 (est.): 221,130; UK 2008: 40,800
Deaths/year	World 2004: 1,375,000 (males: 948,000, females: 427,000); USA 2011 (est.): 156,940; UK 2008: 35,260
Risk factors	Male, over 60 years of age, living in an industrialised area. Smoking causes over 90% of lung cancers. In addition to asbestos, exposure through working conditions to any of the following has been shown to carry an increased risk: arsenic, chromium, iron oxide, coal and petroleum products, and radiation. A family history of lung cancer in a first-degree relative doubles the risk.
Symptoms	None in the earliest stages: as the cancer develops can cause cough, haemoptysis (coughing up blood), chest pain, weight loss, breathlessness and tiredness.
Classification	As shown above, there are two main categories of lung cancer, SCLC and NSCLC. SCLC is classified as 'limited' (confined to chest) or 'extensive' (metastatic). Three main sub-types of NSCLC have similar prognoses and treatment strategies.
Staging	TNM tumour staging system is used.
Major gene mutations	SCLC: Most tumours have inactivated *P53* and *RB1*. Activating mutations commonly occur in *MYC*, *PIK3CA*, *EGFR* and *KRAS*. Whole genome sequencing has identified a set of mutations that commonly arise from carcinogens in tobacco smoke. NSCLC: *EGFR*, *MET* and *PIK3CA* are often mutated; the EML4-ALK fusion protein is present in about 5% of tumours.
Treatment	SCLC: The disease has usually metastasised by time of diagnosis; for this reason <1% of patients receive surgery. Only treatment is chemotherapy (etoposide + cisplatin or carboplatin) combined with radiation therapy. Pravastatin is in Phase 3 trials for SCLC. NSCLC: Surgery if the disease is sufficiently localised (less than 10% of cases). Chemotherapy and radiation therapy. Drugs in use: cisplatin,

carboplatin, docetaxel, gemcitabine, paclitaxel, pemetrexed, irinotecan, vinorelbine. Erlotinib is one of the most promising targeted therapies: patients with EGFR mutation show up to 18 months increased survival. Gefitinib also effective. Drug resistance invariably develops. Novel anti-EGFR drugs are in development. Crizotinib, an inhibitor of ALK, appears promising.

Side effects   Etoposide, cisplatin, carboplatin, docetaxel, gemcitabine, paclitaxel, pemetrexed, irinotecan and vinorelbine generally suppress the immune system and in particular can cause neutropenia (low white cell count). Other side effects can include nausea, hair loss and, for cisplatin, loss of hearing. Erlotinib associated with an itchy, acne-like rash that covers the head and chest.

Prognosis   SCLC: There is usually a good initial response to chemotherapy but this is often rapidly followed by relapse. Two-year survival rate <15%. Overall five-year survival rate 5 to 10%.

NSCLC: The five-year survival rate for Stage IA (T1 N0 M0) is 73%, IB (T2 N0 M0) 55%, IIA (T1 N1 M0) 40%, IIB (T2 N1 M0 or T3 N0 M0) 40%, IIIA (T1–3 N2 M0 or T3 N1 M0) 10–35%, IIIB (Any T4 or any N3 M0) 5%, IV (Any M1) <5%.

Stomach (gastric) cancer

Gastric cancer arises in the stomach but can spread to the oesophagus as well as to the lungs, lymph nodes and liver. There are two main classes: gastric cardia cancer (at the region where the stomach meets the oesophagus) and non-cardia gastric cancer (cancer elsewhere in the stomach). There are about one million new cases every year. It is the fourth most common cancer worldwide and the tenth most common adult cancer in the UK, causing over 700,000 deaths per year worldwide. Over 70% of cases are in developing countries with half the global total being in Eastern Asia.

New cases/year   World 2008: 988,000 (males: 640,000, females: 348,000); USA 2010 (est.): 21,000; UK 2008: 7,610

Deaths/year   World 2008: 736,000 (males: 463,000, females: 273,000); USA 2010 (est.): 10,570; UK 2008: 5,178

Risk factors   Most diagnoses are in those over 65 years of age. More common in men than in women. Most (65–80%) caused by *Helicobacter pylori* infection although cancer only arises in about 2% of infected individuals. Other contributory factors are diet (low in fruit and vegetables, high in salted or preserved foods), smoking and a family history of gastric cancer.

Symptoms   Vague but may include heartburn, abdominal discomfort, bleeding and difficulty in swallowing.

Staging   TNM system is used (see Tumour staging).

Liver cancer

Classification	Over 95% are adenocarcinomas of which there are two major types: intestinal (generally well differentiated) and diffuse (generally poorly differentiated). Gastrointestinal stromal tumours (GIST), lymphoma and neuroendocrine tumours arise infrequently in the stomach.
Major gene mutations	Inherited mutations in E-cadherin (*CDH1*) or *BRCA2* promote hereditary diffuse gastric cancer. Somatic mutations in *KRAS, BRAF* and *ARID1A*, which encodes a member of the SWI-SNF chromatin remodelling family. Mutations in *P53* or *PLCE1* associated with gastric cardia cancer.
Treatment	Surgery. Chemotherapy commonly ECF (epirubicin, cisplatin and 5-fluorouracil) or ECX (epirubicin, cisplatin and capecitabine). Radiotherapy not routinely used.
Side effects	Nausea, hair loss, diarrhoea, mouth ulcers and general suppression of the immune system and increased infection risk.
Prognosis	Over 50% cure rate for localised (early stage) disease. However, most patients (80%) present with metastases. Depending on the extent to which affected tissue can be resected, five-year survival rates range from 10% to 50%.

Liver cancer

Hepatocellular carcinoma (HCC), also called malignant hepatoma, is a primary malignant cancer of the liver and ranks fifth among worldwide causes of cancer mortality. Although relatively rare in the developed world, there are over one million new cases per year worldwide and in some regions it is the most common cancer. The incidence is rising, mainly due to the spread of the hepatitis C virus.

Eighty-five per cent of primary liver cancers are HCCs. Other types include: cholangiocarcinoma (arising in cells lining the bile duct), angiosarcoma (or hemangiosarcoma, arising in blood vessels) and hepatoblastoma, a rare form generally affecting young children.

Hemangiomas, hepatic adenomas and focal nodular hyperplasia are benign liver tumours.

New cases/year	World 2008: 749,000 (males: 523,000, females: 226,000); USA 2011 (est.): 26,190; UK 2008: 3,594
Deaths/year	World 2008: 695,000 (males: 478,000, females: 217,000); USA 2011 (est.): 19,590; UK 2008: 3,390
Risk factors	Much more common in men than in women. Associated with increasing age. Most cases caused by hepatitis virus infection: HBV in Eastern Asia, HCV in Western Europe. Chronic exposure to environmental toxins, notably aflatoxin B1 that is made by a fungus that infects a variety of cereals, is also a cause of HCC, particularly in China and west Africa. HCC can also develop from cirrhosis (commonly caused

by alcoholism). Where hepatitis is not endemic the usual cause of HCC is metastasis from primary tumours, particularly in the colon and breast. Only a fraction of chronic HBV carriers develop HCC. Smoking and/or alcohol consumption together with hepatitis infection increases risk. Frequently associated with cirrhosis and 5% of patients with cirrhosis eventually develop HCC.

Chewing betel quid and exposure to vinyl chloride (used in the production of the plastic PVC).

Symptoms	Nausea, jaundice (yellow skin colour), increased liver mass. Raised levels of alpha-fetoprotein and/or alkaline phosphatase arise from some HCCs.
Staging	Stage 1: single tumour; no metastasis.
	Stage 2: single or multiple small tumours (less than 5 cm diameter); no metastasis.
	Stage 3A: tumours greater than 5 cm diameter; no metastasis.
	Stage 3B: tumour growth into adjacent blood vessels; no further metastasis.
	Stage 3C: tumour growth confined to area around the liver.
	Stage 4A: spread to lymph nodes.
	Stage 4B: spread to other organs in the body.
Major gene mutations	Alterations affecting the retinoblastoma pathway (loss of $RB1$ and $INK4A$ expression and amplification of cyclin D1) occur in 70 to 100% of cases. Also common are loss of $P53$ (15–50%), β-catenin (20–40%).
Treatment	Surgery or a liver transplant are options for localised tumours. Few means available for advanced disease and their efficacy is limited. Most frequently used drugs are doxorubicin (Adriamycin®) and cisplatin.
Side effects	Nausea, diarrhoea, hair loss, fatigue, general suppression of the immune system and increased infection risk.
Prognosis	Five-year survival rate in developed countries: 11%.

Colorectal cancer (colon cancer, large bowel cancer)

This disease includes cancers in the colon, appendix or rectum. They are mainly adenocarcinomas arising in cells that release mucus. Bowel cancer is the third biggest cause of cancer deaths worldwide but there are large variations in rates. Almost 60% occur in more developed regions, the highest rates being in Europe, North America and Australia/New Zealand, the lowest in Africa and Asia.

New cases/year	World 2008: 1,235,000 (male: 663,000, female: 571,000); USA 2011 (est.): 101,340; UK 2008: 39,991
Deaths/year	World 2008: 609,000 (male: 320,000, female: 288,000); USA 2011 (est.): 49,380; UK 2008: 16,259

Breast cancer

Risk factors	Age (most cases after the age of 50).
	Polyps of the colon.
	Previous cancers (e.g. breast).
	Family history of colon cancer (associated conditions being familial adenomatous polyposis (FAP) and hereditary non-polyposis colorectal cancer (HNPCC, also called Lynch syndrome)).
	Smoking.
	Diet (including alcohol and red meat).
	Physical inactivity.
	Inflammatory bowel disease (chronic ulcerative colitis and Crohn's disease).
	Bacterial infection.
Symptoms	None that are specific. Blood in stools, symptoms of anaemia, change in bowel habit (constipation or diarrhoea), tenesmus (feeling of incomplete emptying on defecation), abdominal pain. The nearer the tumour is to the anus, the more likely there will be early symptoms. Metastasis to the liver may cause jaundice. Screening for blood in the faeces can now give early tumour detection.
Staging	TNM tumour staging system is used.
Major gene mutations	*MYC*, *RAS*, *ERBB2*, *P53*. Widespread hypomethylation.
	Inherited mutations: *APC* (in familial adenomatosis polyposis); *MSH2* (in hereditary non-polyposis colon cancer).
Treatment	Surgery: curative (early tumours that have not spread too far can be resected); palliative (used to reduce tumour size). Keyhole surgery now often used. Radiotherapy is not routinely used. Chemotherapy: adjuvant, neoadjuvant or palliative. Drugs used include the combination of 5-fluorouracil, leucovorin and oxaliplatin (FOLFOX), 5-fluorouracil, leucovorin and irinotecan (FOLFIRI) $+/-$ cetuximab or bevacizumab. Other drugs in use are panitumumab. Bortezomib, oblimersen, gefitinib and erlotinib and topotecan are in clinical trials for treated/untreated metastatic colon cancer.
	Intrinsic and acquired drug resistance limits treatment effectiveness.
Side effects	Fatigue, nausea, diarrhoea and numbness in the extremities: hair loss may also occur. General suppression of the immune system and increased infection risk.
Prognosis	Critically dependent on stage at which the cancer is detected: early stage detection has ~five-fold longer survival rates than for late stage detection. The overall USA five-year relative survival rate is about 60%: for tumours diagnosed when they are still confined to the primary site the rate is 90%; for tumours that have metastasised it is less than 12%.

Breast cancer

Breast cancers usually arise in the inner lining of the small tubes (ducts) that carry milk to the nipple (ductal carcinomas) or in the lobules where milk is made (lobular carcinoma). These tubes and lobules are lined by two types of epithelial cells: luminal (secretory) cells and myoepithelial cells.

Breast cancer is by far the most common cancer in women, with 1.38 million new cases worldwide in 2008: 23% of all cancers. It is over 100 times more common in women than in men and it causes nearly 14% of all female cancer deaths. Over the last 30 years mortality has declined, due to contributions from new drugs and more widespread screening mammography, and over 80% of women now live more than five years after diagnosis. In the UK this X-ray method for the detection of abnormalities is recommended every three years for women between 50 and 70 years of age: from 2012 the age range is to be extended to women between 47 and 73 years. In the USA the recommendation is that women between 40 and 50 are screened every two years with annual examinations after the age of 50.

Treatment is almost always initially by surgery, either the removal of the tumour and a small amount of adjacent tissue ('lumpectomy') or the removal of all the breast tissue ('mastectomy'). If the cancer has spread to the lymph nodes in the armpit (axilla), axillary lymph node dissection (removal) can be carried out at the same time. Prophylactic mastectomy is an option for women at high risk, for example, if they have a family history of the disease and they are carrying a mutation in a pre-disposing gene (e.g. *BRCA1*). Adjuvant therapy usually involves radiotherapy and chemotherapy.

New cases/year	World 2008: 1,384,155; USA 2011 (est.): 230,620 (males: 2,140, females: 230,480); UK 2008: 48,034 (males: 341, females: 47,693)
Deaths/year	World 2008: 458,503; USA 2011 (est.): 39,970 (males: 450, females: 39,520); UK 2008: 12,116 (males: 69, females: 12,047)
Risk factors	Sex (99% female, 1% male) Age: increases with age until menopause when it plateaus. Four in every five cases aged 50 or over. Race: most common in Caucasians, then African Americans, Hispanic, Asian Americans. Family history: about 10% of breast cancers are caused by inherited mutations. Hormonal status: early menarche and late first full-term pregnancy (longer exposure to oestrogen and higher risk). High breast tissue density (visualised in mammograms). Obesity: linked to oestrogen-dependent breast cancer via increased aromatase expression. The AMP-activated protein kinase (AMPK: a master regulator of energy homeostasis that plays an important role in balancing energy expenditure and the requirement for food intake) regulates aromatase expression and reduced AMPK activity occurs in breast cancer. Alcohol also increases the risk and physical activity may decrease it.

Breast cancer

Symptoms Most commonly the first symptom is a lump or thickening in an area of the breast. However, 90% of such abnormalities are benign. Other symptoms may include change in breast size, lumps in the armpit, bleeding or discharge from the nipple, peau d'orange (orange peel skin) on the breast and weight loss.

Staging TNM tumour staging system is used. For breast cancer the designation Tis (sometimes called 'stage zero breast cancer') means carcinoma *in situ*, which includes ductal carcinoma *in situ* (DCIS), lobular carcinoma *in situ* (LCIS: not a true cancer but does carry an increased risk that breast cancer will develop) and Paget's disease of the nipple. The term *in situ* means that cancer cells have not spread from their place of origin. Invasive cancers have spread beyond the lining of the duct or lobule. A sub-set of women with untreated DCIS progress to develop invasive (or infiltrating) ductal carcinoma (IDC).

Example: T2 N0 M0 describes a tumour between 2 and 5 cm in diameter that has not spread to lymph nodes or metastasised to distant sites.

Classification There are four main molecular classes, based on hormonal status (oestrogen receptor (ER) and progesterone receptor (PR)) and ERBB2 expression, listed in order of increasingly poor prognosis.

Luminal-A: mostly ER^+ and/or PR^+, $ERBB2^-$. Histologically low grade, less aggressive, most common subtype, associated with increasing age.

Luminal-B: mostly ER^+ and/or PR^+, $ERBB2^+$. High grade, poorer outcome than Luminal-A.

Basal-like: mostly ER^-, PR^-, $ERBB2^-$ (triple negative): cytokeratin $5/6^+$ and/or $EGFR^+$. 15% of all diagnoses, risk at younger age (<40).

$ERBB2^+$: amplification and high ERBB2 amplicon expression. Highly aggressive, less common, risk at younger age (over 40) greater than luminal subtypes.

Luminal-type cancers are the most common sub-type and tend to have the most favourable long-term survival, whereas the basal-like and ERBB2-positive tumours are most sensitive to chemotherapy but have the worst prognosis overall.

Metastasis to the bone occurs in ~30% of invasive breast cancers. This condition is essentially untreatable (average survival from diagnosis two to three years).

Major gene mutations *EGFR* and *ERBB2* are abnormal in about 25% of invasive breast tumours. Other frequently abnormal genes include *FGFR1* and *MYC*.

Mutations in *BRCA1* or *BRCA2* contribute to ~5% of all breast cancers and to about two thirds of familial breast cancers. Inheritance of a *BRCA1* mutation carries a 60% risk of breast cancer for that individual by age 50 and a 90% lifetime risk. Inherited mutations in *P53* (Li–Fraumeni syndrome) also strongly promote breast cancer and mutations in *PTEN* (Cowden) or *STK11* (Peutz–Jeghers) also increase risk.

Over 20 breast cancer susceptibility variants (SNPs) have been identified, each associated with a modest increase in risk. Taken together these account for about 8% of the excess familial risk of breast cancer.

Treatment   Surgery, radiotherapy and chemotherapy. For cancers that are either ER$^+$, PR$^+$ or both, three different classes of hormonal therapy agents may be used: (1) aromatase inhibitors (anastrozole, exemestane or letrozole); (2) selective oestrogen receptor modulators (SERMs) (tamoxifen, raloxifene or toremifene); and (3) oestrogen receptor downregulators (fulvestrant). In addition goserelin, which blocks the production of oestrogen and/or progesterone by the pituitary gland, is also used in the treatment of breast cancer.

Trastuzumab (better known as Herceptin®) for tumours over-expressing ERBB2: only one third of newly diagnosed tumours show regression with Herceptin® monotherapy and the majority that do have an initial response become resistant to Herceptin® within one year. Herceptin® resistance may be promoted by phosphorylation of EPHA2 and activation of PI3K and MAPK signalling.

For the treatment of metastatic bone disease bisphosphonates, inhibitors of bone resorption (the breakdown of bone), are used and the SRC inhibitor dasatinib is undergoing clinical trials.

Side effects   Herceptin®: 1 to 4% develop congestive heart failure; 10% significant decrease in cardiac function.

Aromatase inhibitors: bone loss, arthralgia, cardiovascular effects and possibly cognitive defects.

General suppression of the immune system and increased infection risk.

Prognosis   Five-year survival rates.

USA: over 98% for tumours than have not spread from the primary; over 83% for tumours that have spread to lymph nodes. For tumours that have metastasised to distant sites the rate is 23%. Metastasis to the bone occurs in ∼30% of invasive breast cancers. This condition is essentially untreatable (average survival from diagnosis two to three years).

UK: overall five-year survival rate is 82%.

Oesophageal cancer

The oesophagus is the food pipe that connects the mouth to the stomach in which two main types of cancer can occur, squamous cell carcinoma and adenocarcinoma. Squamous cell carcinoma begins in the flat cells that line the oesophagus; adenocarcinoma in the cells that make and release mucus and other fluids. Oesophageal squamous cell carcinoma (ESCC) is the sixth highest cause of cancer deaths worldwide, the fourth highest in China and ranks seventh in the UK.

The incidence of oesophageal cancer has risen over the last 35 years, for example, from 5 to 25 cases per 100,000 in the UK. In the USA and western Europe there has also been a shift in the major type with adenocarcinoma now being more prevalent than squamous cell carcinoma. The reasons for these shifts are unknown.

Gastric cardia adenocarcinoma (GCA) and distal gastric adenocarcinoma (DGA) bear similarities to ESCC. The cardia is the border between the oesophagus and the stomach.

Oesophageal cancer

New cases/year	World 2008: 481,000 (males: 326,000, females: 155,000); USA 2008: 16,640; UK 2007: 8,000 (includes adenocarcinoma and squamous cell carcinoma)
Deaths/year	World 2008: 406,800 (males: 276,000, females: 130,000); USA 2010: (est) 14,500; UK 2008: 7,610 (includes adenocarcinoma and squamous cell carcinoma)

Risk factors

Squamous cell carcinoma

Marked worldwide variation. In Europe and North America the main risk factors are heavy smoking and alcohol consumption. In some areas of China, central Asia and southern Africa nutritional deficiencies and consumption of pickled vegetables, nitrosamine-rich or mycotoxin-contaminated foods contribute. Chewing tobacco or betel quid also increase risk. High incidence areas show familial aggregation, i.e. genetic susceptibility, and several loci have been identified in Chinese and Japanese populations.

Age, sex and race: oesophageal cancer is rare in people under the age of 45, more common in men than in women and in African Americans than whites. More than 80% of new cases and deaths occur in developing countries.

Two rare conditions pre-dispose to squamous cell carcinoma: Tylosis, an inherited condition in which skin grows too thickly on the palms of the hands and soles of the feet, which carries a high risk; Plummer–Vinson syndrome, a rare form of anaemia due to a lack of iron, causes blockage of the oesophagus.

Adenocarcinoma

Less clear but one of the strongest risk factors is Barrett's oesophagus, a pre-cancerous condition that often results from gastro-oesophageal reflux disease although it can arise from any damage to the tissue. About one in ten cases of Barrett's oesophagus progress to oesophageal adenocarcinoma.

Achalasia, a defect of the cardiac sphincter (the valve between the stomach and oesophagus) causes blockage of the oesophagus and increases by ten-fold the risk of both main types of oesophageal cancer.

There is evidence that some strains of *Helicobacter pylori* may reduce the risk of Barrett's oesophagus and adenocarcinoma but may increase the risk of squamous cell carcinoma. Prolonged, heavy exposure to soot, metal or silica dust or vehicle exhaust may also increase risk.

Symptoms

Painful or difficult swallowing, weight loss, chest pain, hoarseness and cough. Diagnosis usually by endoscopy (a long thin tube with a light and camera inside).

Staging

TNM tumour staging system is used. The number system may also be used:

Stage 0 Carcinoma *in situ* (CIS) or high grade dysplasia (abnormal cell changes in the lining of the oesophagus). The subsequent stages correspond to the TNM system as shown:

Stage 1A (T1 N0 M0), 1B (T2 N0 M0).

Stage 2A (T3 N0 M0), 2B (T1 or T2 N1 M0).

	Stage 3A (T4a N0 M0), (T3 N1 M0) or (T1 or T2 N2 M0), 3B (T3 N2 M0), 3C (T4a N1 or N2 M0), (T4b N any, M0) or (T any, N3 M0).
	Stage 4 (T any, N any, M1).
Major gene mutations	*P53, MSR1, ASCC1* and *CTHRC1*.
Treatment	Surgery, radiotherapy and chemotherapy. For adenocarcinoma a common drug combination is epirubicin, cisplatin and 5-fluorouracil (ECF). For squamous cell carcinoma a common combination is cisplatin and 5-fluorouracil.
Side effects	May include anaemia, nausea, diarrhoea, hair loss, mouth ulcers and fatigue. General suppression of the immune system and increased infection risk.
Prognosis	Five-year survival rate after surgery: 30 to 50% (stage II disease), 10 to 25% (stage III disease).

Cervical cancer

Cancer of the cervix – the neck of the womb that opens into the vagina – arises most frequently (~80% of cases) in the layer of epithelial cells that lines the cervix (squamous cell cervical cancer). The other common form (~15%) is in the glandular cells of the cervix (adenocarcinoma).

Cervical cancer is the eighth most common cancer worldwide, with over half a million new cases and more than 250,000 deaths annually. It is the leading cause of cancer deaths in women in the developing world. Cervical cancer only develops as a result of infection of epithelial cells that form the skin or mucous membranes of the cervix by human papillomavirus (HPV), as indicated by the fact that HPV is detectable in over 99% of cervical tumours. Human papillomavirus infection can also cause cancers of the throat, genitals and anus and it contributes between 3.7% and 5% of the global burden of all cancers.

In 1928, George Papanicolaou discovered that vaginal smears can reveal cervical cancer, which led to the Pap smear test for carcinomas in the female genital tract. Harald zur Hausen won the Nobel Prize in Physiology or Medicine in 2008 for showing the link between cervical cancer and human papillomavirus (HPV) types 16/18. The first HPV vaccine (Gardasil®) entered clinical trials in 2002.

New cases/year	World 2008: 530,000; USA 2011 (est.): 12,710; UK 2008: 2,938
Deaths/year	World 2008: 275,000; USA 2011 (est.): 4,290; UK 2008: 957
Risk factors	Infection by human papillomavirus (HPV): typically transmitted by sexual contact, infecting the anogenital region. Transmission is by skin contact – penetration not required. 50 to 80% of *all* sexually active women have been infected. However, only 10 to 20% of infected women develop cancer.

	Available data shows ~60% of men are infected (though the only dataset is for a mixture of Brazilian, Mexican and Hispanic USA males). Smoking doubles the risk.
Symptoms	Few in the early stages although vaginal bleeding or discharge may occur. Advanced stages may be associated with loss of appetite/weight, pain, heavy vaginal bleeding and fatigue. The most common sign of genital HPV infection are genital or anal warts (condylomata acuminata or venereal warts), 90% of which are caused by HPV 6 and 11.
Staging	TNM system is used (see Tumour staging) or the International Federation of Gynecology and Obstetrics (FIGO) staging system: Stage 0: epithelium with no invasion (carcinoma *in situ*) Stage I: limited to the cervix Stage II: invades beyond the cervix Stage III: extends to pelvic wall or lower third of the vagina Stage IV: extends outside the vagina.
Classification	Smear testing detects abnormal cells (pre-malignant change) that have the potential to progress to cancer. Abnormal smear tests are graded from mild to severe dysplasia depending on the degree that cells have progressed towards a cancerous state. Mild dysplasia often returns to normal within two years. Moderate to severe dysplasia will prompt a colposcopy that can more accurately grade the pre-malignant change using CIN (cervical intraepithelial neoplasia) grading. For pre-malignant changes the CIN (cervical intraepithelial neoplasia) grading defines progression from normal to CIN I (low grade/mild) through CIN II to CIN III (precursor of invasive squamous cell cervical carcinoma).
Major gene mutations	The oncoproteins E6 and E7 of high-risk HPV subtypes inactivate p53 and the retinoblastoma protein, thereby overcoming normal cell cycle control. This leads to high-grade intraepithelial lesions, abnormal growths that are non-invasive and that usually disappear, indicating that HPV infection is not sufficient to cause cervical cancer. Additional factors are therefore required to drive progression to invasive carcinoma. Other than mutations in the *STK11* tumour suppressor gene in some cancers, genetic alterations or other agents that cooperate with HPV have not been identified.
Vaccination	Cervarix® (vs. HPV 16/18) and Gardasil® (vs. HPV 6/11/16/18). A new vaccine is being developed that may be effective against more than 90% of HPV types. Prophylactic efficacy approximately 100%.
Treatment	Surgery may be effective for very early stage disease. Radiotherapy is also used. There are few chemotherapy options.
Side effects	None serious from either Cervarix® or Gardasil®.
Prognosis	The duration of immunity is at least six years.

For further information on cervical cancer, consult the Royal College of Obstetricians and Gynaecologists: www.rcog.org.uk.

Pancreatic cancer

Pancreatic ductal adenocarcinoma is the 13th most commonly diagnosed cancer worldwide and the 8th most common cause of cancer death. In the USA and the UK it is 4th and 5th, respectively, on the death list. It afflicts men and women in almost equal numbers and is extremely difficult to diagnose in its early stages, for which reason over half of patients have distant metastases and more than one quarter have regional spread at the time of diagnosis. The overall five-year survival rate is less than 5%.

New cases/year	World 2008: 278,684 (males: 144,859, females: 133,825); USA 2011 (est.): 44,030; UK 2008: 8,085
Deaths/year	World 2008: 266,669 (males: 138,377, females: 128,292); USA 2011 (est.): 37,660; UK 2008: 7,781
Risk factors	Smoking is thought to contribute to ~20% of cases. Diabetes, long-term inflammation of the pancreas (chronic pancreatitis, 70% of which is caused by heavy alcohol consumption) and stomach ulcers. Diets high in meats and fats, diets low in vegetables and folate. Possible hereditary factors in ~10%. Rare under the age of 40: 63% of new cases present over the age of 70.
Symptoms	Weight loss, stomach or back pain, jaundice (yellow skin colour), development of diabetes.
Staging	TNM system is used (see Tumour staging). Patients almost always present with stage 4.
Classification	CA19-9 in the blood is used as a marker for colon and pancreatic cancer. However, it has low sensitivity and specificity, its level being elevated in some non-malignant diseases. In patients not expressing Lewis antigen (~5% of cases) no increase in CA19-9 levels occurs even with large tumours.
Major gene mutations	80–95% of these tumours have mutations in *KRAS*, so they resemble leukaemias in that one major mutation drives development. Other frequently mutated genes are *P53*, *CDKN2* and *PIK3CA*. Whole genome sequencing of tumour slices has shown that they are made up of mixtures of sub-clones. On average it takes 15 years for metastases to appear from these primary growths. From that point the average lifespan is two years.
Treatment	Because this cancer generally presents late and metastasises early it is usually resistant to treatment. Fewer than 20% of cases are treatable by surgery. Gemcitabine gives partial response in 10% of cases and in 30 to 40% stabilises the disease. Other inhibitors are under development.
Side effects	Gemcitabine has multiple side effects including allergy, nausea and hair loss. General suppression of the immune system and increased infection risk.
Prognosis	Poor: mean overall survival time is six to seven months. The survival rates for stages of exocrine pancreatic cancer range from 37% (Stage IA) to 1% (Stage IV). It is metastatic in more than 50% of patients at the time of diagnosis and has a five-year relative survival rate for all stages of less than 6%.

Prostate cancer

The prostate gland is a male accessory sex gland that stores and secretes an alkaline fluid that makes up part of the semen. Females have an equivalent (the paraurethral gland) but this is located near the lower end of the urethra and prostate cancer cannot occur in women. The gland secretes prostate-specific antigen (PSA), the level of which tends to rise in advanced prostate cancer: the decline in PSA level on treatment is a good prognostic indicator.

The most common non-skin cancer in men, one in six American men will be diagnosed with prostate cancer although more than 65% of them will be over the age of 65. However, about one third of men over the age of 50 would show histological evidence of the disease if their prostates were examined. The fact that in most cases the disease does not become life threatening indicates the complexity of its development. Over 1,000 genes have been shown to be abnormally expressed in prostate cancers and of these 20 were associated with pathological grade. Perhaps unsurprisingly, most of these play a role in controlling progression round the cell cycle.

New cases/year	World 2008: 899,000; USA 2011 (est.): 240,890; UK 2008: 37,051
Deaths/year	World 2008: 258,000; USA 2011 (est.): 33,720; UK 2008: 10,168
Risk factors	Age (rare before age 40), heredity (a father, brother or son with the disease confers double the risk for an individual), race (more common in North America, northwestern Europe, Australia and the Caribbean than in Asia, Africa, and Central and South America). The geographical distribution suggests that diet may affect the risk but there is no unequivocal evidence.
	Benign prostatic hyperplasia (non-cancerous enlargement of the prostate) or prostatitis (infection in the prostate) does not increase the risk of developing prostate cancer.
Symptoms	Urinary difficulties, blood in urine or semen, pelvic, back or hip pain, weight loss.
Staging	TNM system is used (see Tumour staging).
Classification	Conventional methods for identifying prostate tumours are unreliable and may miss up to 30% of significant cancers. A gene signature of over 100 genes has now been identified that predicts tumour status with very high accuracy.
Major gene mutations	Inherited (germ line) mutations that drive prostate cancer can occur in a number of genes (e.g. *BRCA1*, *BRCA2*, *ELAC2*, *MSR1* and *RNASEL*).
	Many other genes can become mutated including *AR*, *P53*, *PTEN*, *RAS* and in mitochondrial DNA. The combination of alterations in *ERBB2* and *PTEN* is associated with poor survival. Chromosome-wide analysis shows widespread loss and some amplification. The extent of disruption reflects the risk of recurrence and suggests that 20% of patients could be nominated for 'watchful waiting'.

Treatment   Surgery. Radiotherapy. Chemotherapy (mitozantrone, doxorubicin,
                         vinblastine, paclitaxel, docetaxel, estramustine phosphate, etoposide).
                         Systemic ablation of testosterone by castration and/or hormone
                         therapy (LHRH agonists, LHRH antagonists or anti-androgens, e.g.
                         abiraterone).
Side effects   Nausea, hair loss, general suppression of the immune system and
                         increased infection risk.
Prognosis   For localised cancer (stages 1 and 2, i.e. T1 or T2) five-year survival rate is
                         99%. For stage 3 (locally advanced cancer) rate is 70 to 80%; for stage 4
                         five-year survival rate is ~30%.

Leukaemia

The term leukaemia refers to a range of cancers in which any one of the types of white blood cell (leucocytes) undergoes progressive over-proliferation in the bone marrow or other haematopoietic organs and circulates throughout the blood and lymphatic systems. This is generally accompanied by suppressed production of other types of leucocyte. Leucocytes are the cells that make up the innate and adaptive immune system and there are five main types: neutrophils, eosinophils and basophils (collectively called granulocytes), monocytes and lymphocytes. Granulocytes and monocytes are collectively called myeloid cells – hence any leucocyte that isn't a lymphocyte is a myeloid cell. There are two major classifications of leukaemia based on the rate of progression of the disease (acute or chronic) and further sub-divisions depending on the type of cell involved (myeloid or lymphoid).

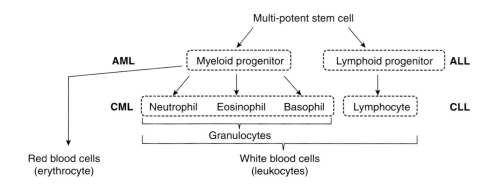

Acute leukaemias are also usually associated with the presence of primitive immature 'blast' cells, while chronic leukaemias are characterised by the presence of increased numbers of mature progeny. Blast cells are immature precursors of either lymphocytes (lymphoblasts), or granulocytes (myeloblasts). They are not normally present in peripheral blood and the appearance of these abnormally large cells often signifies acute leukaemia.

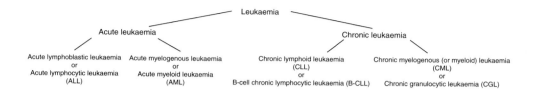

Acute lymphoblastic leukaemia (ALL) is characterised by an excess of lymphoblasts and accounts for about 70% of all cases of childhood leukaemia. Treatment for childhood ALL is very successful but the prognosis for adult ALL is less good.

Acute myeloid leukaemia (AML) is a malignancy of myeloblasts, the precursors of myeloid white blood cells. Although this involves uncontrolled myeloblast growth, there are multiple forms of AML. Acute myeloid leukaemia can now be distinguished from ALL by genome-wide expression profiling, which has also revealed new molecular sub-types.

Chronic lymphoid leukaemia (CLL: alternatively called B-cell chronic lymphocytic leukaemia), also involves excessive production of abnormal lymphocytes but progression is usually slow. Chronic lymphoid leukaemia accounts for 25% of all leukaemia and is the most common form of lymphoid malignancy in Western countries.

Chronic myelogenous leukaemia (CML), also called chronic myeloid leukaemia or chronic granulocytic leukaemia (i.e. granulocytes are the cancerous cells), involves increased production of granulocytes and is characterised by the chromosomal translocation known as the Philadelphia chromosome. A small proportion of patients with a similar myeloproliferative disorder do not have the Philadelphia chromosome. Chronic myelogenous leukaemia is a relatively rare disease, representing about 14% of all leukaemias.

There are several sub-categories within these main groupings. In addition there are other rare forms of leukaemia: hairy cell leukaemia (HCL) and large granular lymphocytic leukaemia, and T-cell prolymphocytic leukaemia (T-PLL, also called T-cell chronic lymphocytic leukaemia), a rare and aggressive adult form. Adult T-cell leukaemia/lymphoma (ATL) is caused by infection of $CD4^+$ T cells by human T-lymphotropic virus (HTLV).

The leukaemias are part of a larger group of diseases called haematological malignancies that includes the four sub-types of Hodgkin's lymphoma and the various forms of non-Hodgkin's lymphoma (including diffuse large B-cell lymphoma (DLBCL), follicular lymphoma (FL), mantle cell lymphoma (MCL), marginal zone lymphoma (MZL), post-transplant lymphoproliferative disorder (PTLD) and B-cell prolymphocytic leukaemia (B-PLL)). The incidence of non-Hodgkin's lymphomas is rising in the western world.

Ten major cancers at a glance

Leukaemias (total)

New cases/year	World 2008: 195,456	USA 2011 (est.): 44,600	UK 2008: 7,659
Deaths/year	World 2008: 143,555	USA 2011 (est.): 21,780	UK 2008: 4,367

Hodgkin's lymphoma

New cases/year	World 2008: 40,265	USA 2011 (est.): 8,830	UK 2008: 1,730
Deaths/year	World 2008: 18,256	USA 2011 (est.): 1,300	UK 2008: 302

Non-Hodgkin's lymphoma

New cases/year	World 2008: 199,736	USA 2011 (est.): 66,360	UK 2008: 11,861
Deaths/year	World 2008: 109,484	USA 2011 (est.): 19,320	UK 2008: 4,438

Acute lymphoblastic leukaemia (ALL)

New cases/year	USA 2011 (est.): 5,730; UK 2008: 654. About 6 in 10 cases occur in children.
Deaths/year	USA 2011 (est.): 1,420; UK 2008: 262
Risk factors	The most common leukaemia in young children: also affects adults, particularly over the age of 65. Exposure to high levels of radiation, previous chemotherapy and chemical carcinogens (e.g. benzine) are possible risk factors.
Symptoms	Vague and non-specific but may include weakness and fatigue, fever, susceptibility to infection, skin rash, bleeding from gums or nose, blood in urine or stools, rash (purpura), enlarged lymph nodes, enlarged spleen, weight loss. Infection by human T-cell leukaemia virus (HTLV-I) may lead to a form of adult T-cell leukaemia.
Sub-types	There are four main sub-types in the FAB classification:

Early precursor B cell (60–65% of cases)

Precursor B cell (20–25%; 70% of childhood ALL)

Mature B cell (2–3%)

T cell (13–15%)

However, more meaningful is the WHO recommended classification into the categories Precursor B cell, Precursor T cell, Burkitt's leukaemia/lymphoma and the rare sub-type acute biphenotypic leukaemia. In Precursor B cell ALL major genetic alterations are:

ETV-RUNX1 (25% of cases)

BCR-ABL1 (25%): the Philadelphia chromosome (present in 95% of CML cases): occurs in 5% of pediatric ALLs.

E2A-PBX1 (5%)

MLL-AF4 (2%): very common in infantile leukaemias, about 80% of which have *MLL* rearrangements and are very aggressive.

MLL-AF10 (mainly in paediatric not adult ALL)

IGH-MYC

TCR-RBTN2

ETV-RUNX1 translocations are associated with a good prognosis: *BCR-ABL1*, *MLL-AF4* and *IGH-MYC* have poor prognoses.

Major gene mutations	In addition to the translocations defining sub-types, mutations in more than 30 other genes have been associated with ALL, including components of the RAS signalling pathway.

Treatment	Over 30 drugs are approved for ALL including abitrexate (methotrexate), clofarabine and imatinib (Gleevec®).
	Childhood ALL: prednisolone, vincristine, asparaginase, daunorubicin. Therapy: cytotoxic drug $+/-$ stem cell transplantation.
Side effects	Vary depending on the drugs used and the individual patient. May include nausea, general suppression of the immune system, hair loss and increased infection risk. Transplantation therapy may induce an immune response that can cause skin problems and diarrhoea. Some children may suffer delayed effects of chemotherapy months or years after treatment. These 'late side effects' may affect development and fertility, and give rise to learning disabilities.
Prognosis	Survival rates are ~80% in children, ~30% in adults. In some groups 20 to 80% of patients relapse or develop cancers resistant to treatment.

Acute myeloid leukaemia (AML)

New cases/year	USA 2010 (est.): 12,330; UK 2008: 2,343
Deaths/year	USA 2010 (est.): 8,950; UK 2008: 2,143
Risk factors	More common in adults than children and in men rather than women. Risk group assigned on basis of karyotype.
Symptoms	Vague and non-specific but may include weakness and fatigue, fever, susceptibility to infection, skin rash, bleeding from gums or nose, blood in urine or stools, enlarged lymph nodes, enlarged spleen and weight loss.
Sub-type	There are two systems for classifying AML: the World Health Organization (WHO) system and the French American British (FAB) system.
	The WHO system is based on cell type and whether (1) there are specific chromosomal changes; (2) abnormalities are detected in more than one type of cell; (3) leukaemia developed from myelodysplasia; and (4) previous cancer treatment has led to the leukaemia ('treatment-related AML').
	The FAB system is based on the morphology (appearance) of the leukaemia cells and is numbered from M0 to M7, depending on cell type:
	AML-M0: acute myeloblastic leukaemia; minimal differentiation (very immature cells).
	AML-M1: a high percentage of blasts in the bone marrow without significant evidence of myeloid maturation. Blasts constitute over 90% of the non-erythroid cells.
	AML-M2: more than 20% blasts in the bone marrow or blood and evidence of maturation to more mature neutrophils. (More than 10% of neutrophils are at different stages of maturation.) Monocytes comprise over 20% of bone marrow cells.
	AML-M3: acute promyelocytic leukaemia (APL; promyelocytes are mature myeloblasts).
	AML-M4: acute myelomonocytic leukaemia.
	AML-M5: acute monocytic leukaemia; over 20% blasts in the marrow, 80% of which are of the monocytic lineage.
	AML-M5a: over 80% monoblasts.

AML-M5b: mixture of monoblasts (80%) and promonocytes.

AML-M6: acute erythroleukaemia (affecting precursor red cells).

AML-M7: acute megakaryoblastic leukaemia (affecting megakaryocytes).

Major gene mutations	Numerous chromosome translocations generate fusion proteins. These include RUNX1–RUNX1T1, NUP98-HOXA9 and MOZ-TIF2. The ten most common occur in about 15% of AMLs. About 50% do not have translocations but have mutations in a large number of genes (e.g. *NPM1*, *FLT3*, *KIT*, *RAS*, *MLL*). Mutations in isocitrate dehydrogenases (IDH1/2) can cause widespread changes in gene expression.
	Both translocation and mutation patterns can give rise to a gene-expression signature indicating whether the prognosis is poor or good.
Treatment	Over 30 drugs are approved for AML. The mainstays of therapy are anthracyclines such as daunorubicin and cytarabine. Gemtuzumab ozogmicin (Myelotarg®), a monoclonal antibody that targets a cytotoxic antibiotic to leukaemic blast cells via binding to the surface protein CD33, can improve survival figures in 10% of patients in specific sub-groups.
Side effects	Commonly shivering, fever, nausea. Can cause myelosuppression and liver damage. General suppression of the immune system and increased infection risk.
Prognosis	Five-year survival rate in adults: 33% (below 65 years), 4% over 65 years. Long-term survival rate 21%, children 55%.

Chronic lymphoid leukaemia (CLL)

New cases/year	USA 2010 (est.): 14,990; UK 2008: 2,798
Deaths/year	USA 2010 (est.): 4,390; UK 2008: 1,129
Risk factors	The most common type of leukaemia in adults, usually occurring after the age of 55. Rare in children. Two-thirds of those affected are men. Despite a strong familial basis to CLL, with risks in first-degree relatives of cases being increased seven-fold, to date the inherited genetic basis of the disease is largely unknown.
Symptoms	Usually without symptoms and discovered in a blood test. It can, however, cause swollen lymph nodes, abdominal pain, tiredness and weight loss.
Staging	Stage 0: high lymphocyte levels in blood but no other symptoms: indolent (slow growing).
	Stage I: high lymphocyte levels in blood; enlarged lymph nodes.
	Stage II: high lymphocyte levels in blood; liver, spleen or lymph nodes may be enlarged.
	Stage III: high lymphocyte levels in blood; low red cell count; liver, spleen or lymph nodes may be enlarged.
	Stage IV: as stage III but with reduced platelet count.
Sub-types	B-cell prolymphocytic leukaemia is an aggressive form of CLL.
Major gene mutations	Four main types of genetic damage are associated with CLL and these reflect the probable course of the disease:

1. Loss of *P53* (5–10% of CLLs): very poor prognosis.

2. Loss of *ATM* (5–10%): poor prognosis.

3. An extra copy of chromosome 12 (20–25%): intermediate prognosis.

4. Loss of specific micro RNAs (50%): good prognosis; may not require any treatment.

Treatment	No treatment ('watchful waiting') may be required at first diagnosis if there are no symptoms. Otherwise, chemotherapy (e.g. fludarabine with cyclophosphamide or chlorambucil) with or without radiotherapy. Therapy with rituximab (an anti-CD20 monoclonal antibody) appears to improve outcome in CLL.
Side effects	Lowered blood cell count; fatigue. General suppression of the immune system and increased infection risk.
Prognosis	Five-year survival rate 75%.

Chronic myelogenous leukaemia (CML)

New cases/year	USA 2010 (est.): 4,870; UK 2008: 613
Deaths/year	USA 2010 (est.): 440; UK 2008: 243
Risk factors	Occurs mainly in adults between 40 and 60 years; rare in children. Radiation (e.g. radiotherapy) and chemical carcinogens (e.g. benzine).
Symptoms	85% of patients present in the chronic phase: there may be no symptoms or patients may have shortness of breath (due to anaemia), abdominal discomfort (due to massive enlargement of the spleen), weight loss, fever, sweating, headache and (rarely) bruising or bleeding. Symptoms may cause pallor, swollen lymph nodes and spleen, and bleeding into the retina of the eye.
Course	There are three phases: *chronic phase* followed over several years in the untreated patient by the *accelerated phase* (weeks to months) and ultimately *blast crisis* (also called the acute phase), which is clinically similar to acute leukaemia. Imatinib and other tyrosine kinase inhibitors can alter the normal natural history of the disease, giving a marked reduction in progression.
Cellular effect	In the chronic phase <10% of the cells in blood and bone marrow are immature (blast) cells and the levels of myeloid progenitor cells and mature granulocytes are increased.
Major gene mutations	The genetic signature is the Philadelphia chromosome seen in 90 to 95% of CML cases. This chromosomal translocation joins part of the BCR gene (from chromosome 22) to the *ABL1* gene on chromosome 9. Additional genetic defects are acquired over time and are required for the progression of the disease.
Treatment	First-line for chronic phase: imatinib; gives a complete haematologic response of 96% at one year. Acute phase patients usually have only a brief response to imatinib. Other drugs (e.g. cyclophosphamide, cytarabine) may also be used. Blast crisis, which may progress to ALL or AML, is similarly treated. Radiotherapy may be used as part of stem cell or bone marrow transplant therapy. Resistance occurs in many patients first treated with imatinib during the accelerated phase. Novel drugs designed to solve this problem are currently in clinical trials where they have shown efficacy and are currently being compared to imatinib for the treatment of newly diagnosed patients.

Side effects	Imatinib: well tolerated but can cause nausea, headaches, rashes and cytopenia.
Prognosis	The US five-year survival rate for 2001–2007 was between 50 and 60%. Chronic phase patients treated with imatinib: five-year survival rate 90%. Six per cent of chronic phase patients relapse having developed resistance to imatinib.

For further information consult the Leukemia & Lymphoma Society (LLS) website: www.lls.org/diseaseinformation/leukemia/

GLOSSARY AND ABBREVIATIONS

Acoustic neuroma   See Schwannoma.

Action potential   If Na^+ permeability is increased, Na^+ moves into the cell down its concentration gradient and the membrane becomes depolarised in the region near the open channel. In contrast, if Cl^- permeability is increased, the membrane becomes hyperpolarised. If the depolarisation is large enough an action potential is generated. The resting membrane potential is restored and maintained by activation of the Na^+-K^+-ATPase pump that actively extrudes Na^+ from the cell.

Acute erythroleukaemia   See leukaemias.

Adenocarcinomas   Malignant tumours of glandular epithelial tissue that arise at the sites of the most common cancers including those of the breast, cervix, colon, lung, prostate, pancreas, oesophagus and stomach (see cancer tables in Appendix E).

Adenoma   Benign tumour derived from (secretory) epithelial cells.

Adjuvant   An agent that modifies the effect of other agents (for example incomplete Freund's adjuvant, which is an emulsion of mineral oil in water). In cancer chemotherapy, adjuvant chemotherapy is the use of drugs post-operatively to inhibit recurrence. Neoadjuvant chemotherapy is a pre-operative drug treatment to reduce the size of a primary tumour to facilitate subsequent surgery or radiotherapy.

Age Standardised Rate (ASR)   The Age Standardised Rate is an adjustment of the total mortality figure to take account of the range of ages in that population. This permits comparison of cancer rates between countries ensuring, for example, that a higher rate does not simply reflect a greater proportion of older people.

AIDS   Acquired immune deficiency syndrome. Clinical syndrome caused by infection with human immunodeficiency virus. Diagnostic criteria include opportunistic infections, abnormal T helper/T suppressor ratio and malignancies including Kaposi's sarcoma and lymphoma.

ALL   See leukaemias.

Allostery   Allosteric regulation of an enzyme or protein means modulation of its activity by the binding of an effector molecule at a site other than the active site of the protein.

Alu **sequences**   Transcribed, non-protein coding mobile DNA elements of $\sim$300 bp present at $\sim$1 million sites in the human genome and accounting for $\sim$10% of genomic DNA. Many partial *Alu* sequences (as short as 10 bp) are also scattered between genes and within introns in the human genome. *Alu* sequences are homologous to the 7SL RNA component of the signal recognition particle. *Alu* sequences (containing a recognition site for *AluI*) constitute the most abundant type of human SINE (SINES are short interspersed elements), the second major class of moderately repeated DNA in mammals. The major class, LINES (long interspersed elements), are $\sim$6 to 7 kb long. Each class may be present in thousands of copies that may not be exact repeats.

Glossary and abbreviations

Aminopterin Immunosuppressive drug, closely related to methotrexate, that competes with folate in binding to dihydrofolate reductase.

AML See leukaemias.

Amplification Increase in gene copy number. Enhanced expression of the gene is a general consequence, but the increase in mRNA concentration is not always proportional to the number of gene copies.

AMV Avian myeloblastosis virus: family *Retroviridae*, sub-family *Oncovirinae*, genus Type C oncovirus. Causes myeloblastosis, osteopetrosis, lymphoid leukosis and nephroblastosis in chickens.

Aneuploidy A condition in which the number of chromosomes is abnormal due to extra or missing chromosomes, in other words, it is a chromosomal state where the number of chromosomes is not a multiple of the haploid set. Normal diploid species have $2n$ chromosomes, where n is the number in the haploid set. Aneuploid individuals would have $2n-1$ chromosomes (monosomy), $2n+1$ chromosomes (trisomy), or some other such arrangement.

ANLL See leukaemias.

Anoikis A form of apoptosis induced by abnormal cell–matrix interactions.

Anti-sense oligodeoxynucleotide (ODN) Synthetic nucleotide sequence (usually 15 or 18 bases) complementary to RNA or DNA sequences used to inhibit specifically the translation or transcription of a gene. In addition to unmodified oligonucleotides, classes of ODN include methylphosphonates, phosphorothioates and α-oligonucleotides. The methylation of the phosphate group in methylphosphonates renders the compound uncharged and without ionisable groups. Phosphorothioate oligodeoxynucleotides (S-ODNs) have one of the non-bridging oxygen atoms in each internucleotide phosphate linkage replaced by a sulphur atom: they are relatively stable to cleavage by nucleases and are readily soluble in aqueous solution.

Anti-sense RNA with sequences complementary to some mRNAs has been detected in prokaryotes and eukaryotes and transcription of regions of the negative strand has also been detected in eukaryotes.

APL See leukaemias.

Apoptosis Two pathways lead to apoptosis (programmed cell death): the extrinsic and the intrinsic.

In the extrinsic pathway cell surface receptors are activated by specific ligands (e.g. Apo2L/TRAIL and CD95L/FasL that bind to DR4/DR5 and CD95/Fas, respectively).

The intrinsic pathway is usually activated by p53 in response to DNA damage. Interactions between pro- and anti-apoptotic members of the BCL2 family lead to the release of cytochrome c from mitochondria and permeabilisation of the mitochondrial membrane.

Both pathways converge on a cascade of caspase enzymes that amplify the signal and implement cell destruction.

Aptamer Oligonucleic acids or peptides that bind with high specificity to target molecules optimised for a particular ligand by cycles of selection from a mixture of random sequences.

Astrocytoma Tumour arising from astrocytes, the major structural brain cell type. Range from benign to highly malignant. They are the most common form of primary brain tumour. Rarely metastasised beyond the central nervous system. Astrocytomas are usually slow growing and may evolve into anaplastic astrocytomas (loss of cellular differentiation, tissue organisation and function).

ASV   Avian sarcoma virus: family *Retroviridae*; causing tumours in chickens (e.g. Rous sarcoma virus, RSV).

Ataxia telangiectasia   An autosomal recessive disorder arising from biallelic mutations in the *ATM* gene and characterised by early-onset progressive cerebellar ataxia (loss of motor coordination), oculo-cutaneous telangiectasia (dilation of small blood vessels in the skin and eyes), hypersensitivity to ionising radiation and an increased incidence of malignancy, particularly leukaemia or lymphoma. Up to 40% of patients with the syndrome develop a malignancy.

ATL   Adult T-cell lymphoma–leukaemia. An aggressive form of T-cell lymphoma–leukaemia characterised by the proliferation of mature CD4+ cells. Associated with HTLV-I infection and occurring particularly in individuals born in the Caribbean, West Africa and Southern Japan.

Autophagosome   An intracellular vesicle that forms during autophagy and fuses with lysosomes to degrade its contents.

Autophagy   A process by which organelles (e.g. mitochondria) and protein complexes are delivered to lysosomes for degradation. It is part of normal cellular homeostasis but is also a mechanism for temporary survival under stress (e.g. nutrient starvation or chemotherapy). The requirement of tumour cells to have a sustained level of mitochondrial function as a component of an autophagic response has been termed 'autophagy addiction'.

Autosomal   Arising from the expression of a gene carried on any chromosome that is not a sex chromosome.

Barrett's oesophagus   A condition associated with increased risk of developing invasive adenocarcinoma, the probable progression being from metaplasia via dysplasia to malignant invasive adenocarcinoma (see Oesophageal cancer table in Appendix E).

Basal-type breast cancer   One of four main classes of breast cancer, based on hormonal status, they comprise a very heterogeneous group that are mostly 'triple negative', i.e. they do not express oestrogen receptor, progesterone receptor and ERBB2 (see Breast cancer table in Appendix E).

Basement membrane   The sheet of fibres that supports and anchors the epithelium and endothelium: made of two thin layers, the basal lamina and the reticular lamina, themselves surrounded by connective tissue.

Connective tissue is one of the four classes of tissues found throughout the body (the others are epithelial, muscle and nervous tissue). It has three main components: cells, fibres and extracellular matrix. Collagen is the predominant protein.

Extracellular matrix, a component of connective tissue, is a macromolecular network occupying the extracellular space in tissues comprising locally secreted proteins and polysaccharides, principally collagen, elastin, fibronectin and laminin, and glycosaminoglycans. Matrix metalloproteases are extracellular proteolytic enzymes that cleave and degrade matrix components.

Basal lamina refers to the flexible, thin mats of specialised extracellular matrix that underlie all epithelial cell sheets and tubes and surround smooth muscle cells, fat cells and Schwann cells. The basal lamina thus separates these cells from underlying connective tissue. It is anchored to connective tissue by type VII collagen fibrils, together referred to as the basement membrane. Most basal laminae contain type IV collagen, laminin, nidogen/entactin and perlecan. Reticular lamina is a fibrous network of type III collagen.

Basic helix–loop–helix   A structural motif in a family of transcription factors comprising two α-helices connected by a loop.

Benign tumour   Such tumours may arise in any tissue and cause local damage by pressure or obstruction but do not spread to other sites or invade adjoining tissues.

Bloom's syndrome   One of three autosomal recessive diseases commonly associated with unrepaired chromosomal breaks (the others are Fanconi's anaemia and ataxia telangiectasia). Bloom's syndrome (BS) is characterised by short stature, immunodeficiency and a greatly elevated frequency of many types of cancer. It is associated with telangiectatic redness of the skin in photo-exposed areas and stunted growth with a propensity to develop acute myeloblastic leukaemia or lymphoma. The gene mutated in BS, *BLM*, encodes an ATP- and Mg^{2+}-dependent DNA helicase.

Burkitt's lymphoma   Human B cell malignancy having 20- to 50-fold higher incidence in tropical Africa than elsewhere. Characterised by chromosomal translocations involving the *MYC* locus. Almost all African forms of Burkitt's lymphoma carry Epstein–Barr virus (EBV) DNA and express at least one viral antigen.

Cachexia   Loss of body weight in someone not trying to lose weight. Also called wasting syndrome; it often occurs in cancer.

Cancer nomenclature   Prefixes refer to the cell type involved:

adeno-   epithelial cells of glands (adenocarcinoma)

chondro-   cartilage (chondroma)

erythro-   red blood cell (erythroblastoma)

fibro-   fibrous connective tissue (fibroma)

glio-   glial tissues (non-neuronal cells that support brain neurons)

hemangio-   blood vessels (hemangiosarcoma)

hepato-   liver (hepatocarcinoma/hepatocellular carcinoma)

leiomyo-   smooth muscle (leiomyosarcoma)

lipo-   fat (lipoma)

lympho- or lympha-   lymphocyte, i.e. white blood cell (lymphoblastoma, lymphangioma)

melano-   pigment cell (melanocytoma)

myelo-   bone marrow (myelosarcoma)

myo-   muscle (myocytoma)

neuro-   nerve tissue (neuroma)

osteo-   bone (osteoma, osteosarcoma)

papillo-   epithelial cells of skin, alimentary tract or bladder (i.e. not glands). Papillomas are benign tumours: the corresponding malignant forms are carcinomas

retino-   retinal cells (retinoblastoma)

rhabdo-   skeletal muscle (benign: rhabdomyoma; malignant: rhabdomyosarcoma)

sarco-   connective tissue (sarcoma).

Suffixes refer to the type cancer:

-aemia   bone marrow cells, the precursors of normal blood cells, e.g. white blood cells, red blood cells, etc.

-blastoma   derived from immature cells or embryonic tissue

-carcinoma   malignant tumours of epithelial cells. Carcinoma *in situ* is a pre-malignant change in which epithelium (particularly of the cervix or skin) shows many malignant changes but does not invade the underlying tissue

-oma   this suffix refers to any swelling or tumour

-sarcoma   connective tissue (bone, cartilage, fat, nerve).

Carcinogen   Any substance that can contribute to the development of cancer. Carcinogens include chemicals, radiation and radionuclides that can damage DNA directly as well as substances that can promote cancer by perturbing metabolism.

Carcinoma   Malignant tumour of epithelial cells. Characterised by ability to invade adjacent tissues and to undergo metastasis. The term includes malignant tumours of glandular epithelial tissue that are termed adenocarcinomas. Carcinoma *in situ* is a pre-malignant change in which epithelium (particularly of the cervix or skin) shows many malignant changes but does not invade the underlying tissue.

Carcinosarcoma   A rare form of malignant tumour containing both epithelial and sarcomatous elements.

Caretaker gene   A gene encoding a protein that contributes to genomic stability, e.g. through DNA repair.

Casein kinase I Mg^{2+}-dependent cytoplasmic (30–37 kDa) and nuclear (25–55 kDa) monomeric enzymes for which glycogen synthase and SV40 T antigen are substrates.

Casein kinase II Ca^{2+} and cAMP-independent serine/threonine kinase ($\alpha2/\beta2$, α 37–44 kDa, β 24–28 kDa) that is widely distributed in eukaryotic cells. Activity is stimulated by growth factors (insulin, TPA, serum) and may be involved in the regulation of cell proliferation. Amino acid substrates must lie within acidic domains. Consensus phosphorylation sequences occur in FOS, JUN, MYC, MYCN, MYCL, MYB, p53, ERBA, v-ETS, adenovirus E1A, SV40 T antigen and HPV E7 proteins. (For a review see Tuazon, P. T. and Traugh, J. A. (1991). Casein kinase I and II – multipotential serine protein kinases: structure, function, and regulation. *Advances in Second Messenger and Phosphoprotein Research* **23**, 123–64.)

Caspase   Caspases form a family of cysteine-*asp*artic prote*ases* or cysteine-dependent aspartate-directed proteases that are essential for apoptosis.

CD   Cluster of differentiation: the CD number identifies specific types of protein on the surface of cells.

Cetuximab   Chimeric monoclonal antibody that inhibits the epidermal growth factor receptor (EGFR).

Choline   Essential nutrient that is incorporated into the headgroups of the phospholipids phosphatidylcholine and sphingomyelin.

Chromothripsis   A single event in which chromosomes are fragmented into many pieces, the majority of which are then reassembled in a seemingly random manner.

Glossary and abbreviations

cis activation Stimulation of gene transcription by a DNA sequence located on the same chromosome, i.e. not by a diffusible agent.

Clinical trials Drug testing in humans comprises three main phases.
  Phase 1: increasing doses are administered to a small group (<80) of either patients or healthy volunteers to determine the maximum tolerated dose.
  Phase 2: involves a larger patient cohort and controls to determine safety and acquire evidence for efficacy.
  Phase 3: on the basis of evidence for efficacy several thousand patients are enrolled in a randomised trial

CLL See leukaemias.

CML See leukaemias.

Colorectal cancer More than 95% of colorectal cancers are adenocarcinomas that originate in the gland cells in the lining of the bowel wall (the bowel comprises the small bowel or small intestine and the large bowel or colon and rectum). The gland cells normally produce mucus. One type of adenocarcinoma is a mucinous or signet-ring tumour in which cancer cells of characteristic appearance often reside in pools of mucus (the mucus pushes the nucleus to one side, giving the appearance of a signet ring under the microscope).
  The remaining 5% of bowel tumours are:
  squamous cell cancers (skin-like cells that make up the bowel lining together with the gland cells);
  carcinoid tumours (an unusual, slow-growing tumour called a neuroendocrine tumour that occurs in hormone-producing tissues, usually in the digestive system. Between 4 and 17% of carcinoid tumours begin in the rectum and 2 to 7% of carcinoid tumours begin in the large bowel);
  sarcomas (cancers of the supporting cells of the body – bone, muscle, etc.). Most sarcomas found in the colon or rectum are leiomyosarcomas that originate in smooth muscle;
  lymphomas (cancers of the lymphatic system. Only about one in a 100 cancers in the colon or rectum are lymphomas).
  (See Colorectal cancer table in Appendix E.)

Connective tissue See Basement membrane.

Contact inhibition Arrest of cell growth upon cell–cell contact.

Cytokine Diverse class of proteins secreted in particular by cells of the nervous and immune systems that are widely involved in cell signalling.

Differentiation Process by which cells give rise to progeny with a more specialised function. Embryonic stem cells are said to be pluripotent because they can differentiate into all the types of cell found in the adult organism. The grade of a tumour is a measure of its state of differentiation, as judged by the morphological appearance of cells. The least aggressive tumours (Grade 1) are often well differentiated, that is, their cells resemble those of their normal counterparts.

De-differentiation (or retro-differentiation) Reversion of cells possessing characteristics of a mature phenotype to a form of immature, precursor cell.

Dominant (allele) An allelic form of a gene expressed in a given cell phenotype as distinct from the other allele carried on the homologous chromosome. Dominant alleles are thus expressed in the homozygous or heterozygous condition; recessive alleles are expressed only when both members of a pair of alleles (or sets of alleles at corresponding loci) are identical.

Double minute Small fragments of DNA that commonly occur in tumour cells that have undergone gene amplification.

Driver mutation Genetic mutation that promotes a stage or stages of tumour development.

Electron volt The energy gained by a single electron when it accelerates through an electric potential difference of one volt. $1\,\text{eV} = 1.602 \times 10^{-19}$ joules.

Endocrine The endocrine system glands that secrete hormones into the bloodstream.

Endothelial cell and endothelium These cells form a layer one cell thick that is the inner lining of all blood vessels, thereby separating the lumen from the rest of the vessel wall. The lining is collectively referred to as the endothelium.

Enhancer Eukaryotic promoter element that increases the transcriptional efficiency of a gene. Often present as short tandem repeats of 50 to 100 bp (e.g. the 72 bp repeat of SV40 virus). Enhancers may be located upstream or downstream of the gene they control at distances of up to several thousand nucleotides from the gene and in some genes (e.g. chicken β globin) the enhancer is located within the gene. Enhancers are equally effective in either orientation.

Enhanson Multiple functional elements that together comprise an enhancer and act synergistically to stimulate transcription from an associated promoter.

Three classes of enhanson have been defined on the basis of the *trans*-activating capacity of repeated sequences of the enhanson, independent of other enhancer elements (proto-enhancer activity). Class A: proto-enhancer activity as a tandem repeat or when associated with a class B enhanson (e.g. Sph-I, Sph-II, GT-IIC). Class B: no proto-enhancer activity alone but cooperate with Class A enhansons (e.g. GT-I). Class C: independent proto-enhancer activity (e.g. TC-II, octamer). Class D enhansons act independently as monomers, e.g. the response elements activated by steroid hormone-receptor complexes.

Each enhanson binds at least one *trans*-activating factor. For example, Sph-I, Sph-II and GT-IIC bind TEF-1, GT-I binds TEF-2, TC-II binds TC-IIA/NF-κB-like and TC-IIB/KBF1/H2TF1-like proteins (that also bind to the κ light chain κB proto-enhancer and the H-2Kb proto-enhancer in the MHC H-2Kb promoter) and the octameric proto-enhancer binds a variety of lymphoid cell-specific factors.

Enhanson sequences: Sph-I: AAGCATGCA; Sph-II: AAGTATGCA; GT-IIC: GTGGAATGT; GT-I: GGGTGTGG; TC-II: GGAAAGTCCCC; H-2Kb: TGGGGATTCCCCA; SV40-P (AP-1-Binding): TTAGTCA; human metallothionein IIA (AP-1-Binding): TGACTCA.

Epithelial cell Cells that line the inner and outer surfaces of the body in continuous sheets, sometimes called epithelial membranes or epithelia. The epithelium is one of the four basic types of animal tissue, along with connective tissue, muscle tissue and nervous tissue. There are four main classes of epithelium: (1) simple squamous; (2) simple cuboidal; (3) simple columnar; (4) pseudostratified. Squamous cells are flat cells that are part of the epithelium as either a single layer (simple squamous epithelium) or multiple layers (stratified squamous epithelium). Pseudostratified cells are a form of column-shaped cells that may also be ciliated, i.e. have fine hair-like extensions. Ciliated epithelium occurs in several places including the nose.

Epstein–Barr virus Burkitt's lymphoma virus: virus of the family *Herpesviridae*, sub-family *Gammaherpesvirinae*. Isolated from the endemic (African) form of Burkitt's lymphoma that primarily afflicts children.

Erythroleukaemia See leukaemias.

Euchromatin The less condensed portions of chromatin visible as cells exit mitosis. Cf. heterochromatin, dark-staining regions of condensed chromatin. Heterochromatin appears most frequently, but not exclusively, at the centromere and telomeres of chromosomes.

Ewing's sarcoma   The second most common primary bone tumour in children and young adults. Cell of origin is unknown. Highly malignant. The t(11;22) (q24;q12) translocation has been identified that results in the fusion of regions of the *ETS* and *EWS* genes.

Expressed sequence tag (EST)   A partial coding sequence isolated at random from a cDNA library, like a sequence-tagged site for mapping total genomic DNA, used for identification and mapping of coding sequences, for discovery of new genes and (by reference to sequence data banks) for the discovery of identities with other genes.

Expression vector   A plasmid-like construct the DNA sequence(s) of which can be transcribed and translated when the vector is taken up by an appropriate host cell. Vectors may include a constitutively active or an inducible promoter.

Extracellular matrix   See Basement membrane.

Familial medullary thyroid carcinoma (FMTC)   Familial recurrence of endocrine tumours associated with mutations in the *RET* gene.

Fatty acid   Fatty acids are unbranched chains of carbon atoms with a methyl group (CH_3-) at one end and a carboxyl group at the other (–COOH). Thus they are a molecule of acetic acid with additional CH_2 groups inserted between the CH_3- and the –COOH ends. The methyl (CH_3- end) carbon of a fatty acid is sometimes called the omega (ω) carbon: an omega-3 (or n-3) fatty acid has its first double bond on the third carbon counting from the ω (methyl) end. Humans cannot make n-3 fatty acids: important nutritional n-3 fatty acids include a-linolenic (18:3), eicosapentaenoic (20:5) and docosahexaenoic (22:6) acids.

In membranes the most common fatty acids have 18 carbon atoms: they are usually unsaturated, that is, contain at least one double bond (that's two carbon atoms joined by two rather than one covalent bond). One double bond makes them mono-unsaturated, more than one and they are poly-unsaturated. Double bonds may be either *cis* or *trans* (hydrogen atoms on the same side or on opposite sides of the double bond, respectively). Naturally occurring fatty acids are rich in the *cis* forms. Unsaturated fats are considered healthier than saturated fats because they lower total cholesterol and LDL cholesterol levels in the blood but *trans* unsaturated fats are to be avoided because the form of the double bond permits a linear structure that may promote plaque formation.

Fibrosarcoma   Malignant tumour of mesenchymal fibrous tissue.

Fluorescence *in situ* hybridisation   A cytogenetic method by which specific DNA sequences are detected by complementary binding of fluorescently labelled probes.

Fluorouracil   Pyrimidine analogue that non-competitively inhibits thymidylate synthase.

Folinic acid   Adjuvant.

Fullerene   Any molecule composed entirely of carbon.

Gain-of-function mutation   Mutation causing increased activity of a gene product or resulting in a protein with a novel function.

Gatekeeper gene   A gene encoding products that normally restrict cell growth, e.g. the retinoblastoma gene *RB1*.

Genetic instability   A cellular state in which mutations occur continuously at an abnormally high rate. Instability may be manifested from the single nucleotide level to gains or losses of whole chromosomes.

Glossary and abbreviations

Glioma   Tumour of the glial tissues (non-neuronal cells present within the CNS) that accounts for ~60% of all primary CNS tumours. Gliomas comprise five distinct types: astrocytomas, oligodendrogliomas, medulloblastomas, ependymomas and spongioblastomas.

Glioblastoma multiforme   A type of tumour that forms from glial (supportive) tissue in the brain. It is highly malignant, grows very quickly and has cells that look quite different from normal glial cells. Early symptoms may include sleepiness, headache and vomiting. Also called a grade IV astrocytoma.

Glycosaminoglycans   Polysaccharide side-chains of proteoglycans made up of >100 repeating disaccharide units of amino sugars, at least one having a negatively charged side-group (carboxylate or sulphate; e.g. heparin). Formerly called mucopolysaccharides.

Glycosylphosphatidylinositol   A glycolipid (a phosphatidylinositol group linked through glucosamine and mannose sugars and an ethanolamine phosphate bridge to the C-terminus of a protein) that forms a GPI anchor for proteins in the cell membrane.

Granulocyte-macrophage colony stimulating factor (GM-CSF)   A cytokine stimulating the formation of granulocyte or macrophage colonies from myeloid stem cells isolated from bone marrow.

Green fluorescent protein   A fluorescent 27 kDa protein first isolated from jellyfish. It is widely used to permit visualisation of intracellular protein location through its expression from a fusion with the gene encoding the protein of interest. It is also used as a reporter when placed under the control of a specific promoter and introduced exogenously into cells. The gene can also be introduced into transgenic organisms under the control of tissue-specific promoters.

Growth factor   A naturally occurring substance regulating cell division and differentiation; mainly protein (peptide) hormones and steroid hormones.

Hairy cell leukaemia   See leukaemias.

Half-life   The time taken for radioactive material to decay to half the original amount.

Haploinsufficiency   A gene dosage effect in a diploid organism in which, following the loss of one allele, the remaining functional gene copy is unable to maintain the normal phenotype.

Haplotype   The set, made up of one allele of each gene, comprising the genotype. Also used to refer to the set of alleles on one chromosome or a part of a chromosome, i.e. one set of alleles of linked genes. Its main current usage is in connection with the linked genes of the major histocompatibility complex.

Hemangioblastoma   Vascular tumour of the CNS.

Heterocyclic amines and polycyclic aromatic hydrocarbons   Heterocyclic: containing a closed ring of more than one kind of atom; polycyclic: three or more closed rings.

Heterozygosity   Possession of one or more dissimilar pairs of alleles of particular genes.

Hirschsprung's disease   Developmental disease, often familial, which affects ~1 in 6,000 people. Characterised by the congenital absence of parasympathetic innervation in the lower intestinal tract. The most severe cases can be fatal. The disease is not known to be associated with cancers but mutations in the *RET* oncogene occur.

Histone deacetylases   HDACs are enzymes that deacetylate histone proteins around which DNA is wrapped. Specifically they remove acetyl groups ($O=C-CH_3$) from lysine amino

Glossary and abbreviations

acids. Histone de-acetylation and acetylation (by histone acetyltransferases) is a critical regulator of gene expression.

HIV Human immunodeficiency virus (HTLV-III, human lymphotropic virus type III, lymphadenopathy-associated virus (LAV)). Family *Retroviridae*, sub-family *Lentivirinae*.

Hodgkin's lymphoma See Lymphoma.

Homologous recombination Recombination at regions of homology between chromosomes by breakage and reunion of DNA allowing direct replacement of the original DNA sequence with an exogenous segment.

HTLV-I, HTLV-II Human T-cell leukaemia/lymphotropic viruses types I and II. Family *Retroviridae*, sub-family *Lentivirinae*. Replication-competent retroviruses. HTLV-I is the apparent causal agent of ATL and is also associated with a variety of non-neoplastic immunological diseases including TSP.

Hypoxia Condition in which all (generalised hypoxia) or parts of the body (tissue hypoxia) experience reduced oxygen supply. Complete oxygen deprivation is referred to as anoxia.

Immune system The organs, tissues, cells and proteins that enable the body to resist infection. When infection occurs there are two main parts to the defensive response: (1) the innate immune system; and (2) the adaptive immune system. The first is an immediate, non-specific response. The adaptive immune response is activated by the innate system and produces improved recognition of the infectious agent that is retained after the cause has been eliminated – an effect called immunological memory. The immune system forms an important defence against cancer because tumour cells express abnormal proteins and thus appear 'foreign', that is, the body responds as it would to an infection.

Initiation site Initiation site (or sequence) is the point at which transcription of a gene begins. In eukaryotic genes the sequence 5′TATAATA3′ (Hogness box) frequently occurs ~30 nucleotides upstream from the initiation site and a sequence (5′GGC/TCAATCT3′) ~40 nucleotides further upstream is also common and acts to promote accurate initiation.

```
 −35   −30                                    −1+1    5
  :     :                                     : :     :
5′ TGGGCATAAAAGGCAGAGCAGGGCAGCTGCTGCTTACACACT 3′ (left-reading strand)
            mRNA:  5′  m⁷GPPACACU3′
```

(The 3′ to 5′ strand, not shown, would be the transcribing strand)

Eukaryotic transcripts often contain additional *leader* sequences at the 5′ P terminus and *trailer* sequences at the 3′ OH terminus: these may or may not be removed by post-transcriptional processing, depending on the type of RNA. Terminator sequences are typically AT- and GC-rich regions near the 3′ end of genes.

Inositol 1,4,5-trisphosphate (IP_3) Second messenger generated by hydrolysis of phosphatidylinositol bisphosphate (PIP_2), together with diacylglycerol (DAG). IP_3 stimulates the release of calcium into the cytoplasm.

Insertional mutagenesis Alteration of a gene as a consequence of inserting nucleotide sequences from viruses or transposons or by transfection or microinjection of DNA.

Irinotecan Topoisomerase 1 inhibitor.

Glossary and abbreviations

Kaposi's sarcoma   Malignant tumour, probably arising from blood vessels. Rare, endemic form is a slowly progressing disease of the elderly. Epidemic Kaposi's sarcoma occurs in patients with HIV infection and pursues a more aggressive course.

Karyotype   The number and appearance of chromosomes.

Kinase   Kinases are enzymes that transfer phosphate groups from donor molecules (particularly ATP) to specific substrates. The process is called phosphorylation and kinases may therefore also be called phosphotransferases. The term kinome refers to the number of kinases encoded by the genome of an organism. The human kinome comprises over 500 genes, one of the largest human gene families.

Landscaper gene   A gene that contributes to tumour growth through its effect on the environment, for example, when expressed in stromal cells.

LCR   Long control region: regulatory region in viral genomes. In human papillomaviruses the LCR encompasses 5 to 12% of the viral genome and contains an intricate network of *cis* responsive elements.

Leucine zipper   A protein sequence of five leucine amino acids each separated by six residues. It mediates dimer formation via a coiled-coil arrangement of parallel α helices and is normally adjacent to a basic, DNA binding domain. The leucine zipper family includes C/EPB, FOS, JUN, CREB and GCN4. In CREB and GCN4 the fifth leucine is substituted by arginine and lysine respectively.

Leucocyte   Any colourless, amoeboid cell mass, including lymphocytes, monocytes and granulocytes (neutrophils, basophils and eosinophils).

Leukaemia   Malignant proliferation of bone marrow cells, which are the precursors of normal elements of the blood, e.g. lymphocytes, erythrocytes, etc. Classified according to lineage of origin and degree of differentiation of the tumour cells, which often predicts clinical behaviour. The major consequences of leukaemia arise from overgrowth of malignant cells in the bone marrow leading to reduction in production of normal bone marrow and blood components. The infiltration of other organs (e.g. spleen, skin) can also occur.

Acute leukaemias have features of cells early in the differentiation pathway and often have an explosive clinical course. Chronic leukaemia cells appear more closely related to fully differentiated blood cells and pursue a more indolent course, sometimes spanning many years.

The major sub-divisions of leukaemias are those with lymphoid features (acute lymphoblastic leukaemia, ALL; chronic lymphocytic leukaemia, CLL) and those with myeloid features (acute myeloblastic leukaemia, AML; acute myelomonocytic leukaemia, AMML; acute promyelocytic leukaemia, APL; chronic myeloid leukaemia, CML). The acute myeloid leukaemias are also termed acute non-lymphoblastic leukaemias (ANLL). (See Leukaemia table in Appendix E.)

Leukosis   Proliferation of leucocyte-forming tissue: includes myelosis and lymphadenosis and forms the basis of leukaemia. In chickens, fowl leukosis refers to the proliferation of immature myeloid or lymphoid cells, including erythroblastosis, granuloblastosis, lymphomatosis and myelocytomatosis.

Linkage disequilibrium   The occurrence of some genes together, more often than would be expected. Thus, in the HLA system of histocompatibility antigens, HLA-A1 is commonly associated with B8 and DR3, and A2 with B7 and DR2, presumably because the combination confers some selective advantage.

Glossary and abbreviations

Lod score (Z)   When studying the segregation of alleles at two loci that could be linked, a series of likelihood ratios is calculated for different values of the recombination fraction (ϑ), ranging from $\vartheta = 0$ to $\vartheta = 0.5$. The likelihood ratio at a given value of ϑ equals the likelihood of the observed data if the loci are linked at the recombination value of ϑ divided by the likelihood of the observed data if the loci are not linked ($\vartheta = 0.5$).

Z = Log to base 10 of this ratio (i.e. Lod(ϑ) = $\log_{10}[L\vartheta/L(0.5)]$).

Thus Z = 4 at a recombination fraction (ϑ) 0.05 means that in the families studied it is 10,000 times more likely that the disease and marker loci are closely linked (i.e. 5 cM apart) than that they are not linked.

A lod score of +3 or higher is generally agreed to indicate linkage.

The recombination fraction (ϑ) is a measure of the distance separating two loci (i.e. the likelihood that a cross-over will occur between them. If two loci are not linked, $\vartheta = 0.5$, as on average genes at unlinked loci will segregate together during 50% of all meioses. If $\vartheta = 0.05$, on average the syntenic alleles will segregate together 19 times out of 20; a cross-over will occur between them on average only 1 in 20 meioses.

LOH/LOCH   Loss of (constitutional) heterozygosity. Complete absence of one allele at a given locus. Any somatic alteration arising in a tumour in the relative abundance of two alleles should be identifiable from a Southern blot comparing normal and tumour DNA samples.

Loss-of-function mutation   Mutation that reduces or completely ablates the activity of a gene product.

LTR   Long terminal repeat. Identical (may include some inverted repeats) DNA sequences of several hundred base pairs ($\sim$270–1,300) of the structure U3-R-U5 (5′ to 3′) at both ends of the unintegrated linear DNA product of reverse transcription, at both ends of integrated proviral DNA and in closed circular retroviral DNA.

Inverted repeats (IR) occur at the ends of LTRs and are perfect (or slightly imperfect) inverted repeats of $\sim$3–25 bp that form a palindrome when two complete LTRs are joined in circular DNA. Integration of proviral DNA results in the loss of 2 bp from the IRs.

Lung cancer   Major sub-divisions are small cell lung cancer and the non-small cell cancers: namely squamous (epidermoid) carcinoma, adenocarcinoma and large cell undifferentiated carcinoma. Tumours with mixed histology may occur. Uncommon lung neoplasms include carcinoid tumours and bronchioalveolar carcinoma (see Lung cancer table in Appendix E).

Lymphocyte   A type of white blood cell that functions in the immune response. The three main classes are B cells, T cells and natural killer cells. They are typically about the same diameter as red blood cells ($\sim$7 μm).

Lymphoid cell   One of the major classes of cells derived from bone marrow stem cells.

Lymphoid tissue   Lymphatic system tissue predominantly arranged in lymph glands but also found in the bowel wall and other sites.

Lymphoma   Malignant proliferation of lymphoid cells. Major divisions of the disease are Hodgkin's disease (Hodgkin's lymphoma) and non-Hodgkin's lymphomas (see Leukaemia table in Appendix E).

Hodgkin's disease is characterised by the presence of binucleate or multinucleate Sternberg–Reed cells, though the cell of origin of Hodgkin's disease is unknown. The disease predominantly presents as lymph node swellings, which may be isolated or, in more extensive disease, involve multiple lymph node sites, liver, spleen, bone marrow and other organs. Hodgkin's disease is sub-divided according to histological appearance into groups that have differing clinical behaviour and prognosis. In the majority of cases Hodgkin's disease is curable with chemotherapy and/or radiotherapy.

The non-Hodgkin's lymphomas present a varied spectrum of histological features and clinical behaviour, ranging from Burkitt's lymphoma, which has a doubling time shorter than any other human tumour, to well differentiated lymphocytic lymphoma, which has a natural history that may span decades. Division of non-Hodgkin's lymphoma into high-grade, intermediate and low-grade disease, based on morphological and surface marker expression, identifies groups with differing prognosis and therapy. Some types of non-Hodgkin's lymphoma are curable with chemotherapy and/or radiotherapy.

Macrophage   White blood cells that function in both innate and adaptive immune responses to engulf and act as scavengers that digest pathogens and cell debris. They arise from the differentiation of monocytes, both cell types being phagocytes. Human macrophages are typically about 20 µm in diameter.

MALDI and TOF   MALDI (matrix-assisted laser desorption ionisation) is a solid-phase ionisation technique in which samples are dissolved in a solution containing a chromophoric matrix. The matrix-sample mixture is dried on a target plate and subjected to laser irradiation. As the matrix absorbs energy from the laser, it desorbs (vaporises) and ionises, and transfers charge to the sample that was desorbed with the matrix. TOF, or time-of-flight, is a mass analysis technique in which ions are separated, based on the time it takes for them to travel over a given distance. In a TOF MS ions of like charge are simultaneously emitted from the source with the same initial kinetic energy. Those with a lower mass will have a higher velocity and reach the detector earlier than ions with a higher mass.

Peptide mass fingerprinting is a process in which a protein sample is digested with trypsin to produce fragments; the fragments are then subjected to MALDI MS. The resulting peaks, which represent the masses of every peptide fragment in the sample, constitute the fingerprint that can be used to determine the identity of the initial protein by comparing it to known patterns in a sequence database.

Malignant tumour   Uncontrolled cell growth characterised by invasion through the basement membrane into surrounding tissue and a propensity to spread (metastasise) by blood or lymphatic routes to other sites.

MAPK   Mitogen activated protein kinases (MAPKs), also known as extracellular signal-regulated kinases (ERKs): highly conserved cellular enzymes that are activated by a variety of trans-membrane receptors in response to extracellular signals through their phosphorylation on serine and tyrosine residues. Each 'MAPK cascade' consists of three enzymes, MAP kinase, MAP kinase kinase (MKK, MEK or MAP2K) and MAP kinase kinase kinase (MKKK, MEKK or MAP3K) that are activated in series.

RAF1 is a MAP3K and in addition there are two other members of the RAF family (ARAF and BRAF) and several other non-RAF members of the MAP3K (or MAPKKK) family. RAF proteins can phosphorylate two distinct MAP2Ks (MAP1K1 and MAP2K2) but there are five other MAP2Ks. Collectively these MAP2Ks can activate at least 15 MAPKs. Members of the MAP2K family are dual-specificity kinases that act both as tyrosine kinases and serine/threonine kinases: they phosphorylate the threonine and tyrosine residues within a TXY motif that is conserved in MAPKs.

Marfan's syndrome   An autosomal dominant condition resulting in a generalised weakness of the supporting tissues of the body. In its classical form it is associated with abnormalities of the eye, the skeletal system and the cardiovascular system. The prevalence of classical Marfan's syndrome is four to six per 100,000 people.

Mass spectrometry (MS)   The measurement of molecular mass that is gained by determining the mass-to-charge ratio (m/z) of ions generated from the target molecule. Mass spectrometers comprise a source for generating the ions from the sample and delivering them into the gas phase; an analyser for separating and sorting the ions; and a detector for sensing the ions as they are sorted. An MS 'run' generates a spectrum that displays ion intensity as a function of m/z.

Mast cell   Secretes histamine and other chemicals that cause allergic and inflammatory responses and are also involved in wound healing.

Melanoma   Malignant tumour of melanocytes (pigmented skin cells). A benign proliferation of such cells is termed a naevus. Melanoma incidence and mortality rates in fair-skinned populations are increasing worldwide. Epidemiological data suggest that high numbers of benign acquired naevi and dysplastic naevi are associated with higher risk of melanoma. Histopathological analysis also indicates that a proportion of melanomas show evidence of a histologically contiguous naevus. Melanocytic naevi can be present at birth (congenital) or arise during an individual's lifetime (common acquired or dysplastic). Conceptually, the premise that all naevi are pre-malignant is controversial, owing in part to the clinical and histopathological heterogeneity of these lesions, but *BRAF* mutations in naevi suggest that the premise is correct.

MEN 1   Multiple endocrine neoplasia type 1 (Wermer's syndrome). Inherited autosomal dominant trait distinguished by the occurrence of tumours of the anterior pituitary and pancreatic islet cells and parathyroid hyperplasia.

MEN 2A/MEN 2B   Multiple endocrine neoplasia type 2A (MEN 2A) is a cancer inherited as an autosomal dominant trait involving malignant tumours of the 'C' cells of the thyroid (medullary carcinoma of the thyroid, MTC) and usually benign tumours of the adrenal medulla (phaeochromocytoma). MEN 2B is similar to MEN 2A but is characterised by earlier tumour onset, ganglioneuromatosis of the intestine and Marfanoid habitus.

Meningioma   Tumour of the meninges, the three membranes that envelope the brain and spinal cord. Meningiomas are typically benign but may compress the brain and can evolve into malignant sarcomas.

6-mercaptopurine   Purine analogue that inhibits purine biosynthesis and DNA replication.

Mesenchymal cell   Mesenchymal stem cells or marrow stromal stem cells (MSCs) are stem cells that can differentiate into osteoblasts, chondrocytes, myocytes, adipocytes, neuronal cells, and, as described lately, into beta-pancreatic islets cells. MSCs differentiate from colony-forming unit-fibroblast (CFU-F). MSCs cultured without serum in the presence of transforming growth factor (TGF) will differentiate into chondrocytes, whereas MSCs cultured in serum with ascorbic acid and dexamethasone will differentiate into osteoblasts. Mesenchymal stem cells have the capability for renewal and differentiation into various lineages of mesenchymal tissues. These features of MSCs attract a lot of attention from investigators in the context of cell-based therapies of several human diseases. Despite the fact that bone marrow represents the main available source of MSCs, the use of bone-marrow-derived cells is not always acceptable due to the high degree of viral infection and the significant drop in cell number and proliferative/differentiation capacity with age. Thus, the search for possible alternative MSC sources remains to be validated. Umbilical cord blood is a rich source of haematopoietic stem/progenitor cells, and does not contain mesenchymal progenitors. However, MSCs circulate in the blood of pre-term fetuses and may be successfully isolated and expanded. Where these cells home at the end of gestation is not clear.

Mesothelioma   Benign or malignant tumour arising from the mesothelial lining of one of the coelomic cavities, usually the pleura or peritoneum, and consisting of epithelioid and spindle cells (see Lung cancer table in Appendix E).

Metastasis   Spread of malignant cells from the site of origin to other sites. Cells usually migrate via the bloodstream or lymphatics but spread can also occur directly (within the abdominal cavity) or via the cerebrospinal fluid from one site in the central nervous system to another.

Metastatic signature   A metastatic signature describes a sub-set of genes whose expression distinguishes primary tumours from metastatic tumours. Up-regulated and down-regulated sub-sets are part of such signatures. Signatures may confer a poor prognosis, an example being over-expression of components of the WNT signalling pathway in early lung metastases.

Metastasis suppressor   Metastasis suppressors are proteins that retard or inhibit metastasis with no effect on primary tumours. Over 20 metastasis suppressor genes have been identified in a wide variety of tumours and their encoded proteins exert a broad range of cellular effects. A number have been associated with good prognosis of specific cancers including NM23 (breast, gastric, melanoma), RECK and PEBP1 (colon, prostate), KAI1 (breast, colon, ovarian, non-small cell lung), KISS1 (ovarian, non-small cell lung), MAP2K4 (prostate) and CTFG (colon, non-small cell lung).

Methotrexate   A competitive inhibitor of dihydrofolate reductase.

MHC   Major histocompatibility complex: three classes of cell surface molecule that mediate the actions of cells of the immune system.

Micro/minisatellite DNA   Microsatellites (or simple repeated sequences, SRSs) are short, repetitive sequences of DNA (up to 6 bp with between 10 and 50 copies) that are stably inherited, vary from individual to individual and have a relatively low inherent mutation rate. A mutator mechanism for cancer arises from ubiquitous somatic mutations (USMs) at SRSs caused by failures of the strand-specific mismatch repair system to recognise and/or repair replication errors due to slippage by strand misalignment. Microsatellite length heterogeneity is present in colon tumours, classified as replication error positive (RER+), showing that somatic genomic instability (SGI) is a very early event in the development of these tumours.

Minisatellites are tandem arrays of a locus-specific consensus sequence that varies between 14 and 100 bp in length. They are often polymorphic in the number of tandem repeats (hence referred to as variable number of tandem repeats (VNTRs) or variable tandem repetitions (VTRs). Minisatellites are dispersed throughout the genome and often occur just upstream or downstream of genes (or within introns). Many VNTR loci display dozens of alleles.

Minimal inverted repeat   Inverted repeat, or palindromic, sequences of nucleic acids in which the code reads the same from each end on complementary strands.

Mitogen   A chemical promoting cell division.

MNNG   N-methyl-N′-nitrosoguanidine: chemical mutagen.

Modifier gene   A gene the action of which modifies the penetrance of the phenotype arising from mutation or loss of function of another gene.

Glossary and abbreviations

Monoclonal antibody   Usually obtained from spleen cells from a mouse that has been immunised with the required antigen fused with myeloma cells. The antibodies are produced from identical immune cells cloned from a unique parent cell and are therefore monospecific.

Monocyte   Mononuclear blood phagocytes that migrate into tissues and differentiate into macrophages.

Multiple myeloma (myelomatosis)   Malignant proliferation of plasma cells.

Myeloblasts   Bone marrow cells that are the precursors of myelocytes.

Myelocytes   Bone marrow cells that develop into granular leucocytes (polymorphonuclear leucocytes) in the blood and are present in certain forms of leukaemia.

Myelocytomatosis   Leukosis involving myelocytes; in fowl: tumours composed of myeloid cells.

Myeloid cell   One of the two classes of cells derived from bone marrow stem cells (the other being lymphoid cells) and comprising megakaryocytes, erythroid precursors, monocytes and polymorphonuclear leucocytes (granulocytes).

Myeloid tissue   Erythrocyte- and leucocyte-producing tissue in fetal liver and spleen and in the bone marrow.

Myoblast   Muscle cell (myocyte) precursor.

Myofibroblast   A sort of cross between a fibroblast and a smooth muscle cell that can release proteins that signal to other cells and helps with wound repair.

Natural killer cell   A type of lymphocyte that kills cells identified by the immune system as being 'foreign'.

NCAM   Neural cell adhesion molecule.
Cell surface glycoprotein: anti-NCAM antibody inhibits cell–cell adhesion.

Neoplasm   New or abnormal cell growth, e.g. a tumour.

Neural crest   Transient embryonic structure of the neural epithelium that is converted to a mesenchymal state: these cells subsequently migrate throughout the embryo to give rise to many derivatives including most of the peripheral nervous system and melanocytes.

Neuroblastoma   Malignant tumour derived from primitive ganglion cells. Together with nephroblastoma (Wilms' tumour), one of the most common solid tumours of childhood. Have the highest spontaneous regression rate of all tumours.

Neurofibromatosis Type 1 (Von Recklinghausen's disease or multiple neuroma)   Familial condition caused by germ line mutations at the *NF1* locus that give rise to developmental changes in the nervous system, muscle, bones and skin. Tumours of nerve sheaths (neurofibromas) may occur in multiple sites. Occurs in one in 30,000 births.

Next-generation or second-generation sequencing   (See Box 8.1.)

NF-κB   Generic name for the nuclear factor κB (NF-κB) transcription factor family that includes *NFKB1*, *NFKB2* and the *REL* and *IKB* families. They play central roles in inflammation and immunity.

NGF   Nerve growth factor: 118 amino acid polypeptide (13 kDa) with chemotropic properties for sympathetic and sensory neurons. In peripheral tissues it attracts neurites to form synapses.

Non-Hodgkin's lymphoma   See Lymphoma.

Glossary and abbreviations

Oncogene A proto-oncogene that has acquired a mutation that confers the capacity to contribute to cancer development. Oncogenes exert dominant effects, i.e. only a single allele need undergo mutation to produce a phenotype.

Oncogene addiction Refers to the reliance of a cancer cell for growth and survival on a single oncogene, which may thus offer a therapeutic target.

ORF Open reading frame: a sequence of nucleotides in DNA that can be read as codons, does not contain a termination codon and potentially encodes a polypeptide. *ORF* also denotes a highly conserved gene present in the $3'$ LTR of mouse mammary tumour viruses.

Ornithine decarboxylase The earliest immediate early response gene to be activated when quiescent cells are stimulated to enter the cell cycle. Catalyses the reaction: L-ornithine = putrescine + CO_2. Spermidine is formed from putrescine and S-adenosylmethionine and the reaction is repeated to produce spermine. Spermidine $[(CH_2)_{10}NH_2^+)_4]$ and spermine $[(CH_2)_7NH_2^+)_3]$ are polyamines found in all bacteria and most animal cells. They are growth factors for some microorganisms and stabilise the structure of ribosomes, some viruses and the DNA of many organisms. ODC is essential for cell proliferation and differentiation and its expression is responsive to a wide variety of growth-promoting stimuli. Changes in ODC expression resulting in polyamine accumulation have been reported during cell transformation and carcinogenesis, and increased levels of ODC have been particularly associated with colon carcinoma. *ODC* is a direct *trans*-activating target of MYC, although this does not explain its very rapid activation upon the stimulation of quiescent cells. An increase in putrescine after central nervous system injury appears to be involved in neuronal death.

Osteosarcoma Malignant tumour of bone, particularly occurring in childhood and early adulthood.

Oxaliplatin Platinum compound that inhibits DNA synthesis.

Papilloma Benign tumour of epithelial cells of the skin, alimentary tract or bladder. The corresponding malignant forms are carcinomas.

Passenger mutation Genetic mutation in a tumour that is not essential for its development.

PDGF Platelet derived growth factor. Dimer of A and/or B or C or D polypeptide chains. PDGF B chain is almost identical to the v-SIS oncoprotein. The PDGF A gene is 60% similar to the B chain gene. AA, AB and BB dimers occur in normal and transformed cells. There are two distinct types of receptor: PDGFα and PDGFβ. PDGFβ receptors bind only PDGF-BB. PDGFα receptors bind PDGF-AA, PDGF-AB or PDGF-BB with high affinity. PDGFαβ receptors bind PDGF-AB or PDGF-BB. PDGF-CC binds strongly to PDGFRα and PDGF-DD is a specific PDGFRβ agonist.

Phaeochromocytoma See MEN 2.

Phosphatidylinositol Negatively charged membrane lipid comprising the alcohol inositol ($C_6H_{12}O_6$) attached to phosphatidic acid.

Phosphatidylinositol 3-kinases (PI3-kinases or PI3Ks) A family of enzymes that regulate diverse cellular functions by phosphorylating the inositol ring of phosphatidylinositol and its derivatives in the 3 position hydroxyl group.

Phosphatidylserine A phospholipid mainly located on the cytosolic side of membranes that becomes exposed on the outer surface during apoptosis.

Glossary and abbreviations

Phosphoinositide Phosphoinositides are phosphorylated derivatives of phosphatidylinositol (PI) including phosphatidylinositol phosphate (PIP), phosphatidylinositol bisphosphate (PIP_2) and phosphatidylinositol trisphosphate (PIP_3).

Phosphorylation The covalent addition of a phosphate (PO_4) group to a molecule, most commonly a protein in a reaction catalysed by a kinase.

Plasmacytoma Localised malignant tumour of plasma cells (i.e. B lymphocytes). Multiple myeloma is a disseminated form of plasmacytoma. Plasmacytoma in rodents is caused by injection of complete Freund's adjuvant. The hybridoma cells from which monoclonal antibodies are obtained are produced by fusion of plasmacytoma cells and primed lymphocytes.

Plasmid Small, circular, double-stranded DNA (up to 200 kb) that can replicate independently and be transferred from one organism to another. Widely used as carriers of cloned genes.

Plasposon Plasposons are mini-transposons containing an origin of replication between the inverted repeats. Thus the plasmid portion of the molecule moves with the transposon to a new location. These artificially derived plasmid transposons are dubbed plasposons. Once a mutant has been constructed with a plasposon, the region around the insertion point can be rapidly cloned and sequenced. The insertion point can be mapped by PFGE using the rare restriction enzyme cutting sites built into the molecule. Several plasposon insertions can be pooled to construct a genomic library of the target strain.

Platelet Also called thrombocytes, platelets are small cell fragments derived from the fragmentation of megakaryocytes. Involved in the formation of blood clots, they secrete a range of growth factors including platelet-derived growth factor (PDGF) and transforming growth factor beta (TGFβ).

Polymorphonuclear leucocyte (granulocyte) Mammalian blood leucocytes of the myeloid lineage. The sub-classes of PMNLs are neutrophils (short-lived phagocytic cells), eosinophils (poorly phagocytic) and basophils (non-phagocytic; histamine containing).

Positron emission tomography (PET) A method that uses radioactive tracers to produce a three-dimensional picture of tissues in the body, particularly used to detect tumours.

Prednisone Corticosteroid used as an immunosuppressant.

Promoter A gene promoter is a region of DNA that regulates transcription. Typically they are in the regions immediately 5′ to the transcription start site.

Proviral integration The acquisition of virally encoded DNA by a host cell genome. Genes activated or mutated by proviral integration may be identified by cloning somatic proviral DNA-host cellular DNA junction fragments from retrovirally induced tumours. DNA probes for the regions flanking the provirus are then used to screen DNA from other tumours.

Proviral tagging Use of proviral DNA as a probe to isolate flanking DNA from tumour cell libraries. Activated cellular proto-oncogenes and previously unknown genes have been detected by this means. Analogous to transposon tagging.

Provirus Viral DNA that has become integrated into a host cell's chromosomal DNA. The RNA genome of *Retroviridae* must first be transcribed to DNA by the action of reverse transcriptase. Proviral genes may be expressed or latent. Proviral integration of oncogenic viruses may cause cell transformation.

Glossary and abbreviations

Pseudogene Non-functional DNA sequences closely similar to those of expressed genes. May be the result of gene duplications in which loss of promoters or other mutations prevent expression.

Radical Radicals (usually called free radicals) are atoms, molecules or ions that have unpaired electrons (an electron in an orbital, rather than an electron pair). Reactive oxygen species are free radicals.

Reactive oxygen species Chemically reactive species containing oxygen (e.g. superoxide (O_2^-) and peroxides).

Real time PCR The exonuclease-based real time polymerase chain reaction (PCR) exploits 5′ to 3′ exonuclease activity of Taq polymerase and measures PCR product accumulation as the reaction proceeds through a dual-labelled fluorogenic probe.

Renal cell carcinoma (RCC) The most common type of kidney cancer, accounting for 80% of cases.

Retinoblastoma Malignant tumour of the retina composed of primitive retinal cells and usually occurring in children less than five years old. Occurs as sporadic or familial forms. In the familial form a germ line mutation of the *RB* gene is found and retinoblastoma occurs bilaterally in approximately one-third of cases. There is a high incidence of second malignancy in these patients, particularly osteosarcoma.

Retinoic acid Vitamin A precursor. Dietetic deficiency causes visual impairment. Influences cell differentiation. Used therapeutically in some forms of leukaemia and epithelial pre-malignancy.

Retroviruses Family of spherical enveloped viruses divided into three sub-families: *Oncovirinae* (includes all oncogenic and closely related non-oncogenic viruses), *Lentivirinae* (the 'slow' viruses, e.g. visna virus) and *Spumavirinae* ('foamy' viruses that cause persistent infections without any clinical disease). The genome consists of an inverted dimer of linear (+)-sense RNA. All viruses contain an RNA-dependent DNA polymerase activity (reverse transcriptase). Viruses are 80 to 100 nm in diameter with surface glycoprotein projections of 8 nm. The helical ribonucleoprotein is contained in an icosahedral capsid. Transmission is both vertical and horizontal.

A-, B-, C- and D-type virus particles:

Morphologically defined groups of retrovirus particles. The three types of murine retroviruses are classified as:

A-type: double-shelled spherical particles of diameter 65 to 75 nm (outer shell) and 50 nm (inner shell) often found as intracellular particles in tumour cells.

B-type: dense core of 40 to 60 nm diameter within a 90 to 120 nm diameter envelope seen outside mouse mammary carcinoma cells after budding through the cell membrane (e.g. MMTV).

C-type: 90 to 110 nm diameter particles seen in association with leukaemic tissue and with sarcomas (e.g. ALV).

D-type: observed both intracellularly (60 to 90 nm diameter) and extracellularly (100 to 120 nm diameter).

The electron-dense nucleoid of the extracellular form is located eccentrically but the glycoprotein spikes are shorter than those of B-type particles. The nucleoid region is located acentrically in B-type particles and concentrically in C-type particles.

Retrovirus life cycle The genome of a retrovirus (Class VI enveloped viruses) comprises two identical plus strands of RNA. The genome encodes the enzyme reverse transcriptase, which, after infection of a target cell, copies the retroviral RNA into double-stranded DNA that is integrated into the host cell genome. To start the process retroviruses contain a few molecules of reverse transcriptase. The integrated DNA is transcribed by the host cell machinery into (+)RNA that is either translated into viral proteins or packed into new virions.

Retroviral genome (elements)

env *Env*elope: the third coding domain (~2 kb) of a replication competent retroviral genome, the product of which is a polyprotein from which the major structural proteins of the viral envelope are generated.

gag Group-specific *antigen*: the first of the three coding domains of a replication-competent retroviral genome. *gag* (~2 kb) encodes a polyprotein whose products form the major structural proteins of the virus.

L   Untranslated sequence (~250 nucleotides) preceding the coding region of *gag* that may determine the packaging of virion RNA.

onc Generic region found in many oncogenic retroviruses that encodes an oncoprotein. May be 3′ to *env* or replace some or all of the normal retroviral genes resulting in a replication defective virus.

PB+   Undefined site for primer binding for the positive strand.

PBS   Primer binding site for the initiation of negative strand synthesis.

pol RNA-directed DNA *pol*ymerase (reverse transcriptase): the second coding domain of a replication competent retroviral genome (~3 kb). *pol* encodes a polyprotein that generates *gag* peptides as well as reverse transcriptase and RNAse H.

R   A short sequence (20–80 nucleotides) directly repeated at both ends of each retroviral RNA subunit that excludes the cap nucleotide at the 5′ terminus and the poly(A) tract at the 3′ terminus. Present in each LTR in viral DNA.

U3   Untranslated region at the 3′ end of the viral genome (~170–1,200 nucleotides) that contains the viral promoter. Present once in viral RNA and in each LTR of viral DNA.

U5   Sequence (80–100 nucleotides) between R and PBS. Present once in viral RNA and in each LTR of viral DNA.

Retrovirus vector   Retroviruses from which all viral genes have been removed or altered and which contain the foreign gene to be expressed, usually with a selectable marker (typically neomycin phosphotransferase).

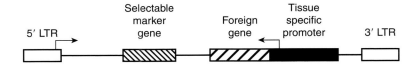

To prepare virus stocks cloned proviral DNA is transfected into a packaging cell that contains an integrated provirus with intact genes but lacking the sequence recognised by the packaging apparatus. Thus packaging cells cannot produce infectious virus but can package RNA transcribed from the transfected vector into virions that are released from the cell and can infect target cells. This virus stock should be helper-free, i.e. lack wild-type replication-competent virions. The non-infectious virions can transfer exogenous genes with high efficiency (30–50%) and precise integration into the DNA of almost all types of eukaryotic cell.

Glossary and abbreviations

Reverse transcriptase   RNA-dependent DNA polymerase: an enzyme of retroviruses that synthesises a single strand of DNA using an RNA template.

Rhabdomyosarcoma   Malignant tumour of striated muscle, particularly occurring in childhood.

RNA editing   Chemical modification in the bases of RNA after it has been transcribed. Examples have been found in mRNA, tRNA, rRNA and micro RNA molecules.

RNA interference (RNAi) or RNA silencing   Modulation of translation from specific mRNAs by one of two types of naturally occurring molecules, micro RNA (miRNA) and small interfering RNA (siRNA, also known as short interfering RNA or silencing RNA). The degradation of target gene mRNA can be induced by short dsRNA molecules (20–25 nucleotides) corresponding to the sequence of the target gene to be silenced (short interfering RNA duplexes (siRNA)). dsRNA causes the degradation of mRNA that is homologous to the trigger dsRNA – the process known as RNA interference.

Sarcoma   Tumours of tissue derived from the mesenchymal layer (connective tissue, bone, cartilage, muscle, fat, blood vessels) as distinct from carcinomas, which are derived from epithelial cells. Often highly malignant.

Scaffold protein   Proteins that regulate cell signalling by having multiple binding sites for other proteins. By tethering into complexes scaffolds localise and focus signal pathway components. Several scaffold proteins participate in the RAS-MAPK pathway including KSR.

Schwannoma   Tumour of the nerve sheath. Termed acoustic neuroma when the tumour occurs on the auditory nerve, the most common site. Also termed neurinoma or neurilemmoma. Benign and usually well demarcated or encapsulated tumour arising from Schwann cells of the neurilemma.

Senescence   In vitro human fibroblasts undergo cellular senescence in two phases: mortality stages 1 and 2.

M1 corresponds to normal replicative senescence and probably occurs when a few telomeres are sufficiently short that the ends are not fully masked from being recognised as requiring double strand break repair. Cell division is then blocked by factors associated with both acute and chronic DNA damage checkpoint activities, including p53, INK4A and RB1. If these checkpoints are blocked by mutations or expression of viral oncogenes, cells divide until M2 (crisis).

M2 probably represents the consequences of terminally short telomeres where end-to-end fusions and chromosome breakage–fusion cycles result in apoptosis.

Single-nucleotide polymorphism (SNP)   A SNP is a variation at a single site in DNA. They occur randomly at an average frequency of one per 1,000 bases and give rise to variation between individuals.

Single nucleus sequencing   A method by which DNA sequence is obtained from single cells isolated from tissues.

Southern blot   A method for the detection of specific DNA sequences: DNA fragments separated by electrophoresis are blotted onto a filter membrane and identified by hybridisation of labelled probes.

Sphingosine 1-phosphate   A sphingolipid, i.e. a lipid derived from the 18 carbon amino alcohol sphingosine.

Glossary and abbreviations

Splice acceptor site (SA)   3′ site at which viral RNA is joined to an SD to form sub-genomic mRNA.

Splice donor site (SD)   Site at which a 5′ portion of the integrated proviral genome is joined to a 3′ region to form spliced, sub-genomic mRNA. The major SD occurs near the 5′ terminus, either within *gag* or in the untranslated region upstream of the first *gag* initiation codon.

Squamous carcinoma   Malignant tumour of squamous epithelial cells occurring in skin and head and neck cancers, and in a minority of bladder, cervix, lung, oesophageal and prostate cancers (see cancer tables in Appendix E).

Stomach (gastric) cancer   More than 95% of stomach cancers are adenocarcinomas, originating in the gland cells in the stomach lining. The other 5% of stomach cancers are either squamous cell cancers, lymphomas, sarcomas or neuroendocrine tumours (see Stomach (gastric) cancer table in Appendix E).

Stroma   Framework of connective tissue that supports organs composed of a variety cells and intercellular material (the extracellular matrix).

Synthetic lethality   Refers to a combination of mutations in two or more genes causing cell death, whereas when only one is mutated the cell remains viable. In cancer it refers to causative mutations, for example, an inactivated tumour suppressor and an activated oncogene driving independent pathways critical for progression. Therapeutic inhibition of the oncoprotein should block cancer development.

Telomerase   Telomeres are synthesised by telomerase, a ribonucleo protein DNA polymerase. Short, tandemly repeated sequences of DNA (TTAGGG in vertebrates) are added to the 3′ termini of chromosomal DNAs, this elongation allowing primer binding and the initiation of synthesis on the other strand, so that chromosome shortening is limited and coding sequences are protected. Telomerase activity is present in many tumours but is largely suppressed in normal tissues.

Telomere   DNA strands can only be made in one direction relative to the parent template DNA (5′ to 3′) and the machinery needs a short sequence (a primer, made of RNA) attached to the template to act as a 'starter'. As double-stranded DNA comes apart during replication, one of the daughter strands has to be made in short lengths, Okazaki fragments, that are then joined together – this is the 'lagging strand'. This system cannot replicate the ends of lagging strands and, to prevent degradation of genes, the ends of chromosomes are capped with telomeres – repetitive DNA sequences.

Teratoma   Tumours comprising tissues that are derived from three germinal layers: the endoderm, mesoderm and ectoderm. They may be solid or cystic and are classified histologically as mature, immature and malignant. The term teratocarcinoma is used to describe tumours in which malignant teratoma tissue co-exists with material that histologically resembles epithelial malignancy.

Terminal differentiation   Regulated progression of eukaryotic cells through successive steps of differentiation and growth inhibition resulting in growth arrest.

6-thioguanine   Purine analogue that inhibits purine biosynthesis and DNA replication.

Thyroid hormones   Thyroxine and tri-iodothyronine: iodinated aromatic amino acid compounds produced and secreted by the thyroid gland.

THR   Thyroid hormone receptor.

trans-activation Modulation of gene transcription by the product of a locus on another chromosome, usually by a regulatory protein.

Transcription factor Proteins that bind to specific DNA sequences to regulate gene expression: the transcription of DNA into RNA. They are therefore sequence-specific DNA-binding factors. Additional factors (e.g. co-activators) may interact with transcription factors to modulate gene expression.

Transcriptome The transcriptional output of a cell (or group of cells) at a given time, i.e. all RNA molecules that have been synthesised (mRNA, rRNA, tRNA, miRNAs, etc.).

Transduction Acquisition of cellular genes by recombination with a viral genome: as a result the cellular genes may become oncogenic.

Transfection Introduction of DNA into a eukaryotic cell and its subsequent integration into that cell's chromosome. Techniques for transfection include precipitation of DNA onto the cell surface by Ca^{2+} ions and electroporation.

Transformation Conversion of a cell to a state of unregulated growth resembling that of a cancer cell. This meaning is distinct from the permanent change in one or more genetic parameters of a cell following the acquisition of novel DNA, e.g. by transfection of DNA.

Transforming growth factors (TGFα, TGFβ) TGFα (50 amino acids) binds competitively to the EGF-R, causing tyrosine auto-phosphorylation and receptor-mediated endocytosis, although there is evidence that the latter process is not identical for the two ligands. In general, TGFα activates similar cell responses to EGF, although it is more potent than EGF in stimulating bone resorption. Induces features of the transformed cell phenotype (growth in semi-solid agar); together with oncogenic *Ras* causes full transformation.

Five distinct TGFβs have been detected in vertebrates: $TGFβ_1$, $TGFβ_2$ and $TGFβ_3$ occur in humans and several other species; $TGFβ_4$ and $TGFβ_5$ have been detected in the chicken and frog, respectively. TGFβs are secreted by many types of cell, regulating proliferation and differentiation. They are potent inhibitors of proliferation of most cell types in vitro. $TGFβ_1$ is a homodimer (two 112 amino acid chains: 25 kDa), the monomers of which are derived by proteolytic cleavage of a 390 amino acid precursor. $TGFβ_1$ stimulates connective tissue formation and is expressed during cellular injury and repair. $TGFβ_2$ and $TGFβ_3$ are expressed during differentiation. Elevated levels of TGFβ1 and $TGFβ_3$ mRNA have been detected in some human tumours.

Transgenic organism Organism transformed by the introduction of novel DNA into its genome, e.g. by the injection of cloned DNA sequences into fertilised eggs. Transgenic mice are usually created by establishing a permanent embryonic stem cell line with cells taken from a blastocyst. These cells are transfected with a vector carrying the gene (or modified gene) to be introduced and cloned cells expressing that gene are injected into a blastocyst for incubation in a foster mother to obtain a chimeric founder. Heterozygous and homozygous recombinants are then generated by breeding with wild-type animals.

Transposons DNA sequences (also called 'jumping genes' or 'mobile genetic elements') that can move between different positions within a cell's genome in a process called transposition. Class I mobile genetic elements (retrotransposons) are transcribed into RNA and reverse transcribed to DNA. Class II transposons are 'cut and pasted' within the genome using the transposase enzyme. About 45% of the human genome comprises transposons and remnants therof.

Tropical spastic paraparesis (TSP)   HTLV-I-linked disease of the CNS, also called HTLV-I-associated myelopathy (HAM). Similar to the chronic progressive form of multiple sclerosis.

Tumour promoter   A tumour promoter is an agent that alone does not cause tumours but increases tumourigenesis caused by a previously applied primary carcinogen. These have been characterised in rodent skin tumours.

Tumour suppressor gene   A gene for which loss or attenuation of function can facilitate cancer progression. 'Classical' tumour suppressor genes (e.g. retinoblastoma, *RB1*) behave according to the 'two-hit hypothesis' in that both alleles must be affected before a phenotype is manifested. Other tumour suppressor genes (e.g. *P53*) can display a phenotype due to loss of just one allele.

Tyrosine kinase   The receptor tyrosine kinase family has been grouped into at least 14 sub-families of which the EPH family is the largest sub-group composed of ~15 members. There are five major classes based on organisation and sequence similarities. Prototypes are the EGF receptor (class I), the insulin receptor (class II), the PDGF-AA, PDGF-BB, CSF-1, KIT and FLT1 receptors (class III), the FGF receptors (class IV), neurotropin receptors (class VI) and the *EPH/ECK*-encoded receptors (Class VII).

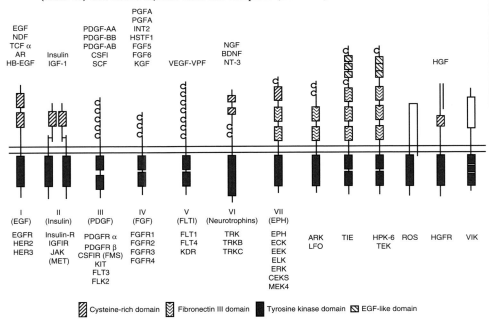

Ubiquitin–proteasome pathway   Mechanism by which proteins are marked for degradation by covalent attachment of ubiquitin proteins through the action of a ubiquitin ligase. (The complete reaction involves the attachment of the 76 amino acid ubiquitin to a ubiquitin-activating enzyme (E1) followed by its transfer to E2 that then binds to one of several ubiquitin ligases (E3s) that transfer ubiquitin from E2 to the protein substrate to be degraded.) Polyubiquitin chains are bound by the large proteasome complex that carries out proteolysis of the attached protein.

Vascular endothelial growth factors (VEGFs)   The family of VEGFs comprises VEGFA, VEGFB, VEGFC, FIGF (FOS induced growth factor, formerly VEGFD)) and placenta growth factor

Glossary and abbreviations

(PGF/PLGF). Five principal, alternatively spliced isoforms of VEGFA are denoted by the number of amino acids (121/145/165/189/206).

VHL   Von Hippel–Lindau disease is a dominantly inherited familial cancer syndrome caused by mutations in the *VHL* gene and named after Eugen von Hippel and Arvid Lindau who first described angiomas in the eye and the cerebellum and spine in 1904 and 1927, respectively. Sporadic mutations in VHL also occur in renal cell carcinoma.

Vincristine   Vinca alkaloid that binds to tubulin and inhibits mitosis.

Wilms' tumour (nephroblastoma)   Childhood solid tumour arising from renal tissue. Most Wilms' tumours are sporadic and unilateral but some are bilateral and approximately 1% are familial.

Xenotropic   Virus that can only replicate in cells of species other than its normal host.

BIBLIOGRAPHY

Chapter 1 Lessons from epidemiology

EUROPE

Ferlay, J., Autier, P., Boniol, M. *et al.* (2007). Estimates of the cancer incidence and mortality in Europe in 2006. *Annals of Oncology* 10, 1–11.

Garin, A. M. (2004). Breast cancer in Russia. *Breast Cancer* 11, 7–9.

Soerjomataram, I., de Vries, E. and Coebergh, J. W. (2009). Did alcohol protect against death from breast cancer in Russia? *Lancet* 374, 975.

Zaridze, D., Brennan, P., Boreham, J. *et al.* (2009). Alcohol and cause-specific mortality in Russia: a retrospective case-control study of 48557 adult deaths. *The Lancet* 373, 2201–14.

CHINA

He, J., Gu, D., Wu, X. *et al.* (2005). Major causes of death among men and women in China. *New England Journal of Medicine* 353, 1124–34.

Yang, L., Parkin, D. M., Li, L. D., Chen, Y. D. and Bray, F. (2004). Estimation and projection of the national profile of cancer mortality in China: 1991–2005. *British Journal of Cancer* 90, 2157–66.

Yang, L., Parkin, D. M., Ferlay, J., Li, L. and Chen, Y. (2005). Estimates of cancer incidence in China for 2000 and projections for 2005. *Cancer Epidemiology, Biomarkers & Prevention* 14, 243–50.

TRENDS

Esserman, L., Wolverton, D. and Hylton, N. (2002). Magnetic resonance imaging for primary breast cancer management: current role and new applications. *Endocrine-Related Cancer* 9, 141–53.

Miniño, A. M., Heron, M. P. and Smith, B. L. (2006). Deaths: preliminary data for 2004. *National Vital Statistics Reports* 54, 1–49.

Nordling, C. O. (1953). A new theory on the cancer-inducing mechanism. *British Journal of Cancer* 7, 68–72.

Chapter 2 Causes of cancer

IONISING RADIATION

Chernobyl Forum: www-ns.iaea.org/meetings/rw-summaries/chernobyl_forum.asp

Griggs, J., Hesketh, R., Smith, G. A. *et al.* (2001). Combretastatin-A4 disrupts neovascular development in non-neoplastic tissue. *British Journal of Cancer* 84, 832–5.

Bibliography

RADON

Gray, A., Read, S., McGale, P. and Darby, S. (2009). Lung cancer deaths from indoor radon and the cost effectiveness and potential of policies to reduce them. *British Medical Journal* **338**, a3110.

TOBACCO

Brennan, P., Buffler, P. A., Reynolds, P. *et al.* (2004). Secondhand smoke exposure in adulthood and risk of lung cancer among never smokers: a pooled analysis of two large studies. *International Journal of Cancer* **109**, 125–31.

California Environmental Protection Agency. (2005). Appendix III. Proposed identification of environmental tobacco smoke as a toxic air contaminant as approved by the Scientific Review Panel on June 24, 2005. www.arb.ca.gov/regact/ets2006/app3exe.pdf [Accessed 26 May 2012].

Lickint, F. (1929). Tabak und Tabakrauch als ätiologischer Factor des Carcinoms. *Zeitschrift für Krebsforschung* **30**, 349–65.

Lickint, F. (1953). *Aetiologie und Prophylaxe des Lungenkrebses als ein Problem der Gewerbehygiene und des Tabakrauches.* Dresden: Steinkopff.

Takahashi, H., Ogata, H., Nishigaki, R., Broide, D. H. and Karin, M. (2010). Tobacco smoke promotes lung tumourigenesis by triggering IKKbeta and JNK1-dependent inflammation. *Cancer Cell* **17**, 89–97.

US Surgeon General. (2006). 2006 Surgeon General's report. The health consequences of involuntary exposure to tobacco smoke. www.cdc.gov/tobacco/data_statistics/sgr/2006/ [Accessed 27 May 2012].

Winickoff, J. P., Park, E. R., Hipple, B. J. *et al.* (2008). Clinical effort against secondhand smoke exposure: development of framework and intervention. *Pediatrics* **122**, e363–75.

ALCOHOL

Collaborative Group on Hormonal Factors in Breast Cancer. (2002). Alcohol, tobacco and breast cancer: collaborative reanalysis of individual data from 53 epidemiological studies, including 58 515 women with breast cancer and 95 067 women without the disease. *British Journal of Cancer* **87**, 1234–45.

TEA AND COFFEE

Galeone, C., Tavani, A., Pelucchi, C. *et al.* (2010). Coffee and tea intake and risk of head and neck cancer: pooled analysis in the International Head and Neck Cancer Epidemiology Consortium. *Cancer Epidemiology Biomarkers & Prevention* **19**, 1723–36.

DIET: EPIDEMIOLOGICAL STUDIES

American Institute for Cancer Research: www.aicr.org/ [Accessed 27 May 2012].

Bastide, N. M., Pierre, F. H. F. and Corpet, D. E. (2011). Heme iron from meat and risk of colorectal cancer: a meta-analysis and a review of the mechanisms involved. *Cancer Prevention Research* **4**, 177–84.

Brunet, A. (2009). When restriction is good. *Nature* **458**, 713–14.

Cross, A. J., Ferrucci, L. M., Risch, A. *et al.* (2010). A large prospective study of meat consumption and colorectal cancer risk: an investigation of potential mechanisms underlying this association. *Cancer Research* **70**, 2406–14.

Doll, R. and Peto, R. (1981). The causes of cancer: quantitative estimates of avoidable risks of cancer in the United States today. *Journal of the National Cancer Institute* **66**, 1191–308.

Bibliography

European Prospective Investigation into Cancer and Nutrition (EPIC) project: http://epic.iarc.fr/ [Accessed 27 May 2012].

Kalaany, N. Y. and Sabatini, D. M. (2009). Tumours with PI3K activation are resistant to dietary restriction. *Nature* **458**, 725–32.

National Institutes of Health (2000). Questions and answers: the Polyp Prevention Trial and the Wheat Bran Fiber Study. www.nih.gov/news/pr/apr2000/nci-19a.htm [Accessed 27 May 2012].

Nurses' Health Study. www.channing.harvard.edu/nhs/?page_id=70 [Accessed 27 May 2012].

Scientific Advisory Committee on Nutrition (2010). Iron and health. www.sacn.gov.uk/pdfs/sacn_iron_and_health_report_web.pdf [Accessed 4 June 2012].

World Cancer Research Fund: www.wcrf-uk.org/ [Accessed 27 May 2012].

OBESITY

Khasawneh, J., Schulz, M. D., Walch, A. *et al.* (2009). Inflammation and mitochondrial fatty acid beta-oxidation link obesity to early tumour promotion. *Proceedings of the National Academy of Sciences of the USA* **106**, 3354–9.

Park, E. J., Lee, J. H., Yu, G. Y. *et al.* (2010). Dietary and genetic obesity promote liver inflammation and tumourigenesis by enhancing IL-6 and TNF expression. *Cell* **140**, 197–208.

Masoro, E. J. (2005). Overview of caloric restriction and ageing. *Mechanisms of Ageing and Development* **126**, 913–22.

STRESS

Hermes, G. L., Delgado, B., Tretiakov, M. *et al.* (2010). Social isolation dysregulates endocrine and behavioral stress while increasing malignant burden of spontaneous mammary tumours. *Proceedings of the National Academy of Sciences of the USA* **106**, 22393–8.

Ye, J., Kumanova, M., Hart, L. S. *et al.* (2010). The GCN2-ATF4 pathway is critical for tumour cell survival and proliferation in response to nutrient deprivation. *The EMBO Journal* **29**, 2082–96.

Chapter 3 Signalling in normal cells

RECEPTOR TYROSINE KINASES (RTKs)

Burke, D., Wilkes, D., Blundell, T. L. and Malcolm, S. (1998). Fibroblast growth factor receptors: lessons from the genes. *Trends in Biochemical Sciences* **23**, 59–62.

Itoh, N. and Ornitz, D. M. (2004). Evolution of the *Fgf* and *Fgfr* gene families. *Trends in Genetics* **20**, 563–9.

Zhang, H., Berezov, A., Wang, Q. *et al.* (2007). ErbB receptors: from oncogenes to targeted cancer therapies. *Journal of Clinical Investigation* **117**, 2051–8.

INTRACELLULAR SIGNALLING FROM ACTIVATED TYROSINE KINASE RECEPTORS

Lopez-Otin, C. and Hunter, T. (2010). The regulatory crosstalk between kinases and proteases in cancer. *Nature Reviews Cancer* **10**, 278–92.

Östman, A., Hellberg, C. and Böhmer, F. D. (2006). Protein-tyrosine phosphatases and cancer. *Nature Reviews Cancer* **6**, 307–20.

Sun, T., Aceto, N., Meerbrey, K. L. *et al.* (2011). Activation of multiple proto-oncogenic tyrosine kinases in breast cancer via loss of the PTPN12 phosphatase. *Cell* **144**, 703–18.

ACTIVATING RAS AND MAPK

Tian, T., Harding, A., Inder, K. *et al.* (2007). Plasma membrane nanoswitches generate high-fidelity Ras signal transduction. *Nature Cell Biology* **9**, 905–13.

MITOGENIC ACTIVATION OF CELL CYCLE PROGRESSION

Jones, R. B., Gordus, A., Krall, J. A. and MacBeath, G. (2006). A quantitative protein interaction network for the ErbB receptors using protein microarrays. *Nature* **439**, 168–74.

MYC IS A CENTRAL REGULATOR OF CELL GROWTH AND PROLIFERATION

Herold, S., Herkert, B. and Eilers, M. (2009). Facilitating replication under stress: an oncogenic function of MYC? *Nature Reviews Cancer* **9**, 441–3.

Knoepfler, P. S. (2007). Myc goes global: new tricks for an old oncogene. *Cancer Research* **67**, 5061–3.

Murphy, D. J., Junttila, M. R., Pouyet, L. *et al.* (2008). Distinct thresholds govern Myc's biological output in vivo. *Cancer Cell* **14**, 447–57.

Chapter 4 'Cancer genes': mutations and cancer development

THE CANCER GENOMIC LANDSCAPE

DeGregori, J. (2011). Evolved tumor suppression: why are we so good at not getting cancer? *Cancer Research* **71**, 3739–44.

Loeb, L. A. (2011). Human cancers express mutator phenotypes: origin, consequences and targeting. *Nature Reviews Cancer* **11**, 450–7.

Wood, L. D., Parsons, D. W., Jones, S. *et al.* (2007). The genomic landscapes of human breast and colorectal cancers. *Science* **318**, 1108–13.

THE FIRST HUMAN ONCOGENE

Parada, L. F., Tabin, C. J., Shih, C. and Weinberg, R. A. (1982). Human EJ bladder carcinoma oncogene is homologue of Harvey sarcoma virus *ras* gene. *Nature* **297**, 474–8.

WHAT TURNS A PROTO-ONCOGENE INTO AN ONCOGENE?

Meyer, K. B., Maia, A. T., O'Reilly, M. *et al.* (2008). Allele-specific up-regulation of *FGFR2* increases susceptibility to breast cancer. *PLoS Biology* **6**(5): e108 doi:10.1371/journal.pbio.0060108.

TUMOUR PROTEIN 53 (TP53 aka p53)

Abbas, T. and Dutta, A. (2009). p21 in cancer: intricate networks and multiple activities. *Nature Reviews Cancer* **9**, 400–14.

Attardi, L. D. and DePinho, R. A. (2004). Conquering the complexity of p53. *Nature Genetics* **36**, 7–8.

Brooks, C. L. and Gu, W. (2006). p53 ubiquitination: Mdm2 and beyond *Molecular Cell* **21**, 307–14.

Bibliography

Cho, Y., Gorina, S., Jeffrey, P. D. and Pavletich, N. P. (1994). Crystal structure of a p53 tumor suppressor-DNA complex: understanding tumorigenic mutations. *Science* **265**, 346–55.

Eischen, C. M. and Lozano, G. (2009). p53 and MDM2: antagonists or partners in crime? *Cancer Cell* **15**, 161–2.

García-Cao, I., García-Cao, M., Martín-Caballero, J. *et al.* (2002). 'Super p53' mice exhibit enhanced DNA damage response, are tumor resistant and age normally. *EMBO Journal* **21**, 6225–35.

Jeffrey, P. D., Gorina, S. and Pavletich, N. P. (1994). Crystal structure of the tetramerization domain of the p53 tumor suppressor at 1.7 angstroms. *Science* **267**, 1498–502.

Krizhanovsky, V. and Lowe, S. W. (2009). Stem cells: the promises and perils of p53. *Nature* **460**, 1085–6.

Mills, A. A. (2005). p53: link to the past, bridge to the future. *Genes and Development* **19**, 2091–9.

Polager, S. and Ginsberg, D. (2009). p53 and E2f: partners in life and death. *Nature Reviews Cancer* **9**, 738–48.

MICRO RNAS (miRNAs)

Corney, D. C., Flesken-Nikitin, A., Godwin, A. K., Wang, W. and Nikitin, A. Y. (2007). *MicroRNA-34b* and *MicroRNA-34c* are targets of p53 and cooperate in control of cell proliferation and adhesion-independent growth. *Cancer Research* **67**, 8433–8.

Kumar, M. S., Pester, R. E., Chen, C. Y. *et al.* (2009). Dicer1 functions as a haploinsufficient tumor suppressor. *Genes and Development* **23**, 2700–4.

Merritt, W. M., Lin, Y. G., Han, L. Y. *et al.* (2008). Dicer, Drosha, and outcomes in patients with ovarian cancer. *New England Journal of Medicine* **359**, 2641–50.

LEUKAEMIA

Anderson, K., Lutz, C., van Delft, F. W. *et al.* (2011). Genetic variegation of clonal architecture and propagating cells in leukaemia. *Nature* **469**, 356–61.

Hong, D., Gupta, R., Ancliff, P. *et al.* (2008). Initiating and cancer-propagating cells in TEL-AML1-associated childhood leukemia. *Science* **319**, 336–9.

Navin, N., Kendall, J., Troge, J. *et al.* (2011). Tumour evolution inferred by single-cell sequencing. *Nature* **472**, 90–4.

Notta, F., Mullighan, C. G., Wang, J. C. Y. *et al.* (2011). Evolution of human *BCR–ABL1* lymphoblastic leukaemia-initiating cells. *Nature* **469**, 362–7.

Rowley, J. D. (1973). A new consistent chromosomal abnormality in chronic myelogenous leukaemia identified by quinacrine fluorescence and Giemsa staining. *Nature* **243**, 290–3.

Varella-Garcia, M. (2006). Stratification of non-small cell lung cancer patients for therapy with epidermal growth factor receptor inhibitors: the EGFR fluorescence in situ hybridization assay. *Diagnostic Pathology* **1**, 19. doi:10.1186/1746–1596–1–19.

Zhu, J., Chen, Z., Lallemand-Breitenbach, V. and de Thé, H. (2002). How acute promyelocytic leukaemia revived arsenic. *Nature Reviews Cancer* **2**, 705–14.

THE RETINOBLASTOMA (RB1) GENE

Burkhart, D. L. and Sage, J. (2008). Cellular mechanisms of tumour suppression by the retinoblastoma gene. *Nature Reviews Cancer* **8**, 671–82.

Ianari, A., Natale, T., Calo, E. *et al.* (2009). Proapoptotic function of the retinoblastoma tumour suppressor protein. *Cancer Cell* **15**, 184–94.

ANALYSIS OF INDIVIDUAL TUMOUR CELLS

Snuderl, M., Fazlollahi, L., Le, L. P. *et al.* (2011). Mosaic amplification of multiple receptor tyrosine kinase genes in glioblastoma. *Cancer Cell* **20**, 810–17.

Chapter 5 What is a tumour?

RESISTANCE TO INHIBITORY GROWTH SIGNALS

Martin, G. S. (2003). Cell signalling and cancer. *Cancer Cell* **4**, 167–74.

Sherr, C. J. (2000). Cancer cell cycles revisited. *Cancer Research* **60**, 3689–95.

Yang, L., Pang, Y. and Moses, H. L. (2010). TGF-β and immune cells: an important regulatory axis in the tumour microenvironment and progression. *Trends in Immunology* **31**, 220–7.

RESISTANCE TO CELL DEATH

Indran, I. R., Hande, M. P. and Pervaiz, S. (2011). hTERT overexpression alleviates intracellular ROS production, improves mitochondrial function, and inhibits ROS-mediated apoptosis in cancer cells. *Cancer Research* **71**, 266–76.

Okada, F., Tazawa, H., Kobayashi, T. *et al.* (2006). Involvement of reactive nitrogen oxides for acquisition of metastatic properties of benign tumours in a model of inflammation-based tumour progression. *Nitric Oxide Biology and Chemistry* **14**, Special Issue: 122–9.

Quan, L. and Wanli, X. (2006). Detection of the apoptosis of Jurkat cell using an electrorotation chip. *Frontiers of Biology in China* **2**, 208–12.

Tazawa, H., Okada, F., Kobayashi, T. *et al.* (2003). Infiltration of neutrophils is required for acquisition of metastatic phenotype of benign murine fibrosarcoma cells: implication of inflammation-associated carcinogenesis and tumour progression. *American Journal of Pathology* **163**, 2221–32.

Webb, B. A., Chimenti, M., Jacobson, M. P. and Barber, D. L. (2011). Dysregulated pH: a perfect storm for cancer progression. *Nature Reviews Cancer* **11**, 671–7.

UNLIMITED REPLICATIVE CAPACITY

Maida, Y., Yasukawa, M., Furuuchi, M. *et al.* (2009). An RNA-dependent RNA polymerase formed by TERT and the *RMRP* RNA. *Nature* **461**, 230–5.

Masutomi, K., Possemato, R., Wong, J. M. Y. *et al.* (2005). The telomerase reverse transcriptase regulates chromatin state and DNA damage responses. *Proceedings of the National Academy of Sciences of the USA* **102**, 8222–7.

Park, J. -I., Venteicher, A. S., Hong, J. Y. *et al.* (2009). Telomerase modulates Wnt signalling by association with target gene chromatin. *Nature* **460**, 66–72.

INDUCTION OF ANGIOGENESIS

Tumour vascularisation

Griggs, J., Metcalfe, J. C. and Hesketh, R. (2001). Targeting tumour vasculature: the development of combretastatin A4. *The Lancet Oncology* **2**, 82–7.

Raica, M., Cimpean, M. and Ribatti, D. (2009). Angiogenesis in pre-malignant conditions. *European Journal of Cancer* **45**, 1924–34.

Tammela, T. and Alitalo, K. (2010). Lymphangiogenesis: molecular mechanisms and future promise. *Cell* **140**, 460–76.

Bibliography

Angiogenic regulators

Fraisl, P., Aragones, J. and Carmeliet, P. (2009). Inhibition of oxygen sensors as a therapeutic strategy for ischaemic and inflammatory disease. *Nature Reviews Drug Discovery* 8, 139–52.

Vascular mimicry by tumour cells

El Hallani, S., Boisselier, B., Peglion, F. *et al.* (2010). A new alternative mechanism in glioblastoma vascularization: tubular vasculogenic mimicry. *Brain* 133, 973–82.

Hormigoa, A., Dinga, B.-S. and Shahin Rafiia, S. (2011). A target for antiangiogenic therapy: vascular endothelium derived from glioblastoma. *Proceedings of the National Academy of Sciences of the USA* 108, 4271–2.

Ricci-Vitiani, L., Pallini, R., Biffoni, M. *et al.* (2010). Tumour vascularization via endothelial differentiation of glioblastoma stem-like cells. *Nature* 468, 824–8.

Soda, Y., Marumoto, T., Friedmann-Morvinski, D. *et al.* (2011). Transdifferentiation of glioblastoma cells into vascular endothelial cells. *Proceedings of the National Academy of Sciences of the USA* 108, 4274–80.

Wang, R., Chadalavada, K., Wilshire, J. *et al.* (2010). Glioblastoma stem-like cells give rise to tumour endothelium. *Nature* 468, 829–35.

ABNORMAL METABOLISM

HIF gives tumour vascular chaos a helping hand

Imtiyaz, H. Z., Williams, E. P., Hickey, M. M. R. *et al.* (2010). Hypoxia-inducible factor 2α regulates macrophage function in mouse models of acute and tumour inflammation. *Journal of Clinical Investigation* 120, 2699–714.

Lanzen, J., Braun, R. D., Klitzman, B. *et al.* (2006). Direct demonstration of instabilities in oxygen concentrations within the extravascular compartment of an experimental tumour. *Cancer Research* 66, 2219–25.

Matsumoto, S., Yasui, H., Mitchell, J. B. and Krishna, M. C. (2010). Imaging cycling tumour hypoxia. *Cancer Research* 70, 10019–25.

Mazumdar, J., Hickey, M. M., Pant, D. K. *et al.* (2010). HIF-2a deletion promotes Kras-driven lung tumour development. *Proceedings of the National Academy of Sciences of the USA* 107, 14182–7.

SonVeaux, P., Copetti, T., De Saedeleer, C. J. *et al.* (2012). Targeting the lactate transporter MCT1 in endothelial cells inhibits lactate-induced HIF-1 activation and tumor angiogenesis. *PLoS One* 7, e33418.

The return of Otto Warburg

Boidot, R., Végran, F., Meulle, A. *et al.* (2012). Regulation of monocarboxylate transporter MCT1 expression by p53 mediates inward and outward lactate fluxes in tumors. *Cancer Research* 72, 939–48.

Figueroa, M. E., Abdel-Wahab, O., Lu, C. *et al.* (2010). Leukemic IDH1 and IDH2 mutations result in a hypermethylation phenotype, disrupt TET2 function, and impair hematopoietic differentiation. *Cancer Cell* 18, 553–67.

Hirschhaeuser, F., Sattler, U. G. A. and Mueller-Klieser, W. (2011). Lactate: a metabolic key player in cancer. *Cancer Research* 71, 6921–5.

Kennedy, K. M. and Dewhirst, M. W. (2010). Tumor metabolism of lactate: the influence and therapeutic potential for MCT and CD147 regulation. *Future Oncology* 6, 127–48.

Najafov, A. and Alessi, D. R. (2010). Uncoupling the Warburg effect from cancer. *Proceedings of the National Academy of Sciences of the USA* **107**, 19135–6.

Reitman, Z. J. and Yan, H. (2010). Isocitrate dehydrogenase 1 and 2 mutations in cancer: alterations at a crossroads of cellular metabolism. *Journal of the National Cancer Institute* **102**, 932–41.

Sonveaux, P., Vegran, F., Schroeder, T. *et al.* (2008). Targeting lactate-fueled respiration selectively kills hypoxic tumour cells in mice. *Journal of Clinical Investigations* **118**, 3930–42.

Végran, F., Boidot, R., Michiels, C., Sonveaux, P. and Feron, O. (2011). Lactate influx through the endothelial cell monocarboxylate transporter MCT1 supports an NF-κB/IL-8 pathway that drives tumor angiogenesis. *Cancer Research* **71**, 2550–60.

Yang, W., Xia, Y., Ji, H. *et al.* (2011). Nuclear PKM2 regulates β-catenin transactivation upon EGFR activation. *Nature* **480**, 118–22.

Dormant tumours

Almog, N., Ma, L., Raychowdhury, R. *et al.* (2009). Transcriptional switch of dormant tumours to fast-growing angiogenic phenotype. *Cancer Research* **69**, 836–44.

Demicheli, R., Retsky, M. W., Hrushesky, W. J. M., Baum, M. and Gukas, I. D. (2008). The effects of surgery on tumor growth: a century of investigations. *Annals of Oncology* **19**, 1821–8.

Folkman, J. and Kalluri, R. (2004). Cancer without disease. *Nature* **427**, 787.

Indraccolo, S., Stievano, L., Minuzzo, S. *et al.* (2006). Interruption of tumour dormancy by a transient angiogenic burst within the tumour microenvironment. *Proceedings of the National Academy of Sciences of the USA* **103**, 4216–21.

INFLAMMATION AND THE IMMUNE SYSTEM

Angel, P., Imagawa, M., Chiu, R. *et al.* (1987). Phorbol ester-inducible genes contain a common cis element recognized by a TPA-modulated trans-acting factor. *Cell* **49**, 729–39.

Chen, J., Yao, Y., Gong, C. *et al.* (2011). CCL18 from tumor-associated macrophages promotes breast cancer metastasis via PITPNM3. *Cancer Cell* **19**, 541–55.

DeNardo, D. G., Brennan, D. J., Rexhepaj, E. *et al.* (2011). Leukocyte complexity predicts breast cancer survival and functionally regulates response to chemotherapy. *Cancer Discovery* doi:10.1158/2159-8274.CD-10-0028.

Dunn, G. P., Koebel, C. M. and Schreiber, R. D. (2006). Interferons, immunity and cancer immunoediting. *Nature Reviews Immunology* **6**, 836–48.

Greten, F. R., Arkan, M. C., Bollrath, J. *et al.* (2007). NF-kappaB is a negative regulator of IL-1beta secretion as revealed by genetic and pharmacological inhibition of IKKbeta. *Cell* **130**, 918–31.

Koebel, C. M., Vermi, W., Swann, J. B. *et al.* (2007). Adaptive immunity maintains occult cancer in an equilibrium state. *Nature* **450**, 903–7.

Kraman, M., Bambrough, P. J., Arnold, J. N. *et al.* (2010). Suppression of antitumour immunity by stromal cells expressing fibroblast activation protein-α. *Science* **330**, 827–30.

Kuper, H., Adami, H. O. and Trichopoulos, D. (2000). Infections as a major preventable cause of human cancer. *Journal of Internal Medicine* **248**, 171–83.

Mancino, A. and Lawrence, T. (2010). Nuclear factor-κB and tumour-associated macrophages. *Clinical Cancer Research* **16**, 784–9.

Minchinton, A. I. and Tannock, I. F. (2006). Drug penetration in solid tumours. *Nature Reviews Cancer* **6**, 583–92.

Powell, A. E., Anderson, E. C., Davies, P. S. *et al.* (2011). Fusion between intestinal epithelial cells and macrophages in a cancer context results in nuclear reprogramming. *Cancer Research* **71**, 1497–505.

Rodriguez-Vita, J. and Lawrence, T. (2010). The resolution of inflammation and cancer *Cytokine and Growth Factor Reviews* **21**, 61–5.

Bibliography

Sakurai, T., He, G., Matsuzawa, A. *et al.* (2008). Hepatocyte necrosis induced by oxidative stress and IL-1 alpha release mediate carcinogen-induced compensatory proliferation and liver tumourigenesis. *Cancer Cell* **14**, 156–65.

Shankaran, V., Ikeda, H., Bruce, A. T. *et al.* (2001). IFNgamma and lymphocytes prevent primary tumour development and shape tumour immunogenicity. *Nature* **410**, 1107–11.

Sica, A., Allavena, P. and Mantovani, A. (2008). Cancer related inflammation: the macrophage connection. *Cancer Letters* **267**, 204–15.

Smyth, M. J., Dunn, G. P. and Schreiber, R. D. (2006). Cancer immunosurveillance and immunoediting: the roles of immunity in suppressing tumour development and shaping tumour immunogenicity. *Advances in Immunology* **90**, 1–50.

Swann, J. B., Uldrich, A. P., van Dommelen, S. *et al.* (2009). Type I natural killer T cells suppress tumours caused by p53 loss in mice. *Blood* **113**, 6382–5.

Vakkila, J. and Lotze, M. T. (2004). Inflammation and necrosis promote tumour growth. *Nature Reviews Immunology* **4**, 641–8.

Wang, D. and Dubois, R. N. (2006). Prostaglandins and cancer. *Gut* **55**, 115–22.

Zitvogel, L., Apetoh, L., Ghiringhelli, F. and Kroemer, G. (2008). Immunological aspects of cancer chemotherapy. *Nature Reviews Immunology* **8**, 59–75.

METASTASIS AND METASTATIC POTENTIAL

Primary and malignant tumours

Coghlin, C. and Murray, G. I. (2010). Current and emerging concepts in tumour metastasis, *Journal of Pathology* **222**, 1–15.

Duda, D. G., Duyverman, A. M. and Kohno, M. (2010). Malignant cells facilitate lung metastasis by bringing their own soil. *Proceedings of the National Academy of Sciences of the USA* **107**, 21677–82.

Hart, I. R. and Fidler, I. J. (1980). Role of organ selectivity in the determination of metastatic patterns of B16 melanoma. *Cancer Research* **40**, 2281–7.

Récamier, J. C. (1829). *Recherches sur le traitement du cancer sur la compression methodique simple ou combinee et sur l'histoire generale de la meme maladie*, 2nd edn.

Early ideas about metastasis

Talmadge, J. E. and Fidler, I. J. (2010). The biology of cancer metastasis: historical perspective. *Cancer Research* **70**, 5649–69.

How do tumour cells become metastatic?

Yachida, S., Jones, S., Bozic, I. *et al.* (2010). Distant metastasis occurs late during the genetic evolution of pancreatic cancer. *Nature* **467**, 1114–17.

The epithelial to mesenchymal transition

Polyak, K. and Weinberg, R. A. (2009). Transitions between epithelial and mesenchymal states: acquisition of malignant and stem cell traits. *Nature Reviews Cancer* **9**, 265–75.

Thiery, J. P. and Sleeman, J. P. (2006). Complex networks orchestrate epithelial–mesenchymal transitions. *Nature Reviews Molecular Cell Biology* **7**, 131–42.

Stem cells

Diehn, M., Cho, R. W., Lobo, N. A. *et al.* (2009). Association of reactive oxygen species levels and radioresistance in cancer stem cells. *Nature* **458**, 780–5.

Eramo, A., Haas, T. L. and De Maria, R. (2010). Lung cancer stem cells: tools and targets to fight lung cancer. *Oncogene* **29**, 4625–35.

Kim, J., Woo, A. J., Chu, J. *et al.* (2010). A Myc network accounts for similarities between embryonic stem and cancer cell transcription programs. *Cell* **143**, 313–24.

Labelle, M., Begum, S. and Hynes, R. O. (2011). Direct signaling between platelets and cancer cells induces an epithelial–mesenchymal-like transition and promotes metastasis. *Cancer Cell* **20**, 576–90.

Prestegarden, L. and Enger, P. O. (2010). Cancer stem cells in the central nervous system: a critical review. *Cancer Research* **70**, 8255–8.

Quintana, E., Shackleton, M., Sabel, M. S. *et al.* (2008). Efficient tumour formation by single human melanoma cells. *Nature* **456**, 593–8.

Rothenberg, M. E., Clarke, M. F. and Diehn, M. (2010). The Myc connection: ES cells and cancer. *Cell* **143**, 184–6.

How do metastatic tumour cells know where to go?

Friedl, P. and Wolf, K. (2010). Plasticity of cell migration: a multiscale tuning model. *Journal of Cell Biology* **188**, 11–19.

Kaplan, R. N., Riba, R. D., Zacharoulis, S. *et al.* (2005). VEGFR1-positive haematopoietic bone marrow progenitors initiate the pre-metastatic niche. *Nature* **438**, 820–7.

Chapter 6 Cancer signalling networks

TUMOURIGENIC DNA VIRUSES

Neuveut C., Yu, W. and Annick, B. M. (2010). Mechanisms of HBV-related hepatocarcinogenesis. *Journal of Hepatology* **52**, 594–604.

Ronit, S. and Shou-Jiang, G. (2011). Viruses and human cancer: from detection to causality. *Cancer Letters* **305**, 218–27.

Stanley, M. A., Pett, M. R. and Coleman, N. (2007). HPV: from infection to cancer. *Biochemical Society Transactions* **35**, 1456–60.

SIGNALLING PATHWAYS THAT IMPACT ON THE CENTRAL AXIS

Phosphatidylinositol 3-kinase (PI3K): survival and apoptosis signalling

Carretero, J., Shimamura, T., Rikova, K. *et al.* (2010). Integrative genomic and proteomic analyses identify targets for Lkb1-deficient metastatic lung tumors. *Cancer Cell* **17**, 547–59.

Foster, K. G. and Fingar, D. C. (2010). Mammalian target of Rapamycin (mTOR): conducting the cellular signalling symphony. *Journal of Biological Chemistry* **285**, 14071–7.

Guo, J. Y., Chen, H. Y., Mathew, R. *et al.* (2011). Activated Ras requires autophagy to maintain oxidative metabolism and tumorigenesis. *Genes and Development* **25**, 460–70.

Ji, H., Ramsey, M. R., Hayes, D. N. *et al.* (2007). LKB1 modulates lung cancer differentiation and metastasis. *Nature* **448**, 807–11.

Lian, Z. and Di Cristofano, A. (2005). Class reunion: PTEN joins the nuclear crew. *Oncogene* **24**, 7394–400.

Mayo, L. D. and Donner, D. B. (2002). The PTEN, Mdm2, p53 tumour suppressor-oncoprotein network. *Trends in Biochemical Sciences* **27**, 462–7.

Bibliography

Sato, T., Nakashima, A., Guo, L., Coffman, K. and Tamanoi, F. (2010). Single amino-acid changes that confer constitutive activation of mTOR are discovered in human cancer. *Oncogene* **29**, 2746–52.

Scott, K. L., Kabbarah, O., Liang, M. C. *et al.* (2009). GOLPH3 modulates mTOR signalling and rapamycin sensitivity in cancer. *Nature* **459**, 1085–90.

Slack-Davis, J., DaSilva, J. O. and Parsons, S. J. (2010). LKB1 and Src: antagonistic regulators of tumour growth and metastasis. *Cancer Cell* **17**, 527–8.

Takamura, A., Komatsu, M., Hara, T. *et al.* (2011). Autophagy-deficient mice develop multiple liver tumors. *Genes and Development* **25**, 795–800.

Yang, S., Wang, X., Contino, G. *et al.* (2011). Pancreatic cancers require autophagy for tumor growth. *Genes and Development* **25**, 717–29.

The JAK–STAT pathway

Constantinescu, S. N., Girardot, M. and Pecquet, C. (2007). Mining for JAK–STAT mutations in cancer. *Trends in Biochemical Sciences* **33**, 122–31.

Verstovsek, S., Kantarjian, H., Mesa, R. A. *et al.* (2010). Safety and efficacy of INCB018424, a JAK1 and JAK2 inhibitor, in myelofibrosis. *New England Journal of Medicine* **363**, 1117–27.

WNT and GPCR signalling

Barker, N. and Clevers, H. (2000). Catenins, Wnt signalling and cancer. *BioEssays* **22**, 961–5.

Li, F. Q., Mofunanya, A., Harris, K. and Takemaru, K. (2010). Chibby cooperates with 14–3–3 to regulate β-catenin subcellular distribution and signalling activity. *Journal of Cell Biology* **181**, 1141–54.

Ribas, C., Penela, P., Murga, C. *et al.* (2007). The G protein-coupled receptor kinase (GRK) interactome: role of GRKs in GPCR regulation and signalling. *Biochimica et Biophysica Acta* **1768**, 913–22.

Cell adhesion: cadherin signalling

Pylayeva, Y., Gillen, K. M., William, G. *et al.* (2009). Ras- and PI3K-dependent breast tumorigenesis in mice and humans requires focal adhesion kinase signaling. *Journal of Clinical Investigation* **119**, 252–66.

Yagi, T. and Takeichi, M. (2000). Cadherin superfamily genes: functions, genomic organization, and neurologic diversity. *Genes and Development* **14**, 1169–80.

Integrin signalling

Cabodi, S., Camacho-Leal, M., Di Stefano, P. and Defilippi, P. (2010). Integrin signalling adaptors: not only figurants in the cancer story. *Nature Reviews Cancer* **10**, 858–70.

Caswell, P. T., Vadrevu, S. and Norman, J. C. (2009). Integrins: masters and slaves of endocytic transport. *Nature Reviews Molecular and Cell Biology* **10**, 843–53.

Muller, P. A. J., Caswell, P. T., Doyle, B. *et al.* (2009). Mutant p53 drives invasion by promoting integrin recycling. *Cell* **139**, 1327–41.

Sandilands, E., Serrels, B., McEwan, D. G. *et al.* (2012). Autophagic targeting of Src promotes cancer cell survival following reduced FAK signalling. *Nature Cell Biology* **14**, 51–60.

BCR-ABL1

Kolch, W. and Pitt, A. (2010). Functional proteomics to dissect tyrosine kinase signalling pathways in cancer. *Nature Reviews Cancer* **10**, 618–29.

Titz, B., Low, T., Komisopoulou, E. *et al.* (2010). The proximal signalling network of the BCR-ABL1 oncogene shows a modular organization. *Oncogene* **29**, 5895–910.

The Hedgehog pathway and GLI signalling

Das, S., Harris, L. G., Metge, B. J. *et al.* (2009). The Hedgehog pathway transcription factor GLI1 promotes malignant behavior of cancer cells by up-regulating osteopontin. *Journal of Biological Chemistry* **284**, 22888–97.

Scales, S. J. and de Sauvage, F. J. (2009). Mechanisms of Hedgehog pathway activation in cancer and implications for therapy. *Trends in Pharmacological Sciences* **30**, 303–12.

Steg, A. D., Yuan, K., Johnson, M. R. *et al.* (2010). Gli1 promotes cell survival and is predictive of a poor outcome in ERa-negative breast cancer. *Breast Cancer Research and Treatment* **123**, 59–71.

Watson, A., Kent, P., Alam, M. *et al.* (2009). GLI1 genotypes do not predict basal cell carcinoma risk: a case control study. *Molecular Cancer* **8**, 113.

Xu, L., Kwon, Y. -J., Frolova, N. *et al.* (2010). Gli1 promotes cell survival and is predictive of a poor outcome in ERalpha-negative breast cancer. *Breast Cancer Research and Treatment* **123**, 59–71.

Transforming growth factor beta (TGFβ)

Bierie, B. and Moses, H. L. (2006). Tumour microenvironment. TGF β: the molecular Jekyll and Hyde of cancer. *Nature Reviews Cancer* **6**, 506–20.

Derynck, R. and Akhurst, R. J. (2007). Differentiation plasticity regulated by TGF-β family proteins in development and disease. *Nature Cell Biology* **9**, 1000–4.

Heldin, C. -H., Landström, M. and Moustakas, A. (2009). Mechanism of TGF-β signalling to growth arrest, apoptosis, and epithelial–mesenchymal transition. *Current Opinion in Cell Biology* **21**, 166–76.

Vascular endothelial growth factors (VEGFs) and Notch signalling

Gault, C. R. and Obeid, L. M. (2011). Still benched on its way to the bedside: sphingosine kinase 1 as an emerging target in cancer chemotherapy. *Critical Reviews in Biochemistry and Molecular Biology* **46**, 342–51.

Niu, G. and Chen, X. (2010). Vascular endothelial growth factor as an anti-angiogenic target for cancer therapy. *Current Drug Targets* **11**, 1000–17.

Siekmann, A. F., Covassin, L. and Lawson, N. D. (2008). Modulation of VEGF signalling output by the Notch pathway. *BioEssays* **30**, 303–13.

SIGNALLING AND SYSTEMS BIOLOGY

Ergün, A., Lawrence, C. A., Kohanski, M. A., Brennan, T. A. and Collins, J. J. (2007). A network biology approach to prostate cancer. *Molecular Systems Biology* **3**, 82.

Helikar, T., Konvalina, J., Heidel, J. and Rogers, J. A. (2008). Emergent decision-making in biological signal transduction networks. *Proceedings of the National Academy of Sciences of the USA* **105**, 1913–18.

Natarajan, M., Lin, K. -M., Hsueh, R. C., Sternweis, P. C. and Ranganathan, R. (2006). A global analysis of cross-talk in a mammalian cellular signalling network. *Nature Cell Biology* **8**, 571–80.

Pujana, M. A., Han, J. D., Starita, L. M. *et al.* (2007). Network modeling links breast cancer susceptibility and centrosome dysfunction. *Nature Genetics* **39**, 1338–49.

Chapter 7 The future of cancer prevention, diagnosis and treatment

THE DEVELOPMENT OF ANTI-CANCER DRUGS

Del Monte, U. (2009). Does the cell number 10^9 still really fit one gram of tumor tissue? *Cell Cycle* **8**, 505–6.

Peer, P. G., van Dijck, J. A., Hendriks, J. H., Holland, R. and Verbeek, A. L. (1993). Age-dependent growth rate of primary breast cancer. *Cancer* **71**, 3547–51.

CHEMOTHERAPEUTIC STRATEGIES FOR CANCER

Kinase inhibitors

DeMatteo, R. P., Lewis, J. J., Leung, D. *et al.* (2000). Two hundred gastrointestinal stromal tumors: recurrence patterns and prognostic factors for survival. *Annals of Surgery* **231**, 51–8.

Demetri, G. D., von Mehren, M., Blanke, C. D. *et al.* (2002). Efficacy and safety of imatinib mesylate in advanced gastrointestinal stromal tumors. *New England Journal of Medicine* **347**, 472–80.

Druker, B. J., Guilhot, F., O'Brien, S. G. *et al.* (2006). Five-year follow-up of patients receiving imatinib for chronic myeloid leukemia. *New England Journal of Medicine* **355**, 2408–17.

Hirota, S., Isozaki, K., Moriyama, Y. *et al.* (1998). Gain-of-function mutations of c-kit in human gastrointestinal stromal tumors. *Science* **279**, 577–80.

Lynch, T. J., Bell, D. W., Sordella, R. *et al.* (2004). Activating mutations in the epidermal growth factor receptor underlying responsiveness of non-small-cell lung cancer to gefitinib. *New England Journal of Medicine* **350**, 2129–39.

Mok, T. S., Wu, Y.-L., Thongprasert, S. *et al.* (2009). Gefitinib or carboplatin-paclitaxel in pulmonary adenocarcinoma. *New England Journal of Medicine* **361**, 947–57.

Paez, J. G., Jänne, P. A., Lee, J. C. *et al.* (2004). EGFR mutations in lung cancer: correlation with clinical response to gefitinib therapy. *Science* **304**, 1497–500.

DRUG RESISTANCE

Engelman, J. A., Zejnullahu, K., Mitsudomi, T. *et al.* (2007). MET amplification leads to gefitinib resistance in lung cancer by activating ERBB3 signalling. *Science* **316**, 1039–43.

Maheswaran, S., Sequist, L. V., Nagrath, S. *et al.* (2008). Detection of mutations in EGFR in circulating lung-cancer cells. *New England Journal of Medicine* **359**, 366–77.

Zhang, N., Qi, Y., Wadham, C. *et al.* (2010). FTY720 induces necrotic cell death and autophagy in ovarian cancer cells: a protective role of autophagy. *Autophagy* **6**, 1157–67.

MOLECULAR IMAGING

Positron emission tomography (PET)

Nimmagadda, S., Pullambhatla, M. and Pomper, M. G. (2009). Immunoimaging of CXCR4 expression in brain tumour xenografts using SPECT/CT. *Journal of Nuclear Medicine* **50**, 1124–30.

Bibliography

Magnetic resonance imaging (MRI)

Beloueche-Babari, M., Jackson, L. E., Al-Saffar, N. M. S. *et al.* (2005). Magnetic resonance spectroscopy monitoring of mitogen-actived protein kinase signalling inhibition. *Cancer Research* **65**, 3356–63.

Chung, Y. L., Troy, H., Banerji, U. *et al.* (2003). 17-Allylamino, 17-demethoxygeldanamycin (17AAG) in human colon cancer models. *Journal of the National Cancer Institute* **95**, 1624–33.

Damadian, R. V. (1971). Tumour detection by nuclear magnetic resonance. *Science* **171**, 1151–3.

Gillies, R. J. and Morse, D. L. (2005). In vivo magnetic resonance spectroscopy in cancer. *Annual Review of Biomedical Engineering* **7**, 287–326.

Mountford, C. E., Somorjai, R. L., Malycha, P. *et al.* (2001) Diagnosis and prognosis of breast cancer by magnetic resonance spectroscopy of fine-needle aspirates analysed using a statistical classification strategy. *British Journal of Surgery* **88**, 1234–40.

Sinkus, R., Siegmann, K., Xydeas, T. *et al.* (2007). MR elastography of breast lesions: understanding the solid/liquid duality can improve the specificity of contrast-enhanced MR mammography. *Magnetic Resonance in Medicine* **58**, 1135–44.

Smart contrast agents

Krishnan, A. S., Neves, A. A., de Backer, M. M. *et al.* (2008). Detection of cell death in tumours by using MR imaging and a gadolinium-based targeted contrast agent. *Radiology* **246**, 854–62.

^{13}C *hyperpolarisation*

Albers, M. J., Bok, R., Chen, A. P. *et al.* (2008). Hyperpolarized ^{13}C lactate, pyruvate, and alanine: noninvasive biomarkers for prostate cancer detection and grading. *Cancer Research* **68**, 8607–15.

Branca, R. T., Cleveland, Z. I., Fubara, B. *et al.* (2010). Molecular MRI for sensitive and specific detection of lung metastases. *Proceedings of the National Academy of Sciences of the USA* **107**, 3693–7.

Gallagher, F. A., Kettunen, M. I., Hu, D. E. *et al.* (2009). Production of hyperpolarized $[1,4-^{13}C_2]$-malate from $[1,4-^{13}C_2]$fumarate is a marker of cell necrosis and treatment response in tumours. *Proceedings of the National Academy of Sciences of the USA* **106**, 19801–6.

Proteomics

Stephens, A. N., Hannan, N. J., Rainczuk, A. *et al.* (2010). Post-translational modifications and protein-specific isoforms in endometriosis revealed by 2D DIGE. *Journal of Proteome Research* **9**, 2438–49.

Xue, W., Zender, L., Miething, C. *et al.* (2007). Senescence and tumour clearance is triggered by p53 restoration in murine liver carcinomas. *Nature* **445**, 656–60.

Metabolomics

Cheng, L. L., Burns, M. A., Taylor, J. L. *et al.* (2005). Metabolic characterization of human prostate cancer with tissue magnetic resonance spectroscopy. *Cancer Research* **65**, 3030–4.

Lindon, J. C., Holmes, E. and Nicholson, J. K. (2004). Metabonomics and its role in drug development and disease diagnosis. *Expert Review of Molecular Diagnostics* **4**, 89–99.

Mountford, C. E., Somorjai, R. L., Malycha, P. *et al.* (2001). Diagnosis and prognosis of breast cancer by magnetic resonance spectroscopy of fine-needle aspirates analysed using a statistical classification strategy. *British Journal of Surgery* **88**, 1234–40.

Odunsi, K., Wollman, R. M., Ambrosone, C. B. *et al.* (2005). Detection of epithelial ovarian cancer using 1HNMR-based metabonomics. *International Journal of Cancer* **113**, 782–8.

Yang, J., Xu, G., Zheng, Y. *et al.* (2004). Diagnosis of liver cancer using HPLC-based metabonomics avoiding false-positive result from hepatitis and hepatocirrhosis diseases. *Journal of Chromatography. B, Analytical Technologies in the Biomedical and Life Sciences* **813**, 59–65.

Bibliography

Gene expression profiling

Anastassiou, D., Rumjantseva, V., Cheng, W. *et al.* (2011). Human cancer cells express Slug-based epithelial-mesenchymal transition gene expression signature obtained in vivo. *BMC Cancer* **11**, 529 doi:10.1186/1471-2407-11-529.

Bristow, A. R., Agrawal, A., Evans, A. J. *et al.* (2008). Can computerised tomography replace bone scintigraphy in detecting bone metastases from breast cancer? A prospective study. *Breast* **17**, 98–103.

Golub, T. R., Slonim, D. K., Tamayo, P. *et al.* (1999). Molecular classification of cancer: class discovery and class prediction by gene expression monitoring. *Science* **286**, 531–7.

Pichler, B. J., Wehrl, H. F. and Judenhofer, M. S. (2008). Latest advances in molecular imaging instrumentation. *Journal of Nuclear Medicine* **49**, 5S–23S.

Sitter, B., Lundgren, S., Bathen, T. F. *et al.* (2006). Comparison of HR MAS MR spectroscopic profiles of breast cancer tissue with clinical parameters. *NMR in Biomedicine* **19**, 30–40.

Wang, F., Fang, W., Ji, S. D. *et al.* (2007) Technetium-99m labeled synaptotagmin I C2A detection of paclitaxel-induced apoptosis in non-small cell lung cancer. *Chinese Journal of Oncology* **29**, 351–4.

Breast cancer: MammaPrint

van't Veer, L. J., Dai, H., van de Vijver, M. J. *et al.* (2002). Gene expression profiling predicts clinical outcome of breast cancer. *Nature* **415**, 530–6.

Breast cancer: oncotype DX

Lo, S. S., Mumby, P. B., Norton, J. *et al.* (2010). Prospective multicenter study of the impact of the 21-gene recurrence score assay on medical oncologist and patient adjuvant breast cancer treatment selection. *Journal of Clinical Oncology* **28**, 1671–6.

Paik, S., Shak, S., Tang G. *et al.* (2004). A multigene assay to predict recurrence of tamoxifen-treated, node-negative breast cancer. *New England Journal of Medicine* **351**, 2817–26.

Breast cancer: Rotterdam signature

Wang, Y., Klijn, J. G. M., Zhang, Y. *et al.* (2005). Gene-expression profiles to predict distant metastasis of lymph-node-negative primary breast cancer. *Lancet* **365**, 671–9.

Colon cancer

Kerr, D., Gray, R., Quirke, P. D. *et al.* (2009). A quantitative multigene RT-PCR assay for prediction of recurrence in stage II colon cancer: selection of the genes in four large studies and results of the independent, prospectively designed QUASAR validation study. *Journal of Clinical Oncology* **27**, Supplement: 15s (Abstract 4000).

Acute myeloid leukaemia

Bullinger, L., Döhner, K., Kranz, R. *et al.* (2008). An FLT3 gene-expression signature predicts clinical outcome in normal karyotype AML. *Blood* **111**, 4490–5.

Diffuse large-B-cell lymphoma

Rosenwald, A., Wright, G., Chan, W. C. *et al.* (2002). The use of molecular profiling to predict survival after chemotherapy for diffuse large-B-cell lymphoma. *New England Journal of Medicine* **346**, 1937–47.

Burkitt's lymphoma

Dave, S. S., Fu, K., Wright, G. W. *et al.* (2006). Molecular diagnosis of Burkitt's lymphoma. *New England Journal of Medicine* **354**, 2431–42.

Bibliography

Protein imaging

Cohen, A. A., Geva-Zatorsky, N., Eden, E. *et al.* (2008). Dynamic proteomics of individual cancer cells in response to a drug. *Science* 322, 1511–16.

Issaeva, I., Cohen, A. A., Eden, E. *et al.* (2010). Generation of double-labeled reporter cell lines for studying co-dynamics of endogenous proteins in individual human cells. *PLoS ONE* 5, e13524. doi:10.1371/journal.pone.0013524.

Jarvik, J. W., Fisher, G. W., Shi, C. *et al.* (2002). In vivo functional proteomics: mammalian genome annotation using CD-tagging. *Biotechniques* 33, 852–4.

Chapter 8 The future of cancer in the post-genomic era

HUMAN GENOME SEQUENCING

Eid, J., Fehr, A., Gray, J. *et al.* (2009). Real-time DNA sequencing from single polymerase molecules. *Science* 323, 133–8.

Gierhart, B. C., Howitt, D. G., Chen, S. J. *et al.* (2008). Nanopore with transverse nanoelectrodes for electrical characterization and sequencing of DNA. *Sensors and Actuators B: Chemical* 132, 593–600.

Meyerson, M., Gabriel, S. and Getz, G. (2010). Advances in understanding cancer genomes through second-generation sequencing. *Nature Reviews Genetics* 11, 685–96.

Ng, S. B., Buckingham, K. J., Lee, C. *et al.* (2009). Exome sequencing identifies the cause of a mendelian disorder. *Nature Genetics* 42, 30–5.

Prickett, T. D., Agrawal, N. S., Wei, X. *et al.* (2009). Analysis of the tyrosine kinome in melanoma reveals recurrent mutations in ERBB4. *Nature Genetics* 41, 1127–32.

Travers, K. J., Chin, C. -S., Rank, D. R., Eid, J. S. and Turner, S. W. (2010). A flexible and efficient template format for circular consensus sequencing and SNP detection. *Nucleic Acids Research* 38, e159.

Xi, L., Feber, A., Gupta, V. *et al.* (2008). Whole genome exon arrays identify differential expression of alternatively spliced, cancer-related genes in lung cancer. *Nucleic Acids Research* 36, 6535–47.

WHOLE GENOME SEQUENCING (WGS) AND CANCER

Bentley, D. R., Balasubramanian, S., Swerdlow, H. P. *et al.* (2008). Accurate whole human genome sequencing using reversible terminator chemistry. *Nature* 456, 53–9.

Fong, P. C., Boss, D. S., Yap, T. A. *et al.* (2009). Inhibition of poly(ADP-ribose) polymerase in tumours from BRCA mutation carriers. *New England Journal of Medicine* 361, 123–34.

Ley, T. J., Mardis, E. R., Ding, L. *et al.* (2008). DNA sequencing of a cytogenetically normal acute myeloid leukaemia genome. *Nature* 456, 66–72.

McCabe, N., Turner, N. C., Lord, C. J. *et al.* (2006). Deficiency in the repair of DNA damage by homologous recombination and sensitivity to poly(ADP-ribose) polymerase inhibition. *Cancer Research* 66, 8109–15.

Pleasance, E. D., Stephens, P. J., O'Meara, S. *et al.* (2009). A small-cell lung cancer genome with complex signatures of tobacco exposure. *Nature* 463, 184–90.

Shah, S. P., Morin, R. D., Khattra, J. *et al.* (2009). Mutational evolution in a lobular breast tumour profiled at single nucleotide resolution. *Nature* 461, 809–13.

Stephens, P. J., McBride, D. J., Lin, M. -L. *et al.* (2009). Complex landscapes of somatic rearrangement in human breast cancer genomes. *Nature* **462**, 1005–10.

Stephens, P. J., Greenman, C. D., Fu, B. *et al.* (2011). Massive genomic rearrangement acquired in a single catastrophic event during cancer development. *Cell* **144**, 27–40.

Summerer, D., Schracke, N., Wu, H. -G. *et al.* (2010). Targeted high throughput sequencing of a cancer-related exome subset by specific sequence capture with a fully automated microarray platform. *Genomics* **95**, 241–6.

To, M. D., Wong, C. E., Karnezis, A. N. *et al.* (2008). Kras regulatory elements and exon 4A determine mutation specificity in lung cancer. *Nature Genetics* **40**, 1240–4.

Wang, J., Wang, W., Li, R. *et al.* (2008). The diploid genome sequence of an Asian Individual. *Nature* **456**, 60–6.

Wheeler, D. A., Srinivasan, M., Egholm, M. *et al.* (2008). The complete genome of an individual by massively parallel DNA sequencing. *Nature* **452**, 872–6.

GENOMIC PARTITIONING

Jones, S., Zhang, X., Parsons, D. W. *et al.* (2008). Core signaling pathways in human pancreatic cancers revealed by global genomic analyses. *Science* **321**, 1801–6.

TAILORING THERAPY ON THE BASIS OF WHOLE GENOME SEQUENCING

BRAF

Bollag, G., Hirth, P., Tsai, J. *et al.* (2010). Clinical efficacy of a RAF inhibitor needs broad target blockade in BRAF-mutant melanoma. *Nature* **467**, 596–9.

Corcoran, R. B., Settleman, J. and Engelman, J. A. (2011). Potential therapeutic strategies to overcome acquired resistance to BRAF or MEK inhibitors in BRAF mutant cancers. *Oncotarget* **2**, 336–46.

Emery, C. M., Vijayendran, K. G., Zipser, M. C. *et al.* (2009). MEK1 mutations confer resistance to MEK and B-RAF inhibition. *Proceedings of the National Academy of Sciences of the USA* **106**, 20411–16.

Flaherty, K. T., Puzanov, I., Kim, K. B. *et al.* (2010). Inhibition of mutated, activated BRAF in metastatic melanoma. *New England Journal of Medicine* **363**, 809–19.

Garnett, M. J., Rana, S., Paterson, H., Barford, D. and Marais, R. (2005). Wild-type and mutant B-RAF activate C-RAF through distinct mechanisms involving heterodimerization. *Molecular Cell* **20**, 963–9.

Hatzivassilou, G., Song, K., Yen, I. *et al.* (2010). RAF inhibitors prime wild-type RAF to activate the MAPK pathway and enhance growth. *Nature* **464**, 431–5.

Johannessen, C. M., Boehm, J. S., Kim, S. Y. *et al.* (2010). COT drives resistance to RAF inhibition through MAP kinase pathway reactivation. *Nature* **468**, 968–72.

Montagut, C., Sharma, S. V., Shioda, T. *et al.* (2008). Elevated CRAF as a potential mechanism of acquired resistance to BRAF inhibition in melanoma. *Cancer Research* **68**, 4853–61.

Nazarian, R., Shi, H., Wang, Q. *et al.* (2010). Melanomas acquire resistance to B-RAF(V600E) inhibition by RTK or N-RAS upregulation. *Nature* **468**, 973–7.

Poulikakos, P. I., Zhang, C., Bollag, G., Shokat, K. M. and Rosen, N. (2010). RAF inhibitors transactivate RAF dimers and ERK signalling in cells with wild-type BRAF. *Nature* **464**, 427–30.

Poulikakos P. I., Persaud, Y., Janakiraman, M. *et al.* (2011). RAF inhibitor resistance is mediated by dimerization of aberrantly spliced BRAF(V600E). *Nature* **480**, 387–90.

Tsai, J., Lee, J. T., Wang, W. *et al.* (2008). Discovery of a selective inhibitor of oncogenic B-Raf kinase with potent antimelanoma activity. *Proceedings of the National Academy of Sciences of the USA* **105**, 3041–6.

Villanueva, J., Vultur, A., Lee, J. T. *et al.* (2010). Acquired resistance to BRAF inhibitors mediated by a RAF kinase switch in melanoma can be overcome by cotargeting MEK and IGF-1R/PI3K. *Cancer Cell* 18, 683–95.

Wagle, N., Emery, C., Berger, M. F. *et al.* (2011). Dissecting therapeutic resistance to RAF inhibition in melanoma by tumor genomic profiling. *Journal of Clinical Oncology* 29, 3085–96.

MYC

Delmore, J. E., Issa, G. C., Lemieux, M. E. *et al.* (2011). BET bromodomain inhibition as a therapeutic strategy to target c-Myc. *Cell* 146, 904–17.

Mertz, J. A., Conery, A. R., Bryant, B. M. *et al.* (2011). Targeting MYC dependence in cancer by inhibiting BET bromodomains. *Proceedings of the National Academy of Sciences of the USA* 108, 16669–74.

Sodir, N. M., Swigart, L. B., Karnezis, A. N. *et al.* (2011). Endogenous Myc maintains the tumor microenvironment. *Genes and Development* 25, 907–16.

Zuber, J., Shi, J., Wang, E. *et al.* (2011). RNAi screen identifies Brd4 as a therapeutic target in acute myeloid leukaemia. *Nature* 478, 524–8.

p53

Garcia-Cao I., Garcia-Cao, M., Martin-Caballero, J. *et al.* (2002). 'Super p53' mice exhibit enhanced DNA damage response, are tumour resistant and age normally. *EMBO Journal* 21, 6225–35.

Matheu, A., Pantoja, C., Efeyan, A. *et al.* (2004). Increased gene dosage of Ink4a/Arf results in cancer resistance and normal aging. *Genes and Development* 18, 2736–46.

TUMOUR BIOMARKERS

Diehl, F., Schmidt, K., Choti, M. A. *et al.* (2008). Circulating mutant DNA to assess tumor dynamics. *Nature Medicine* 14, 985–90.

Leary, R. J., Kinde, I., Diehl, F. *et al.* (2010). Development of personalized tumor biomarkers using massively parallel sequencing. *Science Translational Medicine* 2, 20ra14.

McBride, D. J., Orpana, A. K., Sotiriou, C. *et al.* (2010). Use of cancer-specific genomic rearrangements to quantify disease burden in plasma from patients with solid tumors. *Genes Chromosomes Cancer* 49, 1062–9.

Press, R. D., Love, Z., Tronnes, A. A. *et al.* (2006). BCR-ABL mRNA levels at and after the time of a complete cytogenetic response (CCR) predict the duration of CCR in imatinib mesylate-treated patients with CML. *Blood* 107, 4250–6.

MICROFLUIDICS AND THE ISOLATION OF CIRCULATING TUMOUR CELLS

Sequist, L. V., Sunitha, N., Toner, M., Haber, D. A. and Lynch, T. J. (2009). The CTC-chip: an exciting new tool to detect circulating tumour cells in lung cancer patients. *Journal of Thoracic Oncology* 4, 281–3.

Stott, S. L., Hsu, C. -H., Tsukrov, D. I. *et al.* (2010). Isolation of circulating tumour cells using a microvortex-generating herringbone-chip. *Proceedings of the National Academy of Sciences of the USA* 107, 18392–7.

DRUG DEVELOPMENT

Gerber, D. E. and Minna, J. D. (2010). ALK inhibition for non-small cell lung cancer: from discovery to therapy in record time. *Cancer Cell* 18, 548–51.

Bibliography

Katayama, R., Khan, T. M., Benes, C. *et al.* (2011). Therapeutic strategies to overcome crizotinib resistance in non-small cell lung cancers harboring the fusion oncogene EML4-ALK. *Proceedings of the National Academy of Sciences of the USA* **108**, 7535–40.

Kwak, E. L., Bang, Y. -J., Camidge, D. R. *et al.* (2010). Anaplastic lymphoma kinase inhibition in non-small-cell lung cancer. *New England Journal of Medicine* **363**, 1693–703.

The expanding field of cancer treatment

Allgood, P. C., Warwick, J., Warren, R. M. L., Day, N. E. and Duffy, S. W. (2008). A case–control study of the impact of the East Anglian breast screening programme on breast cancer mortality. *British Journal of Cancer* **98**, 206–9.

Birkbak, N. J., Eklund, A. C., Li, Q. *et al.* (2011). Paradoxical relationship between chromosomal instability and survival outcome in cancer. *Cancer Research* **71**, 3447–52.

Gotzsche P. C. and Nielsen, M. (2006). Screening for breast cancer with mammography. *Cochrane Database of Systematic Reviews* **4**, CD001877.

Jørgensen, K. J., Zahl, P. -H. and Gøtzsche, P. C. (2010). Breast cancer mortality in organised mammography screening in Denmark: comparative study. *British Medical Journal* **340**, c1241.

Luo, J., Emanuele, M. J., Li, D. *et al.* (2009a). A genome-wide RNAi screen identifies multiple synthetic lethal interactions with the Ras oncogene. *Cell* **137**, 835–48.

Luo, J., Solimini, N. L. and Elledge, S. J. (2009b). Principles of cancer therapy: oncogene and non-oncogene addiction. *Cell* **136**, 823–37.

McPherson, K. (2010). Screening for breast cancer: balancing the debate. *British Medical Journal* **340**, c3106.

Scholl, C., Fröhling, S., Dunn, I. F. *et al.* (2009). Synthetic lethal interaction between oncogenic KRAS dependency and STK33 suppression in human cancer cells. *Cell* **137**, 821–34.

Appendix A

TUMOUR STAGING

Denoix, P. F. (1946). Enquete permanent dans les centres anticancereaux. *Bulletin. Institut National d'Hygiène* **1**, 70–5.

TMN Online: http://onlinelibrary.wiley.com/book/10.1002/9780471420194 [Accessed 29 May 2012].

INDEX

Index